T0318583

# Glyceraldehyde-3-Phosphate Dehydrogenase (GAPDH)

*This work is dedicated to the generations of my family; those that are past, those that are present, and, hopefully, those that are future.*

# Glyceraldehyde-3-Phosphate Dehydrogenase (GAPDH)

The Quintessential Moonlighting Protein in
Normal Cell Function and in Human Disease

**Michael A. Sirover**
Lewis Katz School of Medicine, Temple University
Philadelphia, PA, United States

**ACADEMIC PRESS**

An imprint of Elsevier

Academic Press is an imprint of Elsevier
125 London Wall, London EC2Y 5AS, United Kingdom
525 B Street, Suite 1800, San Diego, CA 92101-4495, United States
50 Hampshire Street, 5th Floor, Cambridge, MA 02139, United States
The Boulevard, Langford Lane, Kidlington, Oxford OX5 1GB, United Kingdom

**Library of Congress Cataloging-in-Publication Data**
A catalog record for this book is available from the Library of Congress

**British Library Cataloguing-in-Publication Data**
A catalogue record for this book is available from the British Library

ISBN: 978-0-12-809852-3

For information on all Academic Press publications visit our website at
https://www.elsevier.com/books-and-journals

Working together
to grow libraries in
developing countries

www.elsevier.com • www.bookaid.org

*Publisher:* Sara Tenney
*Acquisition Editor:* Linda Versteeg-Buschman
*Editorial Project Manager:* Fenton Coulthurst
*Production Project Manager:* Julia Haynes
*Designer:* Matt Limbert

Typeset by TNQ Books and Journals

# Contents

# Section II
# Physiological Stress and GAPDH Functional Diversity

8.  The Significance of Nitric Oxide–Modified GAPDH:
    Regulation of Apoptosis, Cell Signaling, and Heme
    Metabolism

9.  GAPDH and Hypoxia: Mechanisms of Cell Survival
    During Oxygen Deprivation

## 10. Moonlighting GAPDH and Ischemia: Cellular and Molecular Effects of Oxygen Deprivation and Reperfusion

# Section III
# The Pathology of GAPDH Functional Diversity

## 11. GAPDH and Tumorigenesis: Molecular Mechanisms of Cancer Development and Survival

## 12. Moonlighting GAPDH and Age-Related Neurodegenerative Disease: Diversity of Protein Interactions and Complexity of Function

## Section IV
# The Pharmacology of Moonlighting GAPDH

# Section V
# The Unique Role of Sperm-Specific GAPDH

# Section VI
# Discussion                                                  297

# Biography

Michael A. Sirover, PhD, is a professor of Pharmacology, Lewis Katz School of Medicine at Temple University, Philadelphia, PA. He received his BS in Biology from Rensselaer Polytechnic Institute, his PhD from the State University of New York at Stony Brook, and was a postdoctoral fellow at the Fox Chase Cancer Center in Philadelphia. He was an associate editor of the journal *Cancer Research* and, for over a decade, was the chair of a National Cancer Institute Special Advisory Committee on Cancer Prevention.

Dr. Sirover is one of the pioneers in the identification and characterization of multifunctional proteins. His early work on glyceraldehyde-3-phosphate dehydrogenase (GAPDH) helped establish it as the prime example of this new class of cell proteins. His studies focused on its proliferative-dependent regulation, including distinctive changes in its subcellular localization as a function of the cell cycle, its proliferative-dependent transcriptional and translational regulation, its role in DNA repair, the pathology of age-related neurodegenerative disease, and the cellular phenotype of Bloom's syndrome, a cancer protein human genetic disorder. He isolated and characterized anti-GAPDH monoclonal antibodies and the human GAPDH gene, each of which was subsequently used by many other researchers in their individual GAPDH studies. Lastly, he is the author of the definitive reviews of GAPDH structure and function as well as its role in the pathology of human disease.

**Michael A. Sirover**

# Acknowledgments

Words cannot express my gratitude to my wife, Harlene (Lenie), without whose support, every day and every night, I could not have written this work and to my daughter, Jamie, for her assistance in the preparation of each chapter. Work in the author's laboratory was funded by grants from the National Institutes of Health (ES-01735; CA-29414; AG14566; CA119285); the National Science Foundation (77-20183; 8416295); and the W.W. Smith Charitable Trust.

# Introduction

*It's a dangerous business, Frodo, going out your door. You step onto the road, and if you don't keep your feet, there's no knowing where you might be swept off to.*

The Hobbit by J.R.R. Tolkien

It has been approximately three decades since I took my first step on that road and became involved with studies on the functional diversity of glyceraldehyde-3-phosphate dehydrogenase (GAPDH). In that time, the conception of GAPDH has changed from a housekeeping protein of little interest to a moonlighting protein whose structure and function are of importance not only to normal cell function but also to the pathology of human disease. At first, studies on moonlighting GAPDH were met with intellectual puzzlement, curiosity, and, regrettably, sometimes with disdain. However, as time went on, and as study after study demonstrated new and intriguing moonlighting GAPDH activities, the general perception of this protein began to change. Counterintuitively, in science, challenging conventional dogma is difficult, requiring patience, endurance, and the time it takes for us to illuminate Nature's mysteries.

This book is intended to tell the story of that decades-old journey from skepticism to believability. The studies contained in it sum up our current knowledge of the diverse roles of GAPDH, the complex protein–protein and protein–nucleic acid interactions that underlie its moonlighting activities, as well as the distinctive changes that occur in its subcellular localization. The book is divided into three main sections: the first, a consideration of its role in normal cell functions; the second, a discussion of its participation in the etiology of human disease; the third, a "special topics" section in which unique and novel aspects of moonlighting GAPDH are described. In each, salient findings are included and an overview is provided to consider how those results fit into our conception of GAPDH as a moonlighting protein. For that purpose, in many chapters, a model that presents a summation of the studies contained in that chapter is included. I hope that the reader will find this endeavor to be interesting, informative, and intriguing. Any omissions are unintentional, and the interpretation of data is solely that of the author.

Section I

# The Role of Moonlighting GAPDH in Normal Cell Function

Chapter 1

# The Role of Moonlighting GAPDH in Cell Proliferation: The Dynamic Nature of GAPDH Expression and Subcellular Localization

*"If it looks like a duck, if it quacks like a duck, if it walks like a duck, it's a duck"—a humorous term for a form of abductive reasoning*

Wikipedia

A housekeeping protein may be defined as a molecule whose activity, regulation, expression, and subcellular localization remain relatively constant despite changes in cell status (growth, genome expression, environmental stress, etc.). This is reflected in their use as controls in studies quantitating cellular changes that occur in the given situation of interest. In contrast, cellular proteins regulated actively exhibit pronounced changes in such characteristics, which may be of interest in themselves and which would preclude their characterization as a housekeeping gene and protein.

For that reason, it may be suggested that the analyses of GAPDH regulation during cell proliferation may provide perhaps one of the best illustrations not only of its disqualification as a simple, classical housekeeping protein but also its designation as an active, moonlighting protein of considerable significance. In particular, these studies demonstrate pronounced, reversible changes in its intracellular localization as a function of cell growth, cell cycle changes in the transcription of GAPDH mRNA and its translation into protein, its proliferative-dependent physical association with replicating DNA, its requirement for cell cycle transition, and its role in the initiation of cell senescence. In toto, these findings, in accord with those provided in other chapters, cement GAPDH as a moonlighting protein whose function is required not only for normal cell function but also, as discussed in ensuing chapters, its role in the pathology of human disease.

**3**

## 1. SUBCELLULAR LOCALIZATION OF MOONLIGHTING GAPDH DURING CELL PROLIFERATION

The active regulation of GAPDH as a function of cell proliferation was examined initially by immunocytochemical and subcellular fractionation analyses. As illustrated in Fig. 1.1A, immunological determination of GAPDH in confluent, noncycling human fibroblasts revealed its cytosolic, nonnuclear localization (Cool and Sirover, 1989; Sirover, 1997). The recognition of the GAPDH protein was uniform in the former as was its absence in the latter. Analysis of its intracellular localization demonstrated that the overwhelming majority of immunoreactive GAPDH was present not only in the cytosol, membrane, and perinuclear regions as defined both by immunoblot analysis (Fig. 1.1B) and, as indicated in Fig. 1.1C, by comparison with the amount of immunoreactive GAPDH present in a crude cell extract (Mazzola and Sirover, 2005).

In contrast, as illustrated in Fig. 1.2 subsequent to the initiation of cell proliferation, two distinct changes were observed in human fibroblasts: the first was a change in the subcellular distribution of the GAPDH immunoreactive protein; the second was an increase in the level of GAPDH immunofluorescent staining (Cool and Sirover, 1989; Sirover, 1997). With respect to the former, in proliferating human fibroblasts, immunoreactive GAPDH exhibited a perinuclear or nuclear localization. In addition, these intracellular changes in human fibroblast immunoreactive GAPDH appeared to display a defined temporal sequence. As cell growth commenced, there was a cytoplasmic→perinuclear→nuclear movement of immunoreactive GAPDH. In contrast, as cell growth diminished and ultimately stopped, there was a nuclear→perinuclear→cytoplasmic change in immunoreactive GAPDH intracellular localization (Cool and Sirover, 1989). With respect to the latter, as indicated in Fig. 1.2, there was a considerable

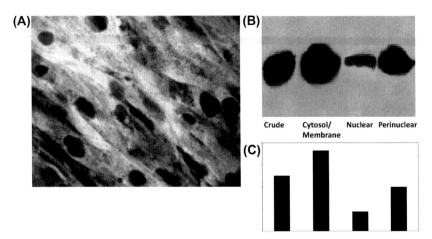

FIGURE 1.1    Subcellular GAPDH localization in noncycling cells. *(A) Reprinted by permission of Wiley and Sons. (B, C) Elsevier.*

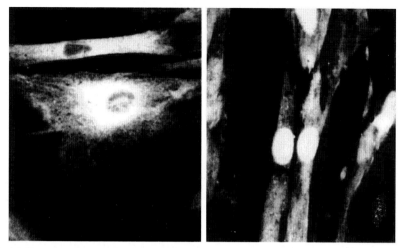

**FIGURE 1.2**   Subcellular GAPDH localization during cell proliferation. *Reprinted by permission of Wiley and Sons.*

increase in GAPDH immunofluorescence in the perinuclear and nuclear regions in proliferating cells as compared with that observed in those regions in non-cycling cells (Fig. 1.1). Further, as cell proliferation diminished, there was a progressive decline in immunofluorescent intensity to that observed in confluent cells (Cool and Sirover, 1989).

As indicated in Table 1.1, these proliferative-dependent changes in GAPDH subcellular distribution and its increased expression appeared to be a general property of growing cells.

Using partial hepatectomy as an experimental paradigm, the proliferative-dependent intracellular localization of GAPDH was determined by immunoblot analysis subsequent to subcellular fractionation (Corbin et al., 2002). In those studies, the nuclear content of GAPDH was increased 3-fold at 24 h, and the level of GAPDH mRNA was increased 1.5-fold following surgery. These two findings suggest that there was an increase in the biosynthesis of the GAPDH protein during hepatocyte cell proliferation. In contrast, the cytoplasmic level of immunoreactive GAPDH protein remained constant. This latter finding would suggest that the newly synthesized GAPDH protein could be selectively located in the nucleus. This will be considered later (Lee and Sirover, 1989).

Previous studies identified a denatured GAPDH isoform, which was antigenically distinct from the native GAPDH isoform (Grigorieva et al., 1999). Recently, using that antibody, which specifically recognized the denatured GAPDH species, its intracellular localization was probed in HeLa cells (Arutyunova et al., 2003, 2013). These studies demonstrated that denatured GAPDH exhibited a nuclear localization during cell proliferation. However, in contrast to native GAPDH, its nuclear localization did not appear to be evenly

**TABLE 1.1** Immunocytochemical Localization of Moonlighting GAPDH[a]

| Organism/ Cell Line | Experimental Paradigm | Change in GAPDH Subcellular Localization | "Unique" Findings | References |
|---|---|---|---|---|
| Normal human fibroblasts | Subculture of confluent cells; asynchronous cell proliferation | Cytoplasmic→perinuclear→nuclear→perinuclear→cytoplasmic | Reversible subcellular translocation of GAPDH | Cool and Sirover (1989) and Sirover (1997) |
| Rat liver | Partial hepatectomy | Cytoplasmic→nuclear | Increase in nuclear GAPDH–no change in cytoplasm | Corbin et al. (2002) |
| HeLa cells | Actively proliferating | Nuclear | Subcellular localization of nonnative GAPDH | Arutyunova et al. (2003, 2013) |
| Human lung adenocarcinoma | $G_1$/S thymidine block | Nuclear→perinuclear→cytoplasmic | S phase, $G_2$/M cell cycle specific nuclear localization | Sundararaj et al. (2004) |

[a]Chronological order.

distributed. This suggested that the denatured form may provide moonlighting functions distinct from those observed with native GAPDH. Evidence was also presented indicating a cytoplasmic colocalization with actin. The significance of this intracellular distribution but could have either a structural or enzymatic function of the GAPDH denatured protein.

Although the studies described above demonstrate proliferative-dependent subcellular changes in immunoreactive GAPDH, they were performed in asynchronously growing cells. Therefore, they do not shed light on the specific phases of the cell cycle in which those changes in intracellular distribution take place. Subsequently, investigations were performed in synchronized cells, which shed light on the cell cycle–dependent changes in immunoreactive GAPDH (Sundararaj et al., 2004).

In these studies, human adenocarcinoma cells were synchronized at the $G_1/S$ border using a thymidine block. Cell cycle stages were defined by flow cytometry. GAPDH localization was determined by immunofluorescence analysis. These studies revealed that GAPDH was localized in the nucleus while the cells were in S phase. Further, in $G_2/M$, GAPDH was still predominantly in the nucleus. This may be related to earlier studies that indicated the role of moonlighting GAPDH as a fusogenic protein required for nuclear membrane assembly (Nakagawa et al., 2003). A perinuclear and cytoplasmic localization was observed as cells began to enter the $G_0/G_1$ phase. As such, these studies not only confirm the subcellular changes in GAPDH localization observed during asynchronous cell growth but they also define, for the first time, the cell cycle localization of immunoreactive GAPDH.

## 2. PROLIFERATIVE-DEPENDENT EXPRESSION OF THE GAPDH GENE

To determine the mechanisms, which may underlie both subcellular changes in GAPDH localization and the observed increase in immunofluorescent intensity, the transcription of the GAPDH gene and the biosynthesis of the GAPDH protein were determined both as a function of cell growth and the cell cycle. In this manner, it would be possible to consider the temporal sequence through which cells regulate GAPDH gene expression in relation to its intracellular distribution.

### 2.1 Transcriptional Regulation of the GAPDH Gene

Following the analysis of immunoreactive GAPDH subcellular localization (Cool and Sirover, 1989), a similar experimental paradigm was used to determine whether cells increased GAPDH mRNA during cell growth (Meyer-Siegler et al., 1992). In particular, in human fibroblasts, northern blot analysis demonstrated a threefold increase in GAPDH mRNA in proliferating cells as compared with that observed in confluent cells. Analysis of its temporal sequence indicated that not only was the highest level of GAPDH mRNA observed at 48 h after plating, the identical interval detected for increases in the rate of DNA synthesis (as defined by $^3$H-thymidine incorporation), but also that the former decreased coordinate with the latter as cell proliferation diminished (96 h after plating). This temporal pattern was identical to that observed for immunoreactive GAPDH, i.e., GAPDH nuclear localization was at its zenith at 48 h, declining thereafter at 96 h. In contrast, in a transformed lymphoblastoid cell line, GAPDH gene expression exhibited a continual increase at each interval. The former study was performed in human cell strains which had a finite life span while the latter result was obtained in an immortal cell line. As GAPDH gene expression is upregulated during tumorigenesis (Chapter 11), the results obtained in the transformed cell line may not be unexpected.

The generality of this finding was suggested by subsequent studies using regenerating rat liver subsequent to partial hepatectomy as an experimental paradigm (Corbin et al., 2002; Iwasaki et al., 2004). In the first study, as described earlier, a 1.5-fold increase in GAPDH mRNA was observed, which appeared to parallel the nuclear localization of immunoreactive GAPDH. In the second study, the increase in GAPDH mRNA appeared to parallel that observed for "clock-related" genes, which are regulated in a circadian manner. Intriguingly, two cycles of GAPDH mRNA expression were observed, i.e., GAPDH mRNA was increased at 12 h after partial hepatectomy, declined thereafter to minimal values at 20 h, increased again at 32 h to the same extent observed at 12 h, and then declined again. The observation that there are sequential increases and decreases in GAPDH mRNA levels demonstrate not only its active regulation but also that mechanisms may exist through which GAPDH mRNA is degraded in a circadian manner. Alternatively, the GAPDH mRNA transcribed during cell growth may be inherently unstable with a significantly low $T_{1/2}$.

The question that remained was to determine both the phase of the cell cycle in which GAPDH mRNA was transcribed and its relationship to DNA replication in S phase. In a further study, GAPDH mRNA regulation was examined using serum depletion as a means to synchronize human fibroblasts (Mansur et al., 1993). Northern blot analysis indicated a dramatic increase in GAPDH mRNA 12–15 h after serum addition followed by a noticeable decline thereafter. Quantitation of DNA replication as defined by ($^3$H)-thymidine incorporation demonstrated that the peak rate of DNA synthesis was observed at 21 h following serum addition. These results suggest that there is a temporal cell cycle–dependent sequence in which GAPDH mRNA precedes DNA replication.

## 2.2 Proliferation-Dependent Biosynthesis of the GAPDH Protein

As the studies described above indicated the proliferative and the cell cycle–dependent transcriptional regulation of moonlighting GAPDH, comparable studies were performed to examine GAPDH biosynthesis. Initial studies were performed in asynchronously growing mammalian cells, with a slight twist, i.e., other investigations using sucrose density gradient analysis indicated the formation of a DNA replicative complex, termed the DNA replitase (Reddy and Pardee, 1980); those subcellular fractionation protocols were utilized to probe whether a physical association may exist between newly synthesized GAPDH and replicating DNA as a function of cell proliferation (Lee and Sirover, 1989).

In those studies, the DNA replitase was isolated by the previously developed sucrose step–gradient analysis (Reddy and Pardee, 1980), and the sedimentation of newly synthesized GAPDH with the DNA replitase was determined in asynchronously proliferating BHK-21 cells. The position of ($^{35}$S)-methionine radiolabeled GAPDH in the sucrose density fractions was identified by western blot analysis of acid-precipitable proteins using an anti-GAPDH monoclonal antibody.

In noncycling cells, radiolabeled GAPDH was detected in sucrose gradient fractions corresponding to low-molecular weight proteins. Little radiolabeled GAPDH was observed in gradient fractions of higher density characteristic of the DNA replitase. In contrast, in cycling cells, the opposite results were obtained, i.e., there was no detectable newly synthesized GAPDH at the lower sucrose density position while significant ($^{35}$S)-methionine radiolabeled GAPDH was detected by immunoblot analysis sedimenting at the higher molecular weight coordinate with that of the DNA replitase as defined by the position of ($^3$H)-thymidine radiolabeled DNA. Further, using equal protein concentrations, there was a noticeable increase in the amount of ($^{35}$S) radiolabeled GAPDH observed at 48 h as compared with that detected at 24 h. As cell proliferation ceased (96 h after initiation of cell growth), there was a demonstrative decrease in ($^{35}$S) radiolabeled GAPDH cosedimenting with the DNA replitase. In contrast at that interval, there was the reappearance of radiolabeled GAPDH at the lower molecular weight sucrose density positions. Further, detergent treatment (2% Tween 80) prior to sucrose density analysis did not affect the cosedimentation of GAPDH with the DNA replitase. In toto, these studies were perhaps the first to indicate the physical association of GAPDH with replicating DNA as a function of cell proliferation.

As the studies described above were performed in asynchronously growing cells, subsequent studies focused on the cell cycle analysis of GAPDH biosynthesis in human fibroblasts (Mansur et al., 1993). In these investigations, as described previously, the serum depletion/readdition experimental paradigm was used to synchronize cells then to stimulate them to enter the cell cycle. The rate of GAPDH biosynthesis was determined by immunoprecipitation/densitometric analyses of ($^{35}$S)-methionine pulsed cells.

In this model system, a specific cell cycle–dependent increase and decrease in the rate of GAPDH biosynthesis was observed. In particular, an approximate sevenfold increase in GAPDH biosynthesis was observed at 18 h following serum stimulation. At subsequent intervals, this rate declined but never reached basal levels observed at 0h. In parallel, the induction of DNA replication was determined by ($^3$H)-thymidine incorporation. Those studies revealed that the rate of DNA synthesis was maximal between 18 and 21h after serum addition. Accordingly, there appeared to be a defined temporal sequence through which human cells regulated GAPDH biosynthesis in relation to DNA replication. In toto, these cumulative studies suggest not only the physical association of GAPDH with replication DNA but also the active transcriptional and translational regulation of moonlighting GAPDH in cell proliferation and in the cell cycle.

## 3. GAPDH—A CELL CYCLE CHECKPOINT IN MAMMALIAN CELLS?

Cell cycle checkpoints represent an important mechanism through which cells control their progression through the complex processes that ultimately result in cell growth and division. Detailed studies have indicated the role of numerous proteins which function as "stop/go" signals as cells traverse each cell cycle stage. Accordingly, those proteins utilized by cells for this purpose are, by definition, of significance and importance.

Recent evidence suggests that moonlighting GAPDH may be one of that set of proteins, which determine cell cycle progression. In an initial study, antisense GAPDH constructs were used to examine the effect of GAPDH depletion in a series of human cervical carcinoma cells (Kim et al., 1999). These studies demonstrated not only a significant inhibition of cell proliferation but also a diminution of colony-formation efficiency. No effect in either experimental paradigm was observed using sense or antisense constructs.

For the most part, studies on moonlighting GAPDH do not normally begin with that intent in mind although such investigations end frequently with the definition of a new GAPDH activity. As indicative of that theme, considering the role of the p21 protein in cell regulation, a study was initiated to define p21 protein–protein interactions. The p21 protein is considered of significance, given its interactions with cyclins and with cyclin-dependent kinases (cdk's). Accordingly, affinity chromatography using GSTp21$^{Cip1}$ as a probe was performed. That analysis detected a 38 kDa binding protein subsequently identified as GAPDH. However, perhaps confusingly, using purified GAPDH, no binding of that protein to the affinity matrix was detected.

As this indicated that additional proteins may be required for GAPDH binding to p21$^{Cip1}$, a GAPDH affinity column was used as a probe. This experiment detected a 39 kDa protein identified as SET, which has been identified as a p21$^{Cip1}$ binding protein (Note: It also detected a series of 13–18 kDa proteins identified as histones, which is also of interest with respect to GAPDH moonlighting functions—explained earlier in this chapter). Control experiments using both immunoblot and coimmunoprecipitation analyses verified this protein–protein interaction.

As p21$^{Cip1}$ and SET may be involved in cell cycle regulation, the GAPDH–SET interaction was examined using the serum starved/addition model first to synchronize cells then to release them to proliferate. Colocalization was defined by immunocytochemistry. These studies revealed preferential colocalization in both S phase and in G$_2$/M. Coimmunoprecipitation protocols were used to verify their in vivo physical association.

The functional significance of the GAPDH–SET interaction was then determined. In particular, SET is known to inhibit cyclin B-cdk1 activity. Dose response analysis demonstrated that this SET activity was diminished by GAPDH. This would have a significant effect in vivo on cell cycle regulation. Subsequently, the interaction of GAPDH and SET with cyclin B was confirmed both by affinity chromatography in vitro and immunoprecipitation in vivo. Accordingly, it appears that a tertiary protein complex may be involved in moonlighting GAPDH function. As indicated in Chapter 8, this may be a common property of GAPDH protein interactions.

The physiological relevance of this series of protein interaction studies was examined using transfection analysis of GAPDH constructs then determining the effect of GAPDH overexpression on cell proliferation. Using synchronized cells, there did not appear to be any effect on progression through S phase as defined by ($^3$H)-thymidine incorporation into DNA. In contrast, using immunocytochemical analysis of the mitotic marker phosphorylated histone H3, it appeared that there was a 50% increase in the number of mitosis in cells overexpressing GAPDH as compared to the control. Using immunoprecipitation coupled with in vitro biochemical assay, it was determined that the cell cycle kinetics of cyclin B-cdk1 activity was altered as a function of GAPDH expression, i.e., its maximal activity was observed earlier than usually detected in synchronously growing cells. From these studies it was suggested that GAPDH may regulate the G$_2$/M transition. This finding, along with those described earlier with respect to the cell cycle regulation of GAPDH mRNA transcription and biosynthesis, indicates further the critical nature of temporal sequence with respect to cell cycle–related moonlighting GAPDH activities.

The significance of moonlighting GAPDH in the control of cell proliferation was indicated again by studies that examined the interrelationship among GAPDH, the cell cycle, and the efficacy of cancer chemotherapeutic agents (Phadke et al., 2009). These investigations utilized human lung and renal carcinoma cell lines, short duplex RNAi and two cancer chemotherapeutic agents, cytarabine, and doxorubicin. Significant findings from this study indicated that GAPDH depletion resulted in the reduction of cell proliferation; accumulation of cells in G$_0$/G$_1$; mechanism underlying this cell cycle block involving p53 and p21; increased resistance to the cancer chemotherapeutic agent cytarabine (araC, cytosine arabinoside).

The reduction of cell growth as a function of GAPDH depletion was monitored in three different cancer cell lines: A549, U031 [both (p53-proficient) and H358 (p53-null)]. Introduction of GAPDH RNAi into A549 and U031 p53-proficient cells eliminated cell proliferation in accord with previous studies (Kim et al., 1999). In contrast, cell proliferation was observed in the H358 p53-null cell line. However, it was reduced approximately 50% as compared with the H358 control. That being said, this study was the first to indicate a role for p53 in the control of cell growth by GAPDH.

Cell cycle analysis in the A549 p53-proficient cell line indicated specific changes in cell cycle distribution as a function of GAPDH depletion. In particular, the percentage of cells in $G_0/G_1$ increased from 57% to 77%; in S phase it decreased from 12% to 6% and in $G_2/M$ from 22% to 11%. This was the first indication that the observed decrease in GAPDH depleted cells was due to a block in $G_0/G_1$.

As a difference in the response to GAPDH depletion was observed primarily in p53-proficient cells, intriguingly, immunoblot analysis demonstrated not only the accumulation of p53 but also the accumulation of p21, a cdk inhibitor. Of note, incubation of the A549 p53-proficient cell line with both siGAPDH and sip21, the level of p21 was lower as compared with proliferation in cells treated with siGAPDH alone. However, the rate of cell growth was reduced. These studies were the first to indicate that the GAPDH-induced $G_0/G_1$ cell block is mediated by changes in the expression of both p53 and p21.

The functional significance of this $G_0/G_1$ cell cycle block was indicated by determining the effect of GAPDH depletion on the efficacy of two, now classical, cancer chemotherapeutic agents, cytarabine (araC, cytosine arabinoside) and doxorubicin. The former is a cell cycle–specific (CCS) agent; the latter a cell cycle–nonspecific (CCNS) drug. GAPDH depletion reduced the sensitivity of treated cells to cytarabine by 50-fold while there was no effect on the cytotoxicity of doxorubicin. Further, the accumulation of cytarabine-induced double strand DNA breaks (DSBs) was lower in GAPDH depleted cells as compared to the control. No difference in the extent of doxorubicin induced DSBs was observed in the former as compared with the latter. In toto, these cumulative studies demonstrate the significance not only of GAPDH as a growth control protein but also its potential effect on the efficacy of cancer chemotherapy.

As described above, these studies suggest that GAPDH depletion–mediated $G_0/G_1$ cell cycle block may be mediated by increased expression of p53 and of cdk-p21. The latter occurs presumably by p53-mediated down regulation of cdk-p21 transcription. Intriguingly, the study described above revealed a GAPDH-mediated cell cycle effect mediated by the interaction of the GAPDH protein not only with the cdk-p21 protein but also the SET protein and cyclin B, each of which is involved with cdk-p21 protein–related cell cycle control mechanisms (Carujo et al., 2006). Thus, in common with the postulated $G_2/M$ role of GAPDH in the cell cycle, these findings with respect to a GAPDH-mediated $G_0/G_1$ block again highlight the complexity of moonlighting GAPDH function.

## 4. MECHANISMS OF GAPDH NUCLEAR TRANSLOCALIZATION AND EXPORT

The demonstrative changes in moonlighting GAPDH subcellular distribution as a function of cell proliferation indicate that mechanisms may exist through which cytosolic GAPDH is transported to the nucleus and, conversely, other mechanisms may be utilized to accomplish its export from the nucleus to the cytoplasm. With respect to the former, although preliminary, sequence analysis indicated the potential presence of a nuclear localization signal (reviewed in Sirover, 1999), there has not been a definitive experimental study to identify such a sequence. As indicated in Table 1.2, a number of studies demonstrate several different pathways through which each transport event may occur. The role of GAPDH as part of the OCA-S coactivator complex, which activates the S histone H2B promoter in S phase (Zheng et al., 2003), and that of the androgen receptor in GAPDH nuclear translocation (Jenster et al., 1993; Harada et al., 2007) has been considered previously in this chapter.

In an initial study, the role of serum, and the factors contained therein, was examined with respect to GAPDH cytoplasmic→nuclear→cytoplasmic movement (Schmitz, 2001). In these studies, endogenous GAPDH or a GFP-tagged GAPDH construct were used. Serum deprivation for 5 days resulted in GAPDH nuclear localization. Similar results were observed for the GFP-GAPDH fusion construct. Readdition of serum or the addition of EGF and PDGF to serum-deprived cells resulted in the relocation of GAPDH to the cytoplasm. Inhibitor analysis indicated that GAPDH nuclear translocation was not dependent on phosphorylation by a cdk.

In contrast, inhibitor studies did indicate the role of phosphatidylinositol 3-kinase (PI3K) in determining GAPDH nuclear export (Schmitz, 2001; Schmitz et al., 2003). A test of inhibitors of protein kinase C, MAPK kinase, and PI3K indicated that only the PI3K inhibitor prevented GAPDH nuclear export. Further, use of an exportin 1, i.e., CRM1, inhibitor demonstrated that the serum addition–induced relocalization of GAPDH from the nucleus to the cytoplasm was independent of exportin 1/CRM1.

In contrast, a subsequent study not only demonstrated the role of exportin 1/CRM1 in GAPDH subcellular distribution but also reported the identification of a GAPDH CRM1 nuclear export domain (Brown et al., 2004). In these studies, genotoxic stress as a consequence of thiopurine exposure was used as the experimental paradigm. The intracellular localization of endogenous and GFP-GAPDH constructs were determined by immunocytochemical analysis.

These studies indicated that a 24–48 h exposure to either 6-mercaptopurine or 6-thiopurine resulted in nuclear localization of both endogenous GAPDH and a full length GFP-GAPDH construct. Antibody mapping studies indicated two overlapping amino acid sequences, which, on further analysis, indicated that the anti-GAPDH monoclonal antibody recognized a domain consisting of GAPDH amino acids 250–269. This was confirmed by affinity chromatography

**TABLE 1.2 Mechanisms of GAPDH Nuclear Translocation and Export[a]**

| Organism/Cell Line | Experimental Paradigm | GAPDH Subcellular Localization | "Unique" Findings | References |
|---|---|---|---|---|
| Mouse fibroblasts | Serum depletion, readdition | Cytoplasmic→nuclear→cytoplasmic | Nuclear export not dependent on exportin 1 | Schmitz (2001) and Schmitz et al. (2003) |
| Human colon carcinoma and osteosarcoma cells | Centrifugal elutriation, aphidicolin | Cytoplasmic→nuclear→cytoplasmic | GAPDH part of OCA-S, coactivator of histone H2B promoter | Zheng et al. (2003) |
| Colon adenocarcinoma, T-derived leukemic cells | Thiopurine exposure | Cytoplasmic→nuclear→cytoplasmic | CRM-1 nuclear export domain[b] | Brown et al. (2004) |
| Human prostate cancer cells, COS-7 cells | Overexpression of androgen receptor; immunochemical analysis | Cytoplasmic→nuclear | GAPDH: androgen receptor nuclear translocation | Harada et al. (2007) |
| Rat insulinoma cells | O-GlcNA[c] effect on subcellular translocation (normal or high glucose) | Cytoplasmic→nuclear | GAPDH[thr227] modification by O-GlcNA required for nuclear translocation | Park et al. (2009) |
| Mouse fibroblasts | Transfection of human GAPDH | Cytoplasmic→nuclear | PCAF[d] acetylation of GAPDH lys[117, 227, 251] | Ventura et al. (2010) |
| Human fibroblasts | Serum depletion/addition | Cytoplasmic→nuclear→cytoplasmic | Nuclear import: AMP-activated protein kinase | Kwon et al. (2010) |
| | | | Nuclear export: PI3K[e] | |
| Lung carcinoma, colon carcinoma cells | Cytarabine-induced genotoxic stress | Cytoplasmic→nuclear | NAD+ binding site determines intranuclear interactions | Phadke et al. (2015) |

[a]Chronological order.
[b]Exportin 1 or chromosomal region maintenance.
[c]O-linked N-acetylglucosamine.
[d]Acetyltransferase P300/CBP-associated factor.
[e]Phosphatidylinositol 3-kinase.

using a GAPDH$^{256-270}$ bound construct. The significance of this finding was that sequence comparison analysis indicated its similarity to an atypical nuclear export signal (NES).

For that reason, GAPDH constructs were prepared with modification of that putative NES signal then tested for their intracellular distribution. A truncated GAPDH mutation (T1), which did not contain the postulated NES sequence, exhibited a predominantly nuclear localization. Mutation of two alanine residues within that domain revealed a cytoplasmic GAPDH localization while mutation of two other alanine residues did not affect GAPDH nuclear localization. Analysis of a K259N mutant resulted in nuclear localization in the absence of thiopurine treatment. This latter finding indicated the role of GADPH$^{lys\,259}$ in the nuclear export of GAPDH.

Coimmunoprecipitation experiments were performed to demonstrate the binding of GAPDH to exportin 1/CRM1. Further, incubation with the exportin 1/CRM1 inhibitor leptomycin B prevented export of GAPDH from the nucleus. This was the same inhibitor used previously to demonstrate that GAPDH redistribution to the cytoplasm was exportin 1 independent (Schmitz et al., 2003). Cumulatively, these findings suggest that there are, at a minimum, two mechanisms through which GAPDH may be exported from the nucleus as a function on an appropriate stimulus.

Another mechanism that may underlie GAPDH subcellular translocations may be that of its posttranslational modification. Two studies indicate that separate posttranslational modifications may affect the subcellular localization of GAPDH. In the first study, the role of O-linked N-acetylglucosamine glycosylation (O-GlcNAc) was examined (Park et al., 2009). The rationale for these investigations was that this modification may effect a cytosolic to nuclear transport (rev. in Duverger et al., 1995; Monsigny et al., 2004).

These studies were performed in rat insulinoma cells, which permitted an analysis of the role of glucose on GAPDH expression, modification, and localization. The intracellular distribution of GAPDH and the extent of its O-GlcNAcetylation were examined in cells treated with low (5 mM) or high glucose (30 mM) for 24 h. Immunoblot analysis indicated an increase in nuclear GAPDH in high glucose treated cells. Further, coimmunoprecipitation with an anti-GAPDH antibody followed by immunoblot analysis with an anti-O-GlcNAc antibody indicated GAPDH posttranslational modification. Intriguingly, the modified GAPDH species was localized primarily in the nucleus. Using a detailed biophysical analysis, the GAPDH O-GlcNAc binding site was identified as thr$^{227}$. Mutational analysis was then utilized to define the significance of that binding site, i.e., the subcellular localization of a GAPDH T227A construct was examined and demonstrated to be cytoplasmic in nature. Accordingly, these studies suggest that O-GlcNAc modification of GAPDH may represent an additional mechanism mediating GAPDH nuclear translocation.

In the second study, the role of GAPDH acetylation was determined (Ventura et al., 2010). The experimental paradigm was to probe acetyltransferase P300/

CBP-associated factor (PCAF) interactions with GAPDH in vitro; its ability to acetylate GAPDH in vivo; to determine the intracellular fate of acetylated GAPDH; and to determine the specific GAPDH amino acids that are acetylated by PCAF. As such, these investigations would shed light on the role of this post-translational modification on GAPDH structure and function.

Initially, the PCAF–GAPDH interaction was examined in vitro using GAPDH-Sepharose affinity analysis to identify this protein–protein interaction. Binding was observed for the PCAF catalytic domain, which also was coimmunoprecipitated with GAPDH when it was transfected as a recombinant construct in vivo. Acetylation of GAPDH by PCAF was determined both by an in vitro biochemical assay and by the use of anti-acetyl-Lys antibodies in immunoblot analysis following GAPDH immunoprecipitation with an anti-GAPDH antibody. Mapping studies were utilized to identify the GAPDH$^{lys}$ residues, which were acetylated by PCAF. This analysis was performed using four GAPDH fragments in the in vitro acetylation reaction. These studies identified lys$^{117}$ and lys$^{251}$ as acetylation sites. One fragment that did not appear to be acetylated was further analyzed by spot mapping. That analysis identified lys$^{227}$ as being acetylated by PCAF.

The physiological significance of these posttranslational modifications as examined by analysis of GAPDH subcellular localization is cells overexpressing inducible PCAF under tetracycline control (termed a Tet-Off system). In uninduced cells, GAPDH exhibited a cytoplasmic localization. In contrast, on PCAF induction, GAPDH was located in the nucleus. Mutational analyses were performed in which both lysine residues were substituted by either arginine or glutamine. Each form exhibited a cytoplasmic localization except when an apoptotic stimulus was involved (discussed in Chapter 8). In toto, these two studies identify separate posttranslation modifications, which may mediate GAPDH nuclear translocation.

As previously described, the serum deprivation/readdition model was used to demonstrate the nuclear accumulation of GAPDH in the former and its cytoplasmic redistribution in the latter (Schmitz, 2001; Schmitz et al., 2003). Those studies indicated also a role for PI3K in the regulation of nuclear GAPDH export. Subsequently, a further study identified an intriguing "shuttle" mechanism through which AMP-activated protein kinase (AMPK) stimulates GAPDH nuclear translocation, which may be coupled to the role of PI3K as the determining factor for GAPDH cytoplasmic relocalization (Kwon et al., 2010).

These studies utilized the serum depletion/readdition model using human diploid fibroblasts (HDFs) as the experimental paradigm. As expected, serum depletion resulted in GAPDH nuclear localization while its readdition resulted in cytoplasmic redistribution. The role of AMPK was examined using its specific activator 5-aminoimidazole-4-carboxamide-1-β-D-ribofuranoside (AICAR). In serum containing HDF cultures, AICAR addition not only stimulated GAPDH nuclear localization but also did this in a dose- and time-dependent manner. Of note, it was estimated that over 90% of GAPDH was detected in the nucleus following treatment with 2 mM AICAR for 4 days. Further, the role of AMPK in AICAR-induced GAPDH nuclear translocation was determined using an AMPK inhibitor, termed compound C. The latter reduced AICAR-induced GAPDH nuclear translocation. In addition, the introduction of AMPK siRNA prevented AICAR-induced GAPDH nuclear translocation. Lastly, the role of PI3K in this "shuttle" mechanism was examined using Leptomycin B (LMB) previously demonstrated to prevent GAPDH export from the nucleus (Schmitz, 2001; Schmitz et al., 2003). LMB prevented GAPDH cytoplasmic redistribution on serum readdition. In toto, these studies suggest an intriguing kinase-mediated pathway controlling cytoplasmic→nuclear→cytoplasmic changes in intracellular GAPDH localization as a function of serum depletion/readdition.

Subsequently, as the studies described earlier indicate the role of protein kinases in GAPDH subcellular translocation, the effect of phosphorylation on GAPDH intracellular distribution was investigated (Phadke et al., 2015). In particular, the role of three such posttranslational modifications was determined, i.e., at GAPDH$^{Y94, S98, T99}$, all of which are in the NAD$^+$ binding domain. Intracellular localization of wt and mutant GAPDH constructs, prepared by site-specific mutagenesis (Y94A, S98A, T99A and T99I), were utilized in lung and colon carcinoma cells (p53$^{+/+}$).

Cytarabine treatment was used to monitor nuclear GAPDH localization. These studies revealed that cells expressing the mutant constructs were characterized by a nuclear GAPDH localization comparable with that of the wt control. However, in situ analysis indicated that the intranuclear distribution of the mutant constructs was significantly different from that observed for the wt species. As protein–protein interactions are a basis for moonlighting GAPDH function, these studies are of significance in that they demonstrate, for the first time, that GAPDH posttranslational phosphorylation may be required for the nuclear protein interactions, which form the basis for its nuclear functions. In toto, as indicated in Table 1.2, there appears to be a diversity of mechanisms, which may be involved in the dynamics of GAPDH subcellular localization. Accordingly, it may be of great interest to determine exactly what role each targeting mechanism plays in GAPDH nuclear translocation and export.

## 5. THE ROLE OF MOONLIGHTING GAPDH IN CELL SENESCENCE

The cessation of cell growth and the assumption of a quiescent state, i.e., cell senescence, require specific changes in gene expression, protein structure, and cell morphology. Recent studies indicate a new, potentially important role for moonlighting GAPDH in this significant cellular alteration. In particular, using human lung carcinoma cells as an experimental paradigm, depletion of GAPDH resulted in the defined changes in cell structure and function, which are characteristics of cell senescence (Phadke et al., 2011).

In these studies, the knockdown of GAPDH using siGAPDH resulted not only in cell growth arrest but also revealed changes in cell morphology characteristic of cell senescence. Further, the appearance of cell senescence markers was also observed. Intriguingly, it was noted that reduction in GAPDH resulted in increases in the phosphorylation of AMPK which, as noted above, was involved in GAPDH nuclear localization. Similarly, increased phosphorylation of p53 was also detected. As noted previously, p53 proficiency was required for GAPDH-induced cell cycle arrest (Phadke et al., 2009). Lastly, overexpression of GAPDH reversed cell senescence.

The functional significance of GAPDH-induced cell senescence was demonstrated by examining its sensitivity to a variety of cancer chemotherapeutic agents (Phadke et al., 2013). Using the model developed as described above, cytotoxicity was determined in lung carcinoma control cells and those induced to senesce by GAPDH depletion. Two groups of cancer chemotherapeutic drugs were used, including a series of antimetabolite CCS drugs and a series of CCNS drugs. GAPDH-induced senescent cells were resistant to the effects of the former but sensitive to the latter. In particular, in CCNS-treated cells, there was a noticeable reduction in cell viability as a function of drug dose, which was not observed in GAPDH-induced senescent cells. Further, significant reduction was observed in the number of apoptotic cells as compared with that of control cells as a function of CCS drug treatment. In contrast, with respect to the CCNS drugs, there was a comparable sensitivity in both control and senescent cells. In toto, these studies suggest a role for moonlighting GAPDH in the regulation of cell senescence.

## 6. SUMMARY

The studies presented in this chapter highlight the dynamic nature of moonlighting GAPDH expression and localization as a function of cell proliferation, the cell cycle, and cell senescence. In particular, these investigations highlight the complex role of GAPDH gene expression, its subcellular distribution, and its macromolecular interactions, which are a priori requirements for the successful initiation and progression of cell growth as well as determining the cessation of cell proliferation. In the former, GAPDH exhibits cytoplasmic to nuclear changes in its intracellular localization, which is coordinate with DNA replication; its cytoplasmic relocalization as cell growth ceases; specific proliferation-dependent increases in GAPDH mRNA transcription and the biosynthesis of the GAPDH protein; and the physical association of GAPDH with replicating DNA. These proliferative-related GAPDH effects are exhibited in diverse cell types, indicating the generality of these moonlighting functions. Further, the diversity of mechanisms, which may be involved in GAPDH nuclear↔cytoplasmic movement, is striking. In addition, GAPDH depletion may represent a signal for cells to enter a senescent phase. These cumulative investigations provide substantial evidence defining the complex role of moonlighting GAPDH in cell proliferation and senescence.

# REFERENCES

Arutyunova, E., Danshina, P., Domnina, L., Pleten, A., Muronet, V., 2003. Oxidation of glyceral dehyde-3-phosphate dehydrogenase enhances its binding to nucleic acids. Biochem. Biophys. Res. Commun. 307, 547–552.

Arutyunova, E., Domnia, L., Chudinova, A., Makshakova, O., et al., 2013. Localization of non-native D-glyceraldehyde-3-phosphate dehydrogenase in growing and apoptotic HeLa cells. Biochemistry (Mosc) 78, 91–95.

Brown, V., Krynetski, E., Krynetskaia, N., Griegeri, D., et al., 2004. A novel CRM-1 mediated nuclear export signal governs nuclear accumulation of glyceraldehyde-3-phosphate dehydroge-nase following genotoxic stress. J. Biol. Chem. 279, 5984–5992.

Carujo, S., Estanyol, J., Ejarque, A., et al., 2006. Glyceraldehyde-3-phosphate dehydrogenase is a SET-binding protein and regulates cyclin B-cdk1 activity. Oncogene 25, 4033–4042.

Cool, B., Sirover, M., 1989. Immunocytochemical localization of the base excision repair enzyme uracil DNA glycosylase in quiescent and proliferating normal human cells. Cancer Res. 49, 3029–3036.

Corbin, I., Gong, Y., Zhang, M., Minuk, G., 2002. Proliferative and nutritional dependent regulation of glyceraldehyde-3-phosphate dehydrogenase in the rat liver. Cell Prolif. 35, 173–182.

Duverger, E., Pellerin-Mendes, C., Mayer, R., Roche, A., Monsigny, M., 1995. Nuclear import of glycoconjugates is distinct from the classical NLS pathway. J. Cell. Sci. 108, 1325–1332.

Grigorieva, J., Dainiak, M., Katrukha, A., Muronetz, V., 1999. Antibodies to the nonnative forms of D-glyceraldehyde-3-phosphate dehydrogenase: identification, purification, and influence on the renaturation of the enzyme. Arch. Biochem. Biophys. 369, 252–260.

Harada, N., Yasunga, R., Higashimura, Y., et al., 2007. Glyceraldehyde-3-phosphate dehydrogenase enhances transcriptional activity of androgen receptor in prostate cancer cells. J. Biol. Chem. 282, 22651–22661.

Iwasaki, T., Nakahama, K., Nagano, M., Fujioka, A., et al., 2004. A partial hepatectomy results in altered expression of clock-related and cyclic glyceraldehyde-3-phosphate dehydrogenase (GAPDH) genes. Life Sci. 74, 3093–3102.

Jenster, G., Trapman, J., Brinkmann, 1993. Nuclear import of the human androgen receptor. Biochem. J. 293, 761–768.

Kim, J., Kim, T., Kim, Y., Kim, Y., et al., 1999. Antisense oligodeoxynucleotide of glyceraldehyde-3-phosphate dehydrogenase gene inhibits cell proliferation and induces apoptosis in human cervical carcinoma cell lines. Antisense Nucleic Acid Drug. Dev. 9, 507–513.

Kwon, H., Rhim, J., Jang, I., et al., 2010. Activation of AMP-activated protein kinase stimulates the nuclear localization of glyceraldehyde-3-phosphate dehydrogenase in human diploid fibroblasts. Exp. Mol. Med. 42, 254–269.

Lee, K., Sirover, M., 1989. Physical association of base excision repair enzymes with parental or replicating DNA in BHK-21 cells. Cancer Res. 49, 3037–3044.

Mansur, N., Meyer-Siegler, K., Wurzer, J., Sirover, M., 1993. Cell cycle regulation of the glyceralde-hyde-3-phosphate dehydrogenase/uracil DNA glycosylase gene in normal human cells. Nucleic Acids Res. 21, 993–998.

Mazzola, J., Sirover, M., 2005. Aging of human glyceraldehyde-3-phosphate dehydrogenase is dependent on its subcellular localization. Biochim. Biophys. Acta 1722, 168–174.

Meyer-Siegler, K., Mansur, N., Wurzer, J., Sirover, M., 1992. Proliferation-dependent regulation of the glyceraldehyde-3-phosphate dehydrogenase/uracil DNA glycosylase gene in human cells. Carcinogenesis 13, 2127–2132.

Monsigny, M., Rondanino, C., Duverger, E., Fajar, I., Roche, A., 2004. Glyco-dependent nuclear import of glycoproteins, glycoplexes and glycosylated plasmids. Biochim. Biophys. Acta 1673, 94–104.

Nakagawa, T., Hirano, Y., Inomata, A., et al., 2003. Participation of a fusogenic protein, glyceraldehyde-3-phosphate dehydrogenase, in nuclear membrane assembly. J. Biol. Chem. 278, 20395–20404.

Park, J., Han, D., Kim, K., Kang, Y., Kim, Y., 2009. O-GlcNAcylation disrupts glyceraldehyde-3-phosphate dehydrogenase homo-tetramer formation and mediates its nuclear translocation. Biochim. Biophys. Acta 1794, 252–262.

Phadke, M., Krynetskaia, N., Mishra, A., Krynetskiy, E., 2009. Glyceraldehyde-3-phosphate dehydrogenase depletion induces cell cycle arrest and resistance to antimetabolites in human carcinoma cell lines. J. Pharmacol. Exp. Ther. 331, 77–86.

Phadke, M., Krynetskaia, N., Mishra, A., Krynetskiy, E., 2011. Accelerated cellular senescence phenotype of GAPDH-depleted human lung carcinoma cells. Biochem. Biophys. Res. Commun. 411, 409–415.

Phadke, M., Krynetskaia, N., Krynetskiy, E., 2013. Cytotoxicity of chemotherapeutic agents in glyceraldehyde-3-phosphate dehydrogenase-depleted human lung carcinoma A549 cells with the accelerated senescence phenotype. Anti-Cancer Drugs 24, 366–374.

Phadke, M., Krynetskaia, N., Mishra, A., Barrero, C., et al., 2015. Disruption of the NAD$^+$ binding site in glyceraldehyde-3-phosphate dehydrogenase affects its intranuclear interactions. World J. Biol. Chem. 6, 366–378.

Reddy, G., Pardee, A., 1980. Multienzyme complex for metabolic channeling in mammalian DNA replication. Proc. Natl. Acad. Sci. U.S.A. 77, 3312–3316.

Schmitz, H., 2001. Reversible nuclear translocation of glyceraldehyde-3-phosphate upon serum stimulation. Eur. J. Cell Biol. 80, 419–427.

Schmitz, H., Dutiné, C., Bereiter-Hahn, J., 2003. Exportin-1 independent nuclear export of GAPDH. Cell Biol. Int. 27, 511–517.

Sirover, M., 1997. Role of the glycolytic protein, glyceraldehyde-3-phosphate dehydrogenase, in normal cell function and in cell pathology. J. Cell. Biochem. 66, 133–140.

Sirover, M., 1999. New insights into an old protein: the functional diversity of mammalian glyceraldehyde-3-phosphate dehydrogenase. Biochim. Biophys. Acta 1432, 159–184.

Sundararaj, K., Wood, R., Ponnusamy, S., et al., 2004. Rapid shortening of telomere length in response to ceramide involves the inhibition of telomere binding activity of nuclear glyceraldehyde-3-phosphate dehydrogenase. J. Biol. Chem. 279, 6152–6162.

Ventura, M., Mateo, F., Serratosa, J., Salaet, I., et al., 2010. Nuclear translocation of glyceraldehyde-3-phosphate is regulated by acetylation. Int. J. Biochem. Cell Biol. 42, 1672–1680.

Zheng, L., Roeder, R., Luo, Y., 2003. S phase activation of the histone H2B promoter by OCA-S, a coactivator complex that contains GAPDH as a key component. Cell 114, 255–266.

## FURTHER READING

Barbini, L., Rodriguez, J., Dominguez, F., Vega, F., 2007. Glyceraldehyde-3-phosphate dehydrogenase exerts different biological activities in apoptotic and proliferating hepatocytes according to its subcellular localization. Mol. Cell. Biochem. 300, 19–28.

# Chapter 2

# Moonlighting GAPDH and the Transcriptional Regulation of Gene Expression: Multiprotein Complex Formation and Mechanisms of Nuclear Translocation

*What's past is prologue*

William Shakespeare

The regulation of gene transcription is, perhaps, one of the greatest challenges faced by any organism. As each cell within it contains the same genotype, it is necessary for the organism to specify which genes in which cells are expressed so that the latter can fulfill their particular phenotypic function. For that reason, as detailed analyses demonstrate, nucleotide sequences contained in upstream regulatory elements as well as those in promoter regions may provide the necessary recognition sites through which cells activate the expression of specific genes. The other requisite components are cellular molecules, which recognize and bind to those sequences, thereby inducing gene expression. That being said, there are numerous genes expressed in all cells. Therefore, common mechanisms must exist to ensure their transcription as well.

It may be axiomatic that both general and specific gene transcription may require complex protein–protein structures through which each may occur. It is also axiomatic that such complexes recognize specific genomic nucleotide sequences. Accordingly, in this chapter, we shall consider those studies which demonstrate not only that moonlighting GAPDH is a necessary component of the protein transcriptional complexes, which mediate gene expression but also may be involved in the protein–nucleic interactions through which phenotypic expression of the cellular genotype is achieved. These may be divided into two categories: the first may be termed a general, nonspecific GAPDH function, i.e., its association with the RNA polymerase II transcription complex and with the intranuclear structure termed the promyelocytic leukemia (PML) body; the

Glyceraldehyde-3-Phosphate Dehydrogenase (GAPDH). http://dx.doi.org/10.1016/B978-0-12-809852-3.00002-9

second is a specific effect of moonlighting GAPDH on individual gene regulation, i.e., its required for the multiprotein complex necessary for histone biosynthesis and its role in the binding to the androgen receptor (AR) required for gene activation in prostate cells. As such, these studies, in common with those presented in Chapter 1, demonstrate the dynamic and complex nature of moonlighting GAPDH function.

## 1. GAPDH AS A DNA-BINDING PROTEIN

Studies on the role of GAPDH in transcriptional regulation have an intriguing similarity to the other investigations presented in Part I of this book. In particular, the subjects discussed in this chapter and in Chapters 3–7, are characterized by early, antecedent studies that hinted at the complex functions of GAPDH, which define now its normal moonlighting activities. As such, these past studies were the forerunner, i.e., the prologue, of the subsequent, detailed analyses, which established the functional diversity of GAPDH (Table 2.1).

As with most studies on GAPDH, the discovery of its multidimensional activities was an unintended consequence of analyses, which focused on other questions. Although seemingly quaint in the early 21st century, using a recently developed new technology, DNA-cellulose column chromatography, particular attention was focused on the proteins involved in DNA replication (Alberts et al., 1968). In this analysis a number of such proteins were detected using both native and denatured DNA-cellulose column chromatography. In such a study interrelating DNA-binding proteins to cell growth, a series of proteins, designated P1–P8, were identified by electrophoresis, following their chromatographic elution on denatured DNA cellulose (Salas and Green, 1971). The former indicated a size differential with P1 exhibiting the highest molecular weight and P8 the lowest. A double label experiment ($^3$H and $^{14}$C-proline) indicated differences in protein levels with the highest level of P8 observed in S phase–arrested cells.

Subsequently, P8 was examined in vitro using both ss and dsDNA cellulose (Tsai and Green, 1973). Of note, P8 was not retained by the latter but did bind to the former. It also did not bind to a phosphocellulose column indicating its repulsion by negative charges. Using purified P8, increasing binding to ssDNA was observed, while no similar increase was detected using dsDNA. Pulse-chase analysis indicated that the $T_{1/2}$ of P8 was approximately 40 h in mouse 3T6 fibroblast cells and approximately 96 h in human SB fibroblast cells. It was also noted that the amount of cellular P8 was considerable, being estimated at c. 2%–3% in the human diploid fibroblast cell strain.

In a further study, the nature of this P1–P8 series of DNA-binding proteins was examined in virus-transformed cell lines as a function of cell growth (Melero et al., 1975). Transformation by either an RNA or a DNA tumor virus did not affect the profile of the P1–P8 proteins. That being said, as a function of cell transformation, there was an appreciable difference in P8 levels in what

**TABLE 2.1 Identification of GAPDH as a DNA-Binding Protein[a,b]**

| System | Experimental Paradigm | Biochemical Analysis | "Unique Finding" | References |
|---|---|---|---|---|
| Mouse fibroblasts | Denatured DNA-cellulose column chromatography | Identification of eight individual DNA-binding proteins (P1–P8) | Differential biosynthesis of DNA-binding proteins in growing versus resting cells | Salas and Green (1971) |
| Mouse fibroblasts | Native and denatured DNA column chromatography | Analysis of the P8 protein | P8 protein levels higher in growing cells; P8 binds to denature DNA | Tsai and Green (1973) |
| Virus-transformed hamster cell lines | Electrophoresis of radiolabeled DNA-binding proteins | Synthesis of P8 protein in growing versus resting cells | P8 protein levels higher in transformed cells[c] | Melero et al. (1975) |
| Transformed hamster fibroblasts | Protein purification | Identification of P8 as GAPDH | GAPDH preferentially binds to single-stranded DNA | Perucho et al. (1977) |
| Calf thymus | Effect of DNA-binding proteins on DNA replication | GAPDH (SSB-37 protein[d]) inhibits DNA replication | Catalytically active GAPDH dimer ($M_r = 80\,kDa$) | Grosse et al. (1986) |
| Calf cerebral cortex | Immunochemical analysis; subcellular analysis | Chromatin association of GAPDH (35K protein) | GAPDH stimulates RNA polymerase II? | Morgenegg et al. (1986) |

[a]In chronological order.
[b]Reviewed in Ronai (1993).
[c]See Chapter 11.
[d]Single-strand DNA-binding protein $M_r = 37\,kDa$.

were termed "dense" versus growing cells. This finding was observed in a series of transformed cell clones indicating the generality of this result. Analysis of P8 synthesis in synchronized cells suggested that its synthesis preceded that observed for DNA replication in both untransformed and transformed cells. In toto, it was concluded that the increased levels of P8, as yet an unidentified protein, was transformation dependent.

Subsequently, these investigators identified the P8 protein as GAPDH (Perucho et al., 1977). The parameters used for that identification included determination of amino acid composition; GAPDH reactivity with anti-P8 rabbit serum as defined by double immunodiffusion analysis; comparison of GAPDH enzymatic activity of P8 and that of purified hamster and rabbit muscle GAPDH; similarities of ssDNA versus dsDNA binding to either P8 or purified hamster muscle GAPDH; and comparable inhibition of ssDNA binding by NAD$^+$. From the latter, it was concluded that DNA bound to GAPDH through its NAD$^+$ binding site and not the glyceraldehyde-3-phosphate catalytic site.

The demonstration that GAPDH (P8) binds specifically to ssDNA as compared with dsDNA coupled with its synthesis in growing cells suggested that there may be a role for GAPDH in DNA replication. This was tested by examining its effect on DNA synthesis in vitro by the DNA polymerase α-primase complex (Grosse et al., 1986). In this study, two single-strand binding (SSB) proteins were isolated from calf thymus. The first was termed SSB-35 based on its $M_r = 35$ kDa. The second was termed SSB-37 based on its $M_r = 37$ kDa. As defined by amino acid composition and enzymatic activity, SSB-35 was identified as lactate dehydrogenase (LDH) and SSB-37 as GAPDH. Intriguingly, GAPDH was detected as a dimer with enzymatic activity, which is in contrast to its usual tetrameric structure. As will be noted in later chapters, this is not an unusual finding.

Biochemical characterization of each protein demonstrated that both preferred binding to ssDNA as compared with dsDNA by 300-fold. Their effect on DNA replication was examined using a purified DNA polymerase α-primase preparation using either activated DNA or M13 ssDNA as template. In either instance, a stimulatory effect of LDH was observed, which was concentration dependent. In contrast, GAPDH inhibited DNA replication in a concentration-dependent manner. This effect was greater at lower GAPDH concentrations when the M13 ssDNA was used as a template. The rationale for this demonstrative inhibition was unclear. However, in another study, it was determined that a monoclonal antibody to DNA polymerase α-inhibited GAPDH enzymatic activity (Seal and Sirover, 1986). Accordingly, both findings suggest the physical association of GAPDH with DNA polymerase α.

In the final early study, calf cerebral cortex was used as a source to probe this new aspect of GAPDH biology (Morgenegg et al., 1986). In accord with the investigations described earlier, a 35-kDa protein was identified through DNA-cellulose column chromatography. Glycerol gradient analysis indicated a coordinate sedimentation of the 35 kDa protein and single-stranded DNA-binding activity. Incubation of the 35-kDa protein with different DNA samples indicated not only its selective binding to single-stranded DNA but also that such binding was proportional to protein concentration. To determine whether the 35-kDa protein was indeed GAPDH, its enzymatic activity was compared with that of rabbit skeletal muscle GAPDH. As peptide mapping studies were also identical, it seemed reasonable to identify the 35-kDa protein as GAPDH.

In a novel approach, the subcellular distribution of the cerebral cortex GAPDH was examined. Initially, using two-dimensional gel electrophoresis, GAPDH was identified as a component of the low mobility group of nonhistone proteins. This nuclear localization was confirmed using immunohistochemical analysis with a rabbit antibody prepared against the 35-kDa protein.

That being said, functional studies did not demonstrate an effect of GAPDH on DNA replication. These studies were performed in vivo in permeabilized Chinese hamster ovary cells in the presence or the absence of the 35-kDa protein. Injection of rabbit skeletal muscle GAPDH enhanced RNA transcription when injected into *Xenopus laevis* oocytes, suggesting an effect on RNA polymerase II. This would be in accord with the preferential binding of GAPDH to single-stranded DNA. Unfortunately, that conclusion was called into question by further studies, which suggested that an impurity in the GAPDH preparation was actually the stimulatory agent. This will be considered later.

In summary, the following conclusions may be drawn from these early studies: Protein P8 isolated by several groups following DNA-cellulose column chromatography is GAPDH; it specifically binds to single-stranded DNA; GAPDH levels are proportional to the degree of cell proliferation; GAPDH levels are a function of cellular transformation by either RNA or DNA tumor viruses; it may or may not be involved in DNA replication or in RNA transcription. As such, these past studies provide the prologue for the investigations that follow.

## 2. MOONLIGHTING GAPDH AND GENE TRANSCRIPTION

As the studies described in Section 1 demonstrate, one of the common characteristics of each report was the observation that GAPDH bound preferentially to either single-stranded or denatured DNA. This structural property indicated that, should GAPDH binding be involved functionally, it would seem most likely that it would be involved in DNA functions, which require it to be accessible readily, i.e., in an "open" conformation as contrasted with its classical double helical structure. As indicated in Table 2.2, those early studies were predicative in nature.

### 2.1 GAPDH–PML Body Association

The initial indication that moonlighting GAPDH may function in transcriptional gene expression was that not only did it bind to the PML but also

**TABLE 2.2 Role of GAPDH in Transcription[a]**

| Organism/Cell Line | Transcriptional Gene Expression | Moonlighting GAPDH Function | "Unique" Findings | References |
|---|---|---|---|---|
| Mouse fibroblasts, human fibroblasts | PML[b] body structure and function in normal cells | GAPDH–PML binding; GAPDH localization in transcriptionally active nuclear domains | Histone genes: PML body association in S phase | Carlile et al. (1998) and Wang et al. (2004) |
| *Schizosaccharomyces pombe* | mRNA synthesis | Identification of GAPDH as part of RNA polymerase transcriptional complex | Binding to RNA polymerase II Rpb7 subunit | Kimura et al. (2002) and Mitsuzawa et al. (2005) |
| Human osteosarcoma cells, HeLa cells | Histone gene transcription | OCA-S[c] contains GAPDH | S phase-dependent GAPDH nuclear translocation | Zheng et al. (2003) |
| Human prostate cancer cell lines | Androgen receptor (AR) mediated gene transcription | GAPDH stimulation of AR transactivation | Cytoplasmic to nuclear GAPDH–AR complex translocation | Harada et al. (2007) |

[a]Chronological order.
[b]Promyelocytic leukemia.
[c]OCA-S: Oct-1 coactivator in S phase.

that this protein–protein interaction was dependent on the presence of RNA (Carlile et al., 1998). In this study, immunocytochemical analysis was used to demonstrate the nuclear colocalization of both proteins. Of interest, it appeared that this colocalization may be increased in cells with low passage numbers. This may be in accord with cell culture studies, demonstrating that aging of GAPDH may be dependent on its intracellular localization (Mazzola and Sirover, 2005).

Second, coimmunoprecipitation studies indicated the physical association of GAPDH and PML, which corroborated the colocalization studies. A kinetic analysis indicated that coimmunoprecipitation appeared to be independent of serum deprivation up to 9 h after serum withdrawal, an interval at which a slight diminution of binding was observed. No differences were observed in non–serum-deprived cultures.

Third, as it was hypothesized that the coimmunoprecipitation results might be due to the presence of a common element, the dependence of GAPDH–PML binding was examined as a function of RNase treatment. These investigations revealed not only a decrease in GAPDH and PML colocalization but also a reduction in their coimmunoprecipitation. There was no difference in GAPDH and PML immunoprecipitation by the corresponding antibodies. Accordingly, it was concluded that the RNA was required for GAPDH–PML binding. As will be discussed in a number of subsequent chapters, the formation of GAPDH tertiary complexes is not unusual with respect to moonlighting GAPDH functions.

The significance of the RNA requirement for GAPDH–PML binding was indicated by a subsequent study, which examined the association between transcriptionally active regions of the genome and PML nuclear bodies (Wang et al., 2004). The latter, which contains the PML protein, is an intranuclear structure implicated both in normal cell function and in cell pathology.

In these investigations, cytogenic protocols were used to interrelate the intranuclear location of genomic regions with that of PML bodies. This was an exhaustive series of studies in which loci from nine different chromosomes were examined as a function of transcription, and the proximity of each transcribed locus to a PML body was calculated. Fluorescence in situ hybridization with RNA probes were used to quantitate regions of genomic transcription Statistical analysis was then utilized to quantitate the localization of actively transcribed genomic regions in relation to that observed for PML bodies.

From this investigation it was concluded not only that PML bodies associated with genomic regions characterized by active transcription but also that there was a specific association of PML bodies with cell cycle–related genes. The latter included the chromosome 6p22 region, which contains histone genomic sequences. Further, it most noted that cells in S phase contained higher numbers of PML bodies than those in $G_0/G_1$ could be distinguished. Given these studies, the requirement of RNA for the GAPDH–PML protein–protein interaction (Carlile et al., 1998) would suggest an interrelationship between the latter and gene transcription.

## 2.2 GAPDH and Histone Gene Transcription

Modulation of gene expression is mediated by transcription factors that recognize upstream regulatory sequences or bind to the promoter region. Two such factors were identified as required for histone gene transcription–octomer binding protein 1 (OCT-1) and the OCT-1 Coactivator in S phase (OCA-S). The former is constitutively expressed while the latter functions as a transcriptional activation cofactor. Detailed analysis of the latter illuminated the role of moonlighting GAPDH in histone 2B (H2B) transcription (Zheng et al., 2003; comment by McKnight, 2003).

Those studies demonstrated that the size of purified OCA-S was c.300 kDa in size and that it was comprised of seven proteins. Peptide sequencing identified each of the seven proteins with the 38-kDa protein in the OCA-S complex being GAPDH. Initial mechanistic studies using in vitro translates of each protein in GST pull-down assays indicated that only the p38/GAPDH protein bound to the transcriptional regulatory domains POU-1 and POU-2. A similar result was observed using homogenous recombinant p38/GAPDH in the assay. These findings suggest a direct interaction of p38/GAPDH with OCT-1. It should be noted that this effect is observed with the p38/GAPDH protein. GAPDH in vivo is normally detected as a tetramer. Given the $M_r = c.300$ kDa of the OCA-S multiprotein complex, tetrameric GAPDH would be precluded as being part of OCA-S. As such, this may be another instance in which monomeric GAPDH exhibits a moonlighting function.

The requirement of p38/GADPH for H2B transcription in vitro was demonstrated by immunodepletion of the protein by an anti-GAPDH antibody. The resultant reduction in the levels of the p38/GAPDH resulted in a fourfold diminution in H2B transcription. Addition of the recombinant p38/GAPDH protein to the reaction restored full transcriptional activity. That being said, addition of commercially available GAPDH did not result in restoration of activity. As that protein is tetrameric in nature, that finding is another indication of the role of monomeric GAPDH in its moonlighting functions.

Complementary studies were performed in vivo using RNA interference protocols to abolish p38/GAPDH synthesis. This resulted not only in a time-dependent decrease in the level of nuclear p38/GAPDH but also a reduction in the extent of H2B mRNA synthesis. Further, this reduction in p38/GAPDH decreased the progression of the treated cells into S phase. As indicated in Chapter 1, immunofluorescent analysis indicated the nuclear localization of GAPDH in actively proliferating cells (Cool and Sirover, 1989).

In this study the cell cycle localization of the p38/GAPDH protein to the H2B promoter was examined. For ChIP analysis, cells were synchronized by centrifugal elutriation. DNA was obtained from antibody-precipitated chromatin, and PCR was performed. The results from this experiment indicated that the p38/GAPDH protein was localized to the H2B promoter only in S phase. In contrast, OCT-1, which is constitutively expressed, was localized to that promoter in all cell cycle phases. As these studies had demonstrated previously that there was a physical association between the p38/GAPDH protein and OCT-1, it was suggested that the S phase recruitment of the p38/GAPDH protein may be central to transcriptional regulation of the H2B gene.

Subsequently, nuclear extracts were prepared to determine the necessity of p38/GAPDH for H2B promoter activation as a function of the cell cycle. The experimental paradigm employed was antibody inhibition analysis of transcription. These studies demonstrated that incubation with the anti-p38/GAPDH antibody inhibited H2B transcription both in unsynchronized cells and cells in S phase. In contrast, the antibody had no effect on H2B transcription in $G_1$ or in $G_2$. As H2B transcription is markedly increased in S phase, it was suggested that there is a specific recruitment of the p38/GAPDH protein to the H2B promoter in S phase.

To complete this elegant examination of p38/GAPDH moonlighting function, the role of $NAD^+$ and NADH in this transcriptional mechanism was determined. Early studies had indicated the stimulatory and inhibitory effects, respectively, of each cofactor in GAPDH enzymatic studies (rev. in Sirover, 1999). In these studies, using the GST pull-down assay, addition of $NAD^+$ increased the binding of p38/GAPDH to POU-1. In contrast, decreased binding was detected when NADH was added to the reaction mixture. Similarly, $NAD^+$ enhanced p38/GAPDH promoter association while NADH inhibited that binding. Further, using a nuclear preparation to directly quantitate H2B transcription, a similar stimulatory/inhibitory effect of $NAD^+$ and NADH was observed, respectively. In toto, this detailed examination demonstrated rigorously, the role of moonlighting GAPDH in the regulation of H2B transcription.

A subsequent study examined the parameters through which the $NAD^+$/NADH ratio modulates H2B (Dai et al., 2008). A nuclear extract transcription system was utilized in which endogenous $NAD^+$ and NADH were removed by column chromatography, and the former was reconstituted. As expected, transcription was stimulated by $NAD^+$ and inhibited by NADH. However, titration of $NAD^+$ indicated a biphasic response, i.e., stimulation at low concentrations and inhibition at higher concentrations. Using microinjection of an H2B promoter-luciferase reporter, p38/GAPDH and $NAD^+$ into *Xenopus* oocytes, a similar biphasic response was observed. As noted, previously, there was some concern with respect to the role of GAPDH in transcription using the *Xenopus* oocytes (Morgenegg et al., 1986). Accordingly, these studies would seem to alleviate those concerns.

Immunodepletion of p38/GAPDH from the nuclear extract–based transcription paradigm indicated that its depletion inhibited binding of the other OCA-S components to the immobilized H2B promoter in a pull-down assay. Further, perturbation of $NAD^+$ biosynthesis by siRNA interference of the responsible enzymes reduced H2B expression. Accordingly, these studies serve to highlight the role both of $NAD^+$ itself or changes in the $NAD^+$/NADH ratio as a means to interrelate cellular redox status, moonlighting p38/GAPDH function, and the transcription regulation of histone synthesis.

## 2.3 GAPDH and the Structure of the Multiprotein RNA Polymerase II Transcription Complex

Detailed studies have indicated the primary role of RNA polymerase II in mRNA transcription, its multiprotein composition, and the identification of associated proteins within the transcriptional RNA polymerase protein complex. With respect to the former, 12 subunits were identified (Rpb1 to Rpb12; rev. in Cramer, 2004). With respect to the latter, GAPDH was identified as an associated protein (Kimura et al., 2002).

In a further examination, the binding of GAPDH to the Rpb7 subunit of RNA polymerase II was determined (Mitsuzawa et al., 2005). Again, the goal of this investigation was not to study GAPDH but to utilize a two-hybrid screening paradigm to detect proteins, which interact with Rpb7. This study identified 10 clones that encoded the C-terminal GAPDH domain as Rpb7-binding partners. To confirm those findings, an independent paradigm was used to determine the physical association of GAPDH with Rpb7. A crude cell extract was absorbed to a recombinant Rpb7 protein affinity column. Proteins that were eluted with 6 M urea were analyzed by silver staining after SDS-PAGE. GAPDH was so identified. Accordingly, using two independent protocols, moonlighting was identified as binding to the RNA polymerase II subunit Rbp7. That being said, the interrelationship between this physical interaction, the association of GAPDH with PML bodies, and its obligatory role as part of the OCA-S complex are currently unknown. However, in toto, these three cumulative studies indicate the complexity of GAPDH function in basic mechanisms of gene transcription.

## 2.4 GAPDH and Androgen Receptor–Mediated Gene Transcription

Identification and characterization of receptor-mediated cell signaling is one of the best examples not only of our understanding of normal cell function but also how that understanding translates into new and effective patient therapies. In that regard, the AR is intimately involved in prostate cell-gene transcription being activated by binding to its testosterone ligand. Recent evidence indicates that moonlighting GAPDH activity is a part of that complex series of cellular pathways linking receptor binding to transcriptional gene activation through its effect of AR transactivation (Harada et al., 2007).

In those studies, the role of GAPDH in AR-mediated prostate gene regulation was examined using both heterologous constructs and endogenous prostate cell transcription. With respect to the former, a prostate cancer cell line was cotransfected with a GAPDH-Myc-His construct and an AR construct linked to a luciferase reporter plasmid. As defined by increases in luciferase activity, increased AR transactivation was proportional to the concentration of the GAPDH construct. No increase in AR activity was observed in the absence of ligand. Further, this effect of GAPDH was detected in both $AR^+$ and in $AR^-$ prostate cancer cells.

An important control in these experiments was the validation of the heterologous construct studies with respect to endogenous genes under AR control, i.e., what was the effect of introducing the GAPDH construct on prostate-specific antigen (PSA) levels? Immunoblot analysis of whole cell lysates indicated increased expression of the PSA protein as a function of GAPDH lconstruct transfection. A further significant study examined the role of endogenous GAPDH on AR transactivation. In these studies, cells were transfected with GAPDH siRNA or control siRNA. Luciferase reporter activity was diminished in the former but not in the latter.

Subsequently, coimmunoprecipitation studies were performed to demonstrate the formation of an endogenous GAPDH–AR complex in vivo. Anti-AR IgG immunoprecipitates were analyzed by SDS-PAGE followed by immunoblot analysis revealing the selective immunoprecipitation of GAPDH. Subcellular fractionation studies indicated the presence of the GAPDH–AR complex in both the cytoplasm and in the nucleus. Although GAPDH does not contain a nuclear localization signal (NLS), an NLS is contained in the AR (Jenster et al., 1993). However, GAPDH does contain a nuclear export signal (Brown et al., 2004). Deletion of that GAPDH nuclear export signal resulted in the nuclear accumulation of GAPDH. Further, mutation of the GAPDH active site cysteine (C151S) abolished GAPDH enzymatic activity but had not effect of GAPDH induced AR transactivation.

An intriguing aspect of AR biology is the presence of CAG triplet repeats located within its N-terminal domain. As will be discussed in Chapter 12, GAPDH binding to expanded triplet repeat sequences are characteristic of a series of age-related neurodegenerative disorders. Recent studies indicate the interrelationship between the extent of AR transactivation, and the length of the CAG sequences within the AR gene and the resultant polyglutamine tracts in the AR protein (Chamberlain et al., 1994). In particular, expansion of the tract length in the AR protein, resulting in a decrease in the extent of transactivation. In contrast, deletion of CAG trinucleotide repeat sequences resulted in an increase in AR activity. This suggests that the CAG domain serves as a regulator of AR activity. Accordingly, it would be of interest to consider the interrelationship between GAPDH-mediated AR transcription activation and that regulated by the length of the triplet repeat sequence.

## 3. SUMMARY

As illustrated in the model presented in Fig. 2.1, the studies described in this chapter demonstrate the active role of moonlighting GAPDH in basic mechanisms of gene transcription. Early studies indicated for the first time the binding of GAPDH to single-stranded or denatured DNA. Later studies presented four instances in which moonlighting GAPDH was involved in transcriptional control. In the first three sections (GAPDH–PML body association; histone gene transcription and binding to the RNA polymerase II transcription complex), each

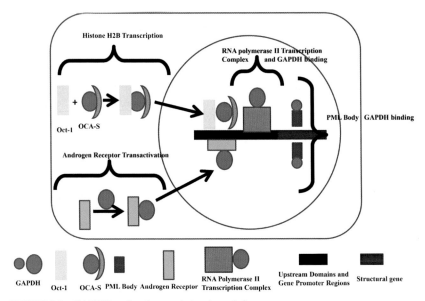

**FIGURE 2.1** GAPDH mediated transcriptional regulation.

study started with the premise that the proteins so identified were localized in the nucleus. This would include GAPDH although the mechanism through which it is transported to the nucleus is unknown. A potential model for GAPDH translocation with respect to histone gene transcription is presented in Fig. 2.1. Other mechanisms, some of which were discussed in Chapter 1, may be involved in other instances in which GAPDH nuclear translocation is required for its role in genomic transcriptional expression. One such mechanism involves the formation of the GADPH–AR complex, which permits the utilization of the NLS contained in the AR receptor to effect GAPDH cytoplasmic→nuclear translocation. This finding would appear to be reflective of a general principle of receptor-mediated cell signaling. In toto, each of the four investigations are not only complementary with those early studies but also, in concert with those investigations, illustrate the complexity of moonlighting GAPDH structure and function in basic mechanisms of gene transcription.

## REFERENCES

Alberts, B., Amodio, F., Jenkins, M., Gutmann, M., Ferris, F., 1968. Studies with DNA-cellulose chromatography. I. DNA-binding proteins from *Escherichia coli*. Cold Spring Harb. Symp. Quant. Biol. 33, 289–305.

Brown, V., Krynetski, E., Krynetskaia, N., et al., 2004. A novel CRM1-mediated nuclear export signal governs nuclear accumulation of glyceraldehyde-3-phosphate dehydrogenase following genotoxic stress. J. Biol. Chem. 279, 5984–5992.

Carlile, G., Tatton, W., Borden, L., 1998. Demonstration of a RNA-dependent nuclear interaction between the promyelocytic leukemia protein and glyceraldehyde-3-phosphate dehydrogenase. Biochem. J. 335, 691–695.

Chamberlain, N., Driver, E., Miesfeld, R., 1994. The length and location of CAG trinucleotide repeats in the androgen receptor N-terminal domain affect transactivation function. Nucleic Acids Res. 22, 3181–3186.

Cool, B., Sirover, M., 1989. Immunocytochemical localization of the base excision repair enzyme uracil DNA glycosylase in quiescent and proliferating normal human cells. Can. Res. 49, 3029–3036.

Cramer, P., 2004. Structure and function of RNA polymerase II. Adv. Protein Chem. 67, 1–42.

Dai, R.-P., Yu, F.-X., Goh, S.-R., et al., 2008. Histone 2B (H2B) expression is confined to a proper NAD⁺/NADH redox status. J. Biol. Chem. 283, 26894–296901.

Grosse, F., Nasheuer, H.-P., Scholtissek, S., Schomburg, U., 1986. Lactate dehydrogenase and glyceraldehyde-3-phosphate dehydrogenase are single-stranded DNA-binding proteins that affect the DNA-polymerase-α-primase complex. Eur J. Biochem. 160, 459–467.

Harada, N., Yasunaga, R., Higashimura, Y., et al., 2007. Glyceraldehyde-3-phosphate dehydroge-nase enhances transcriptional activity of androgen receptor in prostate cancer cells. J. Biol. Chem. 282, 22651–22661.

Jenster, G., Trapman, J., Brinkman, A., 1993. Nuclear import of the human androgen receptor. Biochem. J. 293, 761–768.

Kimura, M., Suzuki, H., Ishihama, A., 2002. Formation of a carboxy-terminal domain phos-phatase (Fcp1)/THIIF/RNA polymerase II (pol II) complex in *Schizosaccharomyces pombe* involves direct interaction between Fcp1 and the Rpb4 subunit of pol II. Mol. Cell. Biol. 22, 1577–1588.

Mazzola, J., Sirover, M., 2005. Aging of human glyceraldehyde-3-phosphate dehydrogenase is dependent on its subcellular localization. Biochim. Biophys. Acta 1722, 168–174.

McKnight, S., 2003. Gene switching by metabolic enzymes-how did you get on the invitation list? Cell 114, 150–152.

Melero, J., Salas, M., Salas, J., Macpherson, I., 1975. Deoxyribonucleic acid-binding proteins in virus-transformed cell lines. J. Biol. Chem. 250, 3683–3689.

Mitsuzawa, H., Kimura, M., Kanda, E., Ishihama, A., 2005. Glyceraldehyde-3-phosphate dehydro-genase and actin associate with RNA polymerase II and interact with its Rpb7 subunit. FEBS Lett. 579, 48–52.

Morgenegg, G., Winkler, G., Hübscher, U., et al., 1986. Glyceraldehyde-3-phosphate dehydrogenase is a nonhistone protein and a possible activator of transcription in neurons. J. Neurochem. 47, 54–62.

Perucho, M., Salas, J., Salas, M., 1977. Identification of the mammalian DNA-binding protein P8 as glyceraldehyde-3-phosphate dehydrogenase. Eur. J. Biochem. 81, 557–562.

Ronai, Z., 1993. Glycolytic enzymes as DNA binding proteins. Int. J. Biochem. 25, 1073–1076.

Salas, J., Green, H., 1971. Proteins binding to DNA and their relation to growth in cultured cells. Nat. New Biol. 229, 165–169.

Seal, G., Sirover, M., 1986. Physical association of the human base excision repair enzyme uracil DNA glycosylase with the 70,000 dalton catalytic subunit of DNA polymerase alpha. Proc. Natl. Acad. Sci. U.S.A. 83, 7608–7612.

Sirover, M., 1999. New insights into an old protein: the functional diversity of mammalian glyceral-dehyde-3-phosphate dehydrogenase. Biochim. Biophys. Acta 1432, 159–184.

Tsai, R., Green, H., 1973. Studies on a mammalian cell protein (P8) with affinity for DNA in vitro. J. Mol. Biol. 73, 307–316.

Wang, J., Shiels, C., Sasieni, P., et al., 2004. Promyelocytic leukemia nuclear bodies associate with transcriptionally active genomic regions. J. Cell Biol 164, 515–526.

Zheng, L., Roeder, R., Luo, Y., 2003. S phase activation of the histone H2B promoter by OCA-S, a coactivator complex that contains GAPDH as a key component. Cell 114, 255–266.

# Chapter 3

# The Diversity of Moonlighting GAPDH Function in Posttranscriptional RNA Regulation: mRNA Stability, tRNA Processing, and Viral Pathogenesis

*Sometimes you can't see the forest for the trees.*

An old idiom

The general theme which characterizes Section I is the surprising and unexpected role of moonlighting GAPDH in multiple processes required for normal cell function. In Chapter 1, evidence was presented to indicate the dynamic changes in GAPDH subcellular localization as an indicator of its moonlighting activity. In Chapter 2, evidence was presented which indicated the physiological significance of GAPDH in basic mechanisms of transcriptional gene expression. In this chapter, this general theme of moonlighting GAPDH activity will be expanded to include the role of GAPDH in posttranscriptional RNA structure and function.

The utilization of posttranscriptional RNA depends on complex cellular regulatory programs through which RNA transcripts are used to transmit coding information from the cellular genome into functional molecules. In particular, protein binding to nucleic acid-binding sites within 5'- and 3'-UTR mRNA sequences provides a basic mechanism to either increase or decrease mRNA stability or to affect translational efficiency. Further, the synthesis of ribosomal RNA and transfer RNA (tRNA) are a priori requirements for the translation of that information into functional proteins.

Recent studies indicate two major roles for moonlighting GAPDH in the structure and function of posttranscriptional RNA. These include its role in tRNA metabolism and in the determination of mRNA stability. The former relates to nuclear GAPDH function in the transport of newly transcribed tRNA molecules. The latter relates to the binding of cytoplasmic GAPDH to 3'-UTR

Glyceraldehyde-3-Phosphate Dehydrogenase (GAPDH). http://dx.doi.org/10.1016/B978-0-12-809852-3.00003-0

**35**

mRNA domains and the subsequent effect of that protein–nucleic acid inter-action on mRNA stability. Finally, there appears to be a significant role for moonlighting GAPDH in viral pathogenesis based on its binding to viral RNA sequences. The physiological significance of each GAPDH activity will be con-sidered. As such, it is hoped that the reader will gain insight into the overall role of moonlighting GAPDH in posttranscriptional RNA regulation, i.e., seeing the forest instead of the trees.

## 1. INITIAL IDENTIFICATION OF GAPDH–RNA INTERACTIONS

As indicated in Chapters 1 and 2, early studies had indicated changes in GAPDH subcellular localization and its capacity to bind to a single-stranded DNA, respectively. Similarly, initial findings indicated that GAPDH possessed an RNA-binding activity. As with the other studies, the initial focus was not on GAPDH but on another goal. In this case, it was to identify and characterize nucleic acid destabilizing proteins (Karpel and Burchard, 1981).

The experimental paradigm was to use sequential affinity chromatography through a series of RNA- and DNA-bound matrices to isolate helix-destabilizing proteins. Purification was monitored by SDS-PAGE. Two proteins of $M_r = 33\,kDa$ and $M_r = 34\,kDa$ were so isolated. Intriguingly, the $M_r$ of the native protein was c.160 kDa. Isoelectrofocusing revealed a $pI = 8.1$, indicating that the protein was moderately basic. Determination of amino acid composition identified the helix-destabilizing protein as GAPDH. This was confirmed by Ouchterlony double diffusion analysis using both the helix-destabilizing protein and commercial GAPDH with an antibody prepared against the helix-destabilizing protein. Enzymatic analysis demonstrated that the helix-destabilizing protein exhibited GAPDH catalytic activity equivalent to that observed for commercial GAPDH. RNA binding was confirmed using radiolabeled poly (U). Mechanistic studies indicated the ability of the protein to depress the $T_m$ of poly (A-U).

The observation of a $pI = 8.1$ suggested that the helix-destabilizing protein was a GAPDH isozyme. Although somatic GAPDH is encoded by a single structural gene (Bruns and Gerald, 1976; Bruns et al., 1979) and is transcribed in mammalian cells as a single transcript with no alternate splicing (Mezquita et al., 1998), many studies have indicated the presence of several isozymes, some of which may exhibit specific moonlighting functions (Susor et al., 1973; Lin and Allen, 1986; Ryzlak and Pietruszko, 1988; Glaser and Gross, 1995; Soukri et al., 1995).

Subsequently, RNA-binding proteins were characterized in rabbit reticulo-cytes (Ryazanov, 1985). In these studies, rRNA was immobilized on Sepharose 4B as an affinity protocol to determine which and how many proteins were bound using a ribosome-free cell extract. Three major RNA-binding proteins were so identified following absorption and elution. The $M_r = 36$-kDa protein was identified as GAPDH by one-dimensional peptide mapping as compared to the pattern observed for commercial GAPDH; by determination of GAPDH catalytic activity and by isoelectric focusing. The latter may be of particular interest as several isozymes were observed, in accord with the earlier study described above.

## 2. ROLE OF MOONLIGHTING GAPDH IN NUCLEAR TRANSFER RNA EXPORT

At times, it seems that tRNA is the "forgotten child" in the family of genetic information molecules. Although it fulfills perhaps the critical step in information transfer, i.e., linking the codons in mRNA to the amino acid which they specify, it, along with ribosomal RNA, tends to be overlooked as one probes the regulation and control of other, more "jazzy" cellular macromolecules. Nevertheless, the 20 odd species of tRNA must be transcribed from their respective genes and the so-produced RNA molecules need to be transported from the nucleus to the cytoplasm. Accordingly, the mechanisms through which those molecules are exported are of significance, and the proteins involved in that transport are of physiological importance.

To consider the latter, Singh and Green (1993) utilized an intriguing experimental paradigm that involved the use of two different tRNA species. The first was wild-type (wt) human initiator methionine tRNA human ($tRNA_i^{met}$); the second was a mutant $tRNA_i^{met}$ that contained a single base substitution (G57U). The significance of that change in base sequence was that it diminished nuclear export of the mutant tRNA species. Accordingly, the experimental approach was to isolate and characterize proteins that bound to the wt tRNA but not to the mutant.

In doing so, they identified a $M_r = 37$-kDa protein that bound to the wt tRNA but not to the mutant species. Tryptic peptide analysis identified the protein as GAPDH. Incubation with an anti-GAPDH antibody removed the tRNA–protein complex. Commercial GAPDH exhibited tRNA-binding activity. Further mutant tRNA studies were used to attempt to localize the GAPDH-binding site on the tRNA molecule. Deletions of several regions diminished binding with the greatest reduction observed in the T stem-loop which contained the G57U mutation.

Again, in what would be in the future a common finding (rev. in, Sirover, 1999, 2011, 2014), incubation with $NAD^+$ inhibited binding in a dose-dependent manner. This indicated the involvement of the GAPDH $NAD^+$ domain and the Rossmann fold contained therein in the GAPDH–tRNA interaction. However, the specificity of GAPDH binding was demonstrated by the inability of lactate dehydrogenase (LDH) to bind tRNA. LDH also contains the Rossmann fold. Therefore, other GAPDH amino acid sequences are required for tRNA binding.

In toto, these findings revealed a specific GAPDH–tRNA protein–nucleic acid interaction, which may be involved in tRNA nuclear export. This is illustrated in Fig. 3.1. Of note, GAPDH contains a unique nuclear export signal (Brown et al., 2004) and has been shown to be involved in nuclear membrane

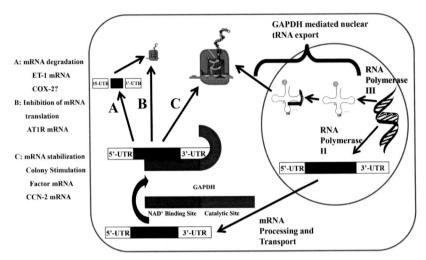

FIGURE 3.1    GAPDH-mediated mRNA regulation and tRNA transport. (A) mRNA degradation ET-1 mRNA COX-2; (B) inhibition of mRNA translation AT1R mRNA; (C) mRNA stabilization colony stimulation factor mRNA CCN2 mRNA.

formation (Nakagawa et al., 2003). Accordingly, the former may be utilized as a means to transport tRNA from the nucleus to the cytoplasm, while the latter activity may facilitate tRNA transport through the nuclear membrane. Further, in the previous chapter, evidence was presented with respect to the role of GAPDH in RNA polymerase II–mediated RNA transcription. As tRNA is transcribed by RNA polymerase III, these studies indicate again the generality of moonlighting GAPDH function.

## 3. ROLE OF MOONLIGHTING GAPDH IN THE DETERMINATION OF mRNA STABILITY

Modulation of mRNA stability and translation may be mediated by protein factors that bind to either the 5′-UTR or to the 3′-UTR base sequences. In particular, AU-rich regions of the latter may be a prime site for regulatory protein binding. As indicated in Table 3.1, a number of studies using diverse cell lineages identified GAPDH not only as a 3′-UTR-binding protein at AU-rich sites but also as a key protein regulating mRNA stability and translation in normal cell function. Studies on the role and mechanisms of GAPDH 3′-UTR colony-stimulating factor-1 in relation to ovarian cancer (Bonafe et al., 2005; Zhou et al., 2008) are discussed in Chapter 11.

### 3.1 GAPDH–Interferon-γ mRNA Binding

As with many GAPDH investigations, the first study was intended to identify those proteins that bound to AU-containing mRNA regions (Nagy and Rigby, 1995).

**TABLE 3.1 GAPDH Regulation of mRNA Expression in Normal Cell Function[a]**

| Organism/Cell Line | mRNA | Characteristics of 3′-UTR GAPDH Binding | "Unique" Finding | References |
|---|---|---|---|---|
| Spleen | GM-CSF[b] c-myc Interferon-γ (IFN-γ) | GAPDH binding strongest to three AUUUA reiterations | Polysomal localization | Nagy and Rigby (1995) |
| Preadipocytes | Glucose transporter-1 (GLUT-1) | Binding to region with highest percentage of A+U | Exposure to TNF[c] increases GAPDH binding | McGowan and Pekala (1996) |
| Umbilical endothelial cells | Endothelin-1 (ET-1) | Facilitation of mRNA destabilization | S-thiolation of GAPDH increases ET-1 mRNA stability | Rodríguez-Pascual et al. (2008) |
| Arterial smooth muscle cells, HEK293 cells | Angiotensin II type 1 receptor (AT1R) | GAPDH binding to secondary structure | GAPDH suppresses AT1R translation | Backlund et al. (2009) |
| Chondrocytes | Connective tissue growth factor (CCN2/CTGF) | Binding to unique 3′-UTR structure (CAESAR[d]) | Hypoxia increases binding | Kondo et al. (2011) |
| Hepatoma | Cyclooxygenase-2 (COX-2) | GAPDH downregulation increases COX-2 mRNA translation | Binding destabilizes mRNA secondary structure | Ikeda et al. (2012) |
| Monocytes | Tumor necrosis factor-α (TNF-α) | Repression of TNF-α protein levels | Metabolism regulates GAPDH–TNF-α ARE binding | Millet et al. (2016) |
| Patient blood samples, HEK293 cells | Sodium channel α-I subunit Naγ 1.1 (SCN1A) | 3′-UTR gene mutation forms a GAPDH-binding site | Binding to mutant reduces SCN1A stability | Zeng et al. (2014) |

[a]*Chronological order (except for Zeng et al., 2014).*
[b]*Granulocyte-macrophage colony-stimulating factor.*
[c]*Tumor necrosis factor.*
[d]*Cis-acting element of structure-anchored repression.*

Using binding to interferon-γ (IFN-γ) 3′-UTR as the experimental probe, a 36-kDa protein was sequentially purified from spleen cytoplasm. N-terminal sequence analysis identified it as GAPDH. This was confirmed using an antihuman placental GAPDH monoclonal antibody (Arenaz and Sirover, 1983). Binding studies indicated that the 36-kDa protein and rabbit muscle GAPDH bound to the 3′-UTR probe with the same extent.

The generality of GAPDH binding to 3′-UTR sequences was demonstrated by using radiolabeled granulocyte-macrophage colony-stimulating factor (GM-CSF) or c-myc probes. Intriguingly, the p36/GAPDH protein bound to each to a greater degree than that detected with IFN-γ. Further, binding assays indicated that the strongest amount of binding was observed when three AUUUA reiterations were present. Comparative analysis revealed that the p36/GAPDH protein bound to a greater extent with the IFN-γ probe than it did to tRNA.

Kinetic studies indicated the both $NAD^+$ and NADH reduced ($^{32}P$)UTP-radiolabeled IFN-γ 3′-UTR binding in a dose-dependent manner. It is of interest that inhibition was observed with both $NAD^+$ and NADH. In GAPDH studies, for the most part, the former is inhibitory, while the latter is not. It would be of interest to determine the $K_i$ for each as a further study indicated that the GAPDH-binding site was located within the GAPDH $NAD^+$ domain.

As described in Chapter 1, the subcellular localization of GAPDH is variable depending on the proliferative state of the cell. In these studies, cell fractionation analysis identified a polysomal localization for the p36/GAPDH protein in non-cycling T lymphocytes. Induction of cell proliferation by phytohemagglutinin resulted in an increase in this subcellular distribution. Studies using a transcriptional inhibitor (5,6-dichlorobenzimide riboside) indicated a further increase in the polysomal localization of the p36/GAPDH protein. Cumulatively, these first studies not only represented the identification of GAPDH as a 3′-UTR-binding protein but also demonstrated a specific translational-related subcellular localization.

## 3.2 GAPDH–GLUT-1 mRNA Binding

In a subsequent study, a general examination of dehydrogenase binding to 3′-UTR mRNA domains was performed (McGowan and Pekala, 1996). These studies used commercially available dehydrogenases and glucose transporter-1 (GLUT-1)–radiolabeled probes in RNA mobility shift assays as the experimental paradigm. Binding was observed with GAPDH as well as with several, but not all, of the other dehydrogenase proteins. The nature of dehydrogenase binding was analyzed further using GLUT-1 transcripts containing specific portions of its AU-rich regions. These studies demonstrated that the region containing the highest AU content was recognized the strongest by GAPDH. Intriguingly, using cell extracts, an increase in GAPDH–3′-UTR GLUT-1 binding was observed after exposure to tumor necrosis factor (TNF).

## 3.3 GAPDH and the Determination of Endothelin-1 mRNA Stability

Structure–function analyses indicated that many mRNA species contain what are termed destabilizing elements in their 3′-UTR sequences. These sequences, when activated, induce mRNA decay. Endothelin-1 (ET-1) mRNA stability was examined in human umbilical endothelial cells (Rodríguez-Pascual et al., 2008). Initially, luciferase reporter plasmids containing the 3′-UTR ET-1 region were used to demonstrate that the latter region controlled ET-1 mRNA stability. Those studies, by measuring decay constants and by mutational analysis, identified three AU-rich regions that mediated ET-1 mRNA decay.

Subsequently, RNA probes containing the AU ET-1 3′-UTR binding domains were used in affinity column chromatography to identify cellular binding proteins. Using this experimental paradigm, a 40-kDa protein was isolated and identified as GAPDH by peptide fingerprinting. Immunoblot analysis with an anti-GAPDH antibody confirmed its identity. Electrophoretic mobility shift assays (EMSAs) using commercial GAPDH demonstrated its binding to the RNA probe. Surprisingly, deletion experiments in which constructs contained only the GAPDH $NAD^+$-binding site or the GAPDH catalytic site demonstrated that neither site exhibited significant 3′-UTR ET-1 binding capacity. In contrast, constructs containing full-length GAPDH bound to the probe, indicating that the full-length protein was required. Knockdown of endogenous GAPDH by transfection of the corresponding siRNA reduced GAPDH levels by 50%. In contrast, ET-1 mRNA expression was increased by 150%. A similar result was observed using luciferase reporter constructs.

GAPDH contains an active site cysteine that can be modified under redox conditions. Chapter 8 considers its relationship to nitric oxide pathology in detail. GAPDH[cys152] is also subject to modification by glutathione. Accordingly, the effect of the active site cysteine on GAPDH–ET-1 binding was examined by two experimental paradigms. The first was to construct a mutant (C152S) and then compare its binding to wt GAPDH. That study demonstrated that such a substitution did not alter GAPDH–ET-1 binding. In contrast, GAPDH[cys152] modification with a nitric oxide donor as well as its S-glutathionylation diminished RNA binding. Studies in vivo using $H_2O_2$ induction of oxidative stress demonstrated an increase in ET-1 3′-UTR expression in HEK293 fibroblasts expressing GAPDH[cys152] but not in cells expressing GAPDH[ser152]. This would indicate that GAPDH S-glutathionylation may provide a mechanism to increase ET-1 expression by increasing ET-1 mRNA stability. As nitric oxide is involved in vascular regulation and as glutathione is also a redox-sensitive agent in vivo, it was suggested that such modification of GAPDH, thereby affecting ET-1 expression, may form a vascular constriction/dilation regulatory pathway.

Mechanistic studies were then utilized to define means through which ET-1 mRNA decay was induced by GAPDH binding to the AU regions of its 3′-UTR region. For these studies, a radiolabeled 3′-UTR ET-1 probe was incubated with cell extracts from siRNA GAPDH-silenced cells and from cells transfected with a siRNA control. As defined by the percentage of radiolabeled RNA remaining, the probe was degraded to a lesser extent in the GAPDH-silenced cells than in the siRNA control. In other words, in cells that contain GAPDH there is a greater degree of RNA loss than in GAPDH-depleted cells.

As it was noted that analysis of GAPDH indicated that it did not possess RNase activity, there would be presumably another mechanism through which GAPDH binding to the 3′-UTR ET-1 binding domain facilitates ET-1 mRNA decay. Analysis of the secondary structure indicated a significant degree of RNA duplex formation in the AU-rich 3′-UTR ET-1 region. This suggested that GAPDH binding would facilitate the unwinding of those duplexes creating susceptible sites for cleavage by cellular RNases (illustrated in Fig. 3.1, Mechanism A).

## 3.4 GAPDH and the Inhibition of AT1R mRNA Translation

The angiotensin II type 1 receptor (AT1R) functions in cell signaling especially with respect to the modulation of angiotensin cardiovascular function. Sequence analysis indicated the presence of an AU-rich 3′-UTR region, which may be involved in its posttranscriptional regulation. Further investigation demonstrated that GAPDH–3′-UTR AT1R binding may provide a mechanism that regulates its expression (Backlund et al., 2009).

In those studies, the role of protein–3′-UTR interactions in the regulation of AT1R mRNA expression was examined. Affinity chromatography was performed using probes containing different regions of the 3′-UTR AT1R domain. Using each of the constructs, a protein with $M_r = 36$ kDa was detected as the major binding protein. Trypsin digestion followed by mass spectroscopic analysis identified the protein as GAPDH. Immunoblot analysis using anti-GAPDH polyclonal antibodies confirmed that identity.

Subsequent studies characterized GAPDH–3′-UTR AT1R binding. The latter was demonstrated by an RNA electromobility shift assay (REMSA) using purified GAPDH (Sigma) and a 1–100 region of the 3′-UTR. Ribonucleoprotein immunoprecipitation analysis demonstrated the binding of endogenous GAPDH and AT1R mRNA as detected using an anti-GAPDH antibody and RT-qPCR with T1R-specific primers. Sequence analysis and computational modeling suggested a complex secondary structure for the 1–100 3′-UTR domain with a duplex AU-rich region that may comprise the GAPDH-binding site.

The functional consequences of that interaction were examined first in HEK293 cells using luciferase constructs, which contained the 3′-UTR AT1R GAPDH-binding site. GAPDH silencing resulted in a 2.5-fold increase in luciferase activity. In those cells, there was no dramatic effect on luciferase mRNA levels, indicating that there was no change in the rate of mRNA decay. That would suggest that GAPDH binding affected rates of translation. Using an in vitro translation paradigm, increasing GAPDH concentration in the reaction mixture, there was a noticeable decrease in luciferase protein synthesis. No change in protein synthesis was observed when a construct that did not contain the 3'-UTR AT1R GAPDH-binding site was used.

Silencing studies were performed also in coronary artery vascular smooth muscle cells to determine the effect on reducing endogenous GAPDH on AT1R expression. Using this protocol, an eightfold decrease in GAPDH mRNA was observed, while a concomitant increase in AT1R protein expression was detected. In toto, as illustrated in Fig. 3.1, these studies suggest a different mechanism through which GAPDH affects posttranscriptional mRNA expression, i.e., its binding inhibits mRNA translation (Fig. 3.1, mechanism B).

## 3.5 GAPDH and the Stability of *ccn2* mRNA

CCN2/GTGF (connective tissue growth factor) is an angiogenesis-related protein whose mRNA regulation is hypoxia dependent and may facilitate tumor cell growth. Hypoxia and the role of GAPDH therein is considered in detail in Chapter 9; the role of GAPDH in tumor development is considered in detail in Chapter 11. For that reason, in this chapter and in this section, the interaction between GAPDH and *ccn2* mRNA will be discussed. In particular, this relates to a unique region in the 3'-UTR of *ccn2* mRNA termed CAESAR (*cis*-acting element of structural-anchored repression), which comprises an 84-nucleotide-long section which is hypoxia responsive.

To consider protein–CAESAR interactions, affinity chromatography was performed using RNA probes that contained CAESAR (Kondo et al., 2011). A 35-kDa protein was identified. Tryptic digestion followed by MALDI-TOF-MS analysis identified the protein as GAPDH. As defined by REMSA, binding was observed when purified human erythrocyte GAPDH was substituted for the 35-kDa protein. Incubation with an antihuman GAPDH antibody abolished REMSA binding.

Subsequently, GAPDH–CAESAR binding was examined as a function of redox status. For this study, GAPDH was treated with either the oxidizing agent diamide or the reducing agent β-mercaptoethanol. Incubation with diamide eliminated GAPDH binding. Treatment with β-mercaptoethanol reversed that effect. Incubation with $NAD^+$ reduced GAPDH–CAESAR binding in a dose-dependent manner, indicating the role of the GAPDH $NAD^+$-binding site in this protein–nucleic acid interaction.

The functional significance of GAPDH–CAESAR binding was examined in chondrosarcoma cells under normal versus hypoxic conditions. Binding of GAPDH was increased in the latter as compared to the former. A model was proposed in which CAESAR regulates rates of *ccn2* mRNA decay. Under normal conditions, *ccn2* mRNA levels are low while they are increased under hypoxic conditions. For that reason, it was postulated that, during hypoxia, GAPDH binding may stabilize *ccn2* mRNA thereby increasing protein production (illustrated in Fig. 3.1, mechanism C).

## 3.6 GAPDH and the Destabilization of Cyclooxygenase-2 mRNA

As the formation of prostaglandin $G_2$ by cyclooxygenase-2 (COX-2) is fundamental to the inflammatory response, posttranscriptional regulation of COX-2 mRNA is, by definition, of importance. Intriguingly, analysis of its 3′-UTR region revealed the presence of seven AUUUA repeats within its initial 60 nucleotides, indicating its potential to serve as a site for mRNA regulation. The role of GAPDH as a 3′-UTR COX-2 mRNA-binding protein and the effect of that binding on COX-2 mRNA stability were examined in hepatoma cells (Ikeda et al., 2012). The experimental paradigm was to examine the effect of GAPDH COX-2 mRNA and protein levels following treatment with lipopolysaccharide (LPS). The latter is considered as the prototype endotoxin, eliciting an inflammatory response.

Initial studies focused on in vivo investigations determining the effect of siRNA GAPDH silencing on COX-2 mRNA and protein levels. These investigations demonstrated that downregulation of GAPDH resulted in a marked increase of both COX-2 mRNA and protein in LPS-treated cells (2.7- and 2.2-fold, respectively). It was suggested that elimination of GAPDH lessens COX-2 mRNA destabilization. RNA-EMSA studies using recombinant GAPDH and the 60-nucleotide COX-2 ARE probe demonstrated in the formation of a GAPDH–COX-2 ARE complex. Competition analysis using $NAD^+$ inhibited complex formation in a dose-dependent manner. This indicated that the GAPDH $NAD^+$-binding site was involved in this RNA interaction. However, the use of a GAPDH$^{C151S}$ mutant revealed that the active site cysteine was not required for the formation of the GAPDH–COX-2 ARE complex.

Accordingly, these studies demonstrate both the functional significance of moonlighting GAPDH in the regulation of COX-2 expression as well as the 60-nucleotide-long region of the COX-2 ARE site and the GAPDH $NAD^+$-binding region as the sites for the formation of this protein–nucleic complex. Molecular modeling studies indicated a high degree of secondary and loop structure in the 60-nucleotide segment. Each contained sufficient AU regions for GAPDH binding. It was hypothesized that such binding would result in GAPDH-mediated unwinding of the RNA permitting its degradation. This is illustrated in Fig. 3.1, mechanism A.

## 3.7 GAPDH and Posttranscriptional Repression of Tumor Necrosis Factor-α mRNA

TNF-α is a well-recognized and studied protein with respect to its role in inflammation. A recent study indicates not only that moonlighting GAPDH binds to its 3′-UTR but also that such binding may be part of a regulatory process, which modulates its synthesis (Millet et al., 2016). In particular, in this investigation the interrelationship between GAPDH–3′-UTR TNF-α mRNA binding to rates of galactose- or glucose-dependent glycolysis. The significance of this nutritional change is that the glycolysis is decreased in galactose-treated cells as compared to rates observed in glucose-treated cells.

Initial studies using a monocyte cell line demonstrated that levels of TNF-α protein but not the corresponding mRNA were decreased in galactose- versus glucose-treated cells. From this result, it was inferred that posttranscriptional repression was responsible for the decrease in protein levels. Subsequently, the binding of GAPDH to TNF-α mRNA was determined by RNA immunoprecipitation in LPS-stimulated galactose- or glucose-treated cells. These studies demonstrated not only that GAPDH bound to TNF-α mRNA but also that a greater extent of binding was observed in the galactose-treated cells.

Knockdown of GAPDH by siRNA silencing demonstrated that there was no effect of GAPDH depletion on TNF-α protein in glucose-treated cells. In contrast, a significant increase in protein levels was observed in galactose-treated cells. Sucrose density gradient analysis suggested that changes in polysomal localization were responsible for this finding, i.e., in glucose-treated cells a greater proportion of TNF-α mRNA was polysome associated as compared to that observed in galactose-treated cells. This observation reinforces the significance of the initial polysomal localization analysis (Nagy and Rigby, 1995).

To define that changes in glycolysis influenced GAPDH–TNF-α binding, glycolysis was blocked by the addition of 2-deoxyglucose, stimulated by the addition of a sirtuin 1 inhibitor or oligomycin and insulin, which increases glucose uptake. Inhibition of glycolysis resulted in an increase in the formation of the GAPDH–TNF-α complex, while its stimulation decreased the formation of the protein–nucleic acid complex. It was suggested that these findings demonstrate the inverse relationship between rates of glycolysis and GAPDH–TNF-α mRNA binding, i.e., lowering glycolysis increases binding, while increasing glycolysis decreases binding.

Mechanistic using constructs containing the 3′-UTR TNF-α region in a luciferase reporter system further indicated the role of metabolism in TNF-α posttranscriptional expression. Studies in primary human monocytes and macrophages indicated greater GAPDH–TNF-α binding in galactose-cultured cells demonstrating the physiological significance of the cell culture studies. In toto, these investigations confirm the hypothesis that reduced levels of glycolysis induce GAPDH binding to the 3′-UTR TNF-α mRNA region thereby preventing its translation. As such, these findings suggest that moonlighting GAPDH functions in a regulatory capacity with respect to TNF-α gene expression, modulating this cellular response to inflammation.

## 3.8 GAPDH and the Regulation of the Human *SCN1A* Gene mRNA

There is an old saying, "You learn more when things go wrong then you do when things go right." Regrettably, nowhere perhaps is that saying more appropriate than in human biology especially with respect to genetic disorders that predispose individuals to the respective affliction. From inborn errors of metabolism to cancer to neurological disorders, etc., analysis of those individuals have yielded insight into the role of mutations, biochemical pathways, and protein structure as the basis for the respective malady. In many cases, this has helped in the development of new treatment protocols to help alleviate the misery these individuals endure.

Dravet syndrome is a neurological disorder in which mutations in the voltage-gated sodium channel α-I subunit gene (SCN1A) appear to underlie this disease. Analysis of such mutations identified a 3′-UTR variant whose expression appeared to create a GAPDH-binding site (Zeng et al., 2014). Using a luciferase reporter gene paradigm, this variant resulted in a marked reduction in luciferase activity (c.70% of control). Transcriptional inhibition studies suggested that this reduction in activity was the result of an effect on mRNA stability. Subsequently, oligonucleotide probes containing the normal 3′-UTR sequence and the variant were used to determine whether there was a difference in protein binding based on that change in structure. Cell extracts from HEK293 cells were used. In the RNA-EMSA assay, a major shift was noted using the mutant construct which was absent using the wt probe. SDS-PAGE of the protein revealed an $M_r = 36.3 \, kDa$ suggesting the protein may be GAPDH. Studies using purified GAPDH yield a similar RNA-EMSA finding. No GAPDH binding was observed using the wt probe.

Knockdown analysis using shRNA resulted in a 60% decrease in GAPDH mRNA. As defined by RNA-EMSA, using HEK293 cell extracts from those cells revealed an 80% reduction in the formation of the GAPDH–variant 3′-UTR RNA complex. Further, analysis of luciferase activity in cotransfected HEK293 cells indicated that the observed reduction in activity was due to GAPDH binding. Using transcriptional inhibitors to quantitate mRNA stability indicated a decrease in the amount of variant mRNA in the cotransfected cells. Taken together, it was concluded that GAPDH was not only a negative regulator of *SCN1A* mRNA expression but also that its binding destabilized the mRNA. Regrettably, this human variant appears to be a "gain of function" through which moonlighting GAPDH regulates inappropriately the posttranscriptional expression of an important human gene (which is why it is italicized in Table 3.1, i.e., it is not a "normal" function).

## 4. ROLE OF MOONLIGHTING GAPDH IN THE TRANSCRIPTIONAL OR THE TRANSLATIONAL CONTROL OF VIRAL PATHOGENESIS

As indicated in the Contents, Section I is concerned with the role of moonlighting GAPDH in normal cell function. In contrast, Section II is intended to consider its role in human pathology. However, as moonlighting GAPDH may function to regulate RNA structure and function in viral pathogenesis through its binding to 3′-UTR, 5′-UTR, or other viral RNA structures, it seemed appropriate to include those studies in this chapter.

As indicated in Table 3.2, a number of studies defined GAPDH–ribonucleic acid interactions as fundamental to viral propagation. Of note, during those

**TABLE 3.2 GAPDH Regulation of RNA Expression in Viral Pathogenesis[a]**

| Virus | GAPDH-Binding Site | Moonlighting GAPDH Activity | "Unique" Finding | References |
|---|---|---|---|---|
| Hepatitis A | 5′-UTR (stem-loop IIIa) | GAPDH binding suppresses translation; colocalized in ribosomal salt wash | Unwinding of double-stranded sequences; competes with PTB[b] for RNA binding | Schultz et al. (1996), Yi et al. (2000) |
| Hepatitis A | 5′-UTR, 3′-UTR | GAPDH binds to three nucleic acid sequences extending into coding regions | Binding site: possible RNA pseudoknot or tRNA-like structure | Kusov et al. (1996), Dollenmaier and Weitz (2003) |
| Parainfluenza virus type 3 (HPIV3) | 3′-UTR | Colocalization with HPIV3 in perinuclear region | Phosphorylated GAPDH incorporated into virion | De et al. (1996), Choudary et al. (2000) |
| Hepatitis B | Posttranscriptional regulatory element (PRE) | GAPDH nuclear→cytoplasmic hepatitis B virus RNA transport | Cytoplasmic GAPDH does not bind to the PRE | Zang et al. (1998) |
| Hepatitis C | 3′-noncoding region (3′-NCR) | Binds to poly (U/UC) tract | Binds to 3′-NCR region located near the replicase complex assembly site | Petrik et al. (1999) |
| Hepatitis delta (HDV) | UC-rich domain (nucleotides 379–414) | GAPDH facilitates HDV ribozyme activity | HDV binding to cytoplasmic GAPDH induces its nuclear translocation | Lin et al. (2000) |
| HIV type 1 | GAPDH incorporated into human immunodeficiency virus type 1 (HIV-1) virions[c] | GAPDH binds to HIV-1 proteins Pr55[gag] and p160[gag-pol] | GAPDH inhibits LysRS[d] and tRNA[Lys3] packaging into virions | Kishimoto et al. (2012) |

[a]Chronological order.
[b]Polypyrimidine tract–binding protein.
[c]No direct RNA binding.
[d]Lysyl-tRNA synthetase.

investigations, it became apparent that the role of RNA secondary structure was not only a most important parameter in determining GAPDH binding but also it was of significance with respect to the consequences of GAPDH–viral ribonucleic complex formation. In addition, the generality of moonlighting GAPDH activity is illustrated by its interrelationship with the pathogenesis of each virus although it has its own unique RNA structure and function.

## 4.1  GAPDH Binding to Hepatitis A Virus RNA

This was illustrated initially in studies examining the role of GAPDH in binding to hepatitis A virus (HAV) 5′-UTR or 3′-UTR mRNA regions. One such investigation focused on the proteins that bound to the IIIa stem-loop structure contained in the HAV 5′-UTR (Schultz et al., 1996) contained within the internal ribosomal entry segment (IRES). Two cell sources were examined: the ribosomal salt wash (RSW) and total cytoplasmic cell extracts. RNA–protein complexes were detected by electrophoresis in nondenaturing gels. Identical results were observed using either protein extract. Gel filtration analysis of the binding proteins indicated an $M_r = 150\,kDa$. In contrast, UV cross-linking studies indicated an $M_r = 30$ and $39\,kDa$. Thus, it was suggested that those proteins were present as high molecular weight forms.

Purification of the protein (monitored by EMSA) resulted in the isolation of the 39-kDa protein. Protein analysis identified it as GAPDH. Confirmation of this finding included immunoblot analysis of the 39-kDa protein and human GAPDH using antihuman placental GAPDH monoclonal antibody 40.10.09 (Arenaz and Sirover, 1983). That study demonstrated equivalent recognition of both proteins by the antibody. Substitution of human GAPDH for the 39-kDa protein in UV cross-linking studies demonstrated that it also bound to stem-loop IIIa RNA. Incubation with the 40.10.09 antibody diminished complex formation. Similarly, $NAD^+$ inhibited GAPDH binding in a dose-dependent manner, indicating the role of the GAPDH $NAD^+$ domain in binding. It should be noted that the presence of GAPDH in the RSW extract is in accord with the previously reported GAPDH polysomal localization (Nagy and Rigby, 1995).

Circular dichroism studies indicated that GAPDH binding to the stem-loop IIIa RNA resulted in a decrease in a reduction in intensity which is indicative of unwinding of the duplex regions contained in the RNA. The magnitude of this effect was GAPDH concentration dependent and resembled that seen on thermal denaturation of the RNA. Stoichiometric studies suggest a 4:1 molar ration of GAPDH monomer required for a half-maximal change in RNA structural perturbation, corresponding to a 1:1 ratio for tetrameric GAPDH.

Further analysis revealed an intriguing dynamic with respect to stem-loop IIIa–GAPDH binding. The polypyrimidine tract (PTB) protein, a potential translation initiation factor, binds to regions similar to that observed for GAPDH. Initial studies indicated that PTB could compete with GAPDH for stem-loop IIIa–GAPDH binding. At equal concentrations, PTB reduced GAPDH binding while, when a 10-fold excess of GAPDH was used, little reduction of PTB binding was detected, suggesting that its affinity for the stem-loop IIIa region was greater than that characteristic of GAPDH.

To define the role of GAPDH in HAV transcriptional regulation, an intriguing construct was prepared that comprised the HAV IRES with a promoter and GAPDH coding sequence upstream and a luciferase reporter gene downstream (Yi et al., 2000). In this dicistronic construct, GAPDH is overexpressed in transfected cells while luciferase activity is a measure of IRES activity. As a control, a GAPDH null mutant was constructed which had a frameshift mutation in GAPDH.

In cells transfected with the wt GAPDH dicistronic construct, luciferase reporter activity was twofold lower than that observed with null mutant transfected cells. This demonstrated that HAV IRES activity was inhibited by GAPDH indicating not only the functional significance of GAPDH–stem-loop IIIa binding but also the unwinding of the latter as the mechanism underlying that inhibitory activity. However, in a cell line characterized by high levels of PTB, IRES repression was reduced. Further, transfection of comparable dicistronic PTB constructs demonstrated a fourfold increase in luciferase activity. Taken together, these findings suggest that IRES activity (and thus HAV propagation) may be controlled by the balance between GAPDH and PTB in virus-infected cells.

In a parallel study, HAV 3′-UTR RNA–protein interactions were examined using a variety of human and monkey cell lines (Kusov et al., 1996). Using radiolabeled RNA probes, a series of proteins were identified using both total cell free extracts and a RSW. The most prevalent was a p38 protein. The latter was observed in all cell lines which were examined, indicating the generality of that observation. Binding of the p38 protein was also observed using 5′-UTR HAV constructs. In these investigations, it was noted, using computer analysis, that the 3′-UTR HAV constructs contained significant secondary structures which included significant double-stranded and single-stranded AU-rich segments.

Subsequently, the p38 protein was identified as GAPDH (Dollenmaier and Weitz, 2003). The experimental paradigm was to use UV cross-linking of radiolabeled RNA probes to either purified human erythrocyte GAPDH or to cytoplasmic cell extracts and then to immunoprecipitate proteins with antihuman placental GAPDH monoclonal antibody 40.10.09 (Arenaz and Sirover, 1983). These studies identified the same p38 protein in the immunoprecipitate. Competition analysis with $NAD^+$ and with ATP demonstrated reduction in GAPDH–RNA binding. The former is in accord with those of previous HAV studies and with those described in Table 3.1.

The RNA-binding domains of the GAPDH protein were identified using a variety of HAV RNA probes. Some of the latter included sequences from the upstream $3D^{pol}$ coding region as well as truncated deletion mutants. These studies identified three nonoverlapping GAPDH-binding domains which were termed α, β, and γ. It was noted that the α and β domains included part of the 3D$^{pol}$ coding region. Further analysis revealed that deletion of the poly (A) tail reduced GAPDH binding. Computer analysis was again used to model the RNA-binding site, which indicated the possibility of a pseudoknot structure.

## 4.2 GAPDH and the Regulation of Human Parainfluenza Virus Type 3 RNA Expression

As with HAV, the 3′-UTR region of human parainfluenza virus type 3 (HPIV3) comprises a regulatory sequence affecting viral propagation. To define proteins which may bind to that region, a cDNA construct was prepared comprising the first 73 nucleotides from the 3′-UTR HPIV3 region under the control of a T7 promoter which was transfected into a monkey kidney cell line, CV-1 (De et al., 1996). Again, computer program analysis indicated a complex structure containing both AU-rich duplex regions and an AU-rich loop sequence. Using REMSA, UV cross-linking, and purification by column chromatography, a 37-kDa protein was purified to near homogeneity. Microsequence analysis identified it as GAPDH. Immunoblot analysis with an anti-GAPDH antibody and the use of a commercially obtained GAPDH in the RNA-binding assay confirmed that identity.

Mechanistic studies demonstrated that an excess of poly (U) inhibited RNA binding as did NAD$^+$ as defined by competition studies. The specificity of GAPDH binding was indicated by the failure of glucose-6-phosphate dehydrogenase and lactate dehydrogenase to bind to the RNA probe. This indicates again that, although the Rossmann fold is involved in the GAPDH–RNA interaction, it, by itself, is insufficient and other portions of the GAPDH protein are necessary.

The in vivo association of GAPDH with HPIV3 was demonstrated by coimmunoprecipitation of both by the anti-GAPDH antibody in infected cells. Further, colocalization studies were performed by double immunofluorescence labeling and confocal microscopy. This indicated their perinuclear association which was in accord with previous studies demonstrating the presence of GAPDH in that intracellular locale (Cool and Sirover, 1989). As a control, the intracellular analysis of tubulin was examined. No colocalization was observed.

In a subsequent investigation, structure–function properties of HPIV3-associated GAPDH were examined (Choudhary et al., 2000). Two-dimensional (2-D) electrophoresis indicated the presence of three GAPDH isoforms (pI 7.6–8.3). In contrast, in CV-1 cells, a single species with a pI of 8.4 was observed. To determine the localization and nature of these GAPDH species, CV-1 cells were infected with HPIV3 and the cells were radiolabeled with ($^{32}$P)orthophosphate. Released virions were harvested and analyzed by 2-D gel electrophoresis.

Analysis of GAPDH within the virion indicated the presence of the three GAPDH isoenzymes. Intriguingly, autoradiographic analysis indicated the presence of multiple $^{32}$P-labeled GAPDH isozymes. Analysis of rabbit muscle

GAPDH phosphorylation in vitro by rat brain protein kinase C indicated multiple $^{32}$P-labeled species as defined by 2-D gel electrophoresis. Purification of phosphorylated cell cytoplasmic GAPDH indicated that it inhibited viral transcription in a dose-dependent manner. Similar results were observed using $^{32}$P-radiolabeled recombinant GAPDH. These results were the initial report not only that a specific GAPDH isozyme may be involved in the recognition and regulation of HPIV3 RNA but also that GAPDH was packaged within the virion itself.

## 4.3 GAPDH Binding to the Posttranscriptional Regulatory Element of Hepatitis B Virus

The hepatitis B virus (HBV) posttranscriptional regulatory element (PRE) is a rather unique structure that appears to serve a transport function for the nuclear to cytoplasmic export of viral mRNA. As with many viruses which contain limited genetic information, cellular factors need to be recruited to facilitate propagation. Using PRE constructs, REMSA followed by UV cross-linking was used to isolated two such nuclear proteins with an $M_r = 30$ and to 45 kDa, respectively, which bound to the PRE (Huang et al., 1996). These proteins were named PRE-interacting protein-1 (PIP-1) and PRE-interacting protein-2 (PIP-2).

Intriguingly, PIP-1 and PIP-2 appeared to be localized solely in the nucleus as no binding was observed using cytoplasmic extracts. Analysis using five different segments of the 560-nucleotide-long PRE prepared by restriction enzyme digestion indicated that binding was strongest using PRE fragment III. However, binding, although reduced, was observed with other fragments as well, suggesting the presence of multiple protein binding sites within the PRE.

The physiological significance of this cellular protein–HMV nucleic acid interaction was examined using a chloramphenicol acetyltransferase (CAT) reporter construct into which these discrete PRE segments were inserted. The most significant CAT expression was observed with the insertion of six fragment III segments, again suggesting the presence of multiple protein-binding sites. Insertion of the fragment into a plasmid containing the entire HBV-transcribed region demonstrated the cytoplasmic localization of the HBV transcripts in transfected cells, thereby illustrating the transport function of PRE.

Further analysis identified PIP-1 as GAPDH (Zang et al., 1998). Using COS cell extracts, PIP-1 was purified by standard chromatographic protocols using REMSA as the identification mechanism. N-terminal sequencing of the purified protein indicated its homology to GAPDH. This identification was confirmed using commercial GAPDH in the fragment III PRE-binding assay. Competition analysis using NADH demonstrated reduced GAPDH–PRE binding, which occurred in an NADH dose-dependent manner. Immunofluorescence analysis demonstrated the nuclear localization of GAPDH. Finally, as further confirmation, nuclear extracts UV cross-linked to radiolabeled fragment III were immunoprecipitated by an anti-GAPDH antibody. The immunoprecipitated protein comigrated with commercial GAPDH UV cross-linked to fragment III. In toto, these studies not only identify PIP-1 as GAPDH but also demonstrate that cytoplasmic GAPDH is unable to bind to the HBV PRE. This suggests that a unique isozyme may be present in cell nuclei.

## 4.4 GAPDH Binding to the 3′-Noncoding Region of the Hepatitis C Virus

As with HBV, hepatitis C virus (HCV) contains its own unique structure, a three-part 3′ noncoding region (NCR) containing a poly (U/UC) tract. To determine cellular factors that bind to that region, an immobilized ribonucleotide $r(U)_{25}$ was used as a probe to isolate liver cell proteins that were candidate HCV 3′-NCR-binding protein (Petrik et al., 1999).

This analysis identified six proteins present in both control and HCV-positive liver explants. A major 36 kDa binding protein was examined further. It appeared to be specific to poly (U) although a 50-fold excess of other polyribonucleotides could eliminate binding. N-terminal sequence analysis identified the protein as GAPDH. Its HCV-binding site was determined using a series of constructs containing variable regions of the HCV 3′-NCR. Those studies indicated a selective binding to what was termed 3′-$NCR_2$, which was confirmed by recognition of the protein by an anti-GAPDH antibody. It was noted that the 3′-$NCR_2$ region was located in direct proximity to the replicase complex assembly site. Attention was directed to the previously reported GAPDH RNA helix-destabilizing activity suggesting that its binding to the 3′-$NCR_2$ region could result in the facilitation of replication based on the activity.

## 4.5 GAPDH Enhancement of Ribozyme Catalysis in the Hepatitis Delta Virus

As indicated in the Introduction to this section, each of the viruses included in Table 3.2 has its own unique RNA characteristics, which distinguish each from the other. Hepatitis D virus (HDV) is no exception as it contains its own ribozyme activity. Yet they have one thing in common, i.e., moonlighting GAPDH is involved in their pathogenesis.

Studies in HeLa cells identified several HDV-binding proteins as determined by UV cross-linking with full-length HDV RNA (Lin et al., 2000). A p36 protein was noted as a major binding protein. It was purified by standard chromatographic protocols using EMSA analysis as the detection method. Protein sequence analysis identified it as GAPDH. Utilization of commercial GAPDH as well as recognition of the p36 protein by an anti-GAPDH monoclonal antibody confirmed its identity. Subsequently, deletion mapping analysis and RNase footprinting studies identified the GAPDH-binding site as a 79-nucleotide-long highly UC-rich nucleotide sequence. Competition experiments indicated that complex formation could be reduced by poly (U) and by poly (A). In situ hybridization/immunofluorescence studies indicated the nuclear colocalization of dimeric HDV cDNA with GAPDH. These studies also suggested a recruitment of cytoplasmic GAPDH into the nucleus by the HDV construct. As previously mentioned in Chapter 1, GAPDH lacks a nuclear localization signal and, therefore, other mechanisms must exist to permit its nuclear translocation. As such, these studies may represent the first example of a viral mechanism for GAPDH redistribution.

Functional analysis indicated that GAPDH increased HDV RNA ribozyme activity. In particular, incubation with GAPDH increased cleavage activity of the HDV ribozyme by twofold. Previous studies demonstrated that GAPDH increased the cleavage rates of hammerhead ribozymes based on its unfolding activity (Sioud and Jespersen, 1996). In this instance, the HDV RNA structure is a double pseudoknot in nature. Presumably, GAPDH can affect that secondary configuration as well.

## 4.6 GAPDH and the Regulation of Human Immunodeficiency Virus Type 1 Infection

As described above, with each of the five viruses, GAPDH has a defined interaction with viral RNA. As such, its role in human immunodeficiency virus type 1 (HIV-1) infection may be the "exception which proves the rule," i.e., GAPDH does not interact with viral RNA but does affect viral pathogenesis through other interactions (Kishimoto et al., 2012).

In this study, analysis of the HIV-1 virion indicated not only that GAPDH was packaged within the virion but also that the virion contained multiple GAPDH isozymes. In particular, five GAPDH species were detected by 2-D electrophoresis and immunoblot analysis. The pIs ranged from 7.6 to 9.02. In comparison, it was noted that six cellular isozymes were detected with an additional species with a pI of 8.68. As such, the virion studies are in agreement with those previously discussed with respect to distinct GAPDH isozymes contained within the HPIV3 virion (Choudhary et al., 2000).

The role of virion-associated GAPDH was examined using siRNA protocols. Such virions did not display a decrease in viral release from infected cells nor did they demonstrate any decrease in the incorporation of viral genomic RNA into viral particles. In contrast, surprisingly, GAPDH-deficient HIV-1 virions exhibited an increase in infectivity, suggesting an inhibitory GAPDH function. This did not appear to be based on the inhibition of reverse transcriptase activity as its quantitation in vitro was not affected by recombinant GAPDH.

Analysis of components packaged into the HIV-1 virion indicated a 1.5-fold enhancement of lysine tRNA synthetase (lysRS) and a fourfold increase of tRNA$^{Lys3}$ into GAPDH-deficient virions. Each is normally packaged into the virion and each is necessary for pathogenesis. The former requires the viral Pr55$^{gag}$ protein for packaging, while the latter requires p160$^{gag-pol}$. Immunoprecipitation studies with an anti-GAPDH antibody demonstrated the physical interaction of GAPDH with both viral proteins. No interaction of GAPDH with lysRS was detected. Overexpression of GAPDH increased virion GAPDH levels, decreased tRNA$^{Lys3}$ packaging, and decreased viral infectivity. In toto, these findings suggest that GAPDH–HIV-1 protein interactions regulate lysRS and tRNA$^{Lys3}$ incorporation in virions, thereby determining the levels of HIV-1 infectivity.

## 5. SUMMARY

The studies presented in this chapter demonstrate the complexity and diversity of GAPDH–RNA interactions in posttranscriptional RNA regulation. They also demonstrate the significance of RNA secondary structure changes as a basic mechanism of GAPDH activity. These investigations indicate the physiological significance of GAPDH as a regulator of posttranscriptional expression in normal cell function. The GAPDH–viral RNA investigations are indicative of the importance of moonlighting GAPDH–RNA interactions in relation to human pathology. As such, each provides evidence with respect to the fundamental role of GAPDH in RNA structure and function.

## REFERENCES

Arenaz, P., Sirover, M., 1983. Isolation and characterization of monoclonal antibodies directed against the DNA repair enzyme uracil DNA glycosylase from human placenta. Proc. Natl. Acad. Sci. U.S.A. 80, 5822–5826.

Backlund, M., Paukku, K., Daviet, L., et al., 2009. Posttranscriptional regulation of angiotensin II type 1 receptor expression by glyceraldehyde-3-phosphate dehydrogenase. Nucleic Acids. Res. 26, 2346–2358.

Bonafe, N., Gilmore-Hebert, M., Folk, N., Azodi, M., Zhou, Y., et al., 2005. Glyceraldehyde-3-phosphate dehydrogenase binds to the AU-rich 3′ untranslated region of colony-stimulating factor-1 (CSF-1) messenger RNA in human ovarian cancer cells: possible role of CSF-1 posttranscriptional regulation and tumor phenotype. Cancer Res. 65, 3762–3771.

Brown, V., Krynetski, E., Krynetskaia, N., Griegeri, D., et al., 2004. A novel CRM-1 mediated nuclear export signal governs nuclear accumulation of glyceraldehyde-3-phosphate dehydrogenase following genotoxic stress. J. Biol. Chem. 279, 5984–5992.

Bruns, G., Gerald, P., 1976. Human glyceraldehyde-3-phosphate dehydrogenase in man-rodent somatic cell hybrids. Science 192, 54–56.

Bruns, G., Gerald, P., Lalley, P., Francke, U., Minna, J., 1979. Gene mapping of the mouse by somatic cell hybridization. Cytogenetics 22, 139.

Choudhary, S., De, B., Banerjee, A., 2000. Specific phosphorylated forms of glyceraldehyde-3-phosphate dehydrogenase associate with human parainfluenza virus type 3 and inhibit viral transcription in vitro. J. Virol. 74, 3634–3641.

Cool, B., Sirover, M., 1989. Immunocytochemical analysis of the base excision repair enzyme uracil DNA glycosylase in quiescent and proliferating normal human cells. Cancer Res. 49, 3029–3036.

De, B., Gupta, S., Zhao, H., Drazba, J., Banerjee, A., 1996. Specific interactions in vitro and in vivo of glyceraldehyde-3-phosphate dehydrogenase and LA protein with cis-acting RNAs of human parainfluenza virus type 3. J. Biol. Chem. 271, 24728–24735.

Dollenmaier, G., Weitz, M., 2003. Interaction of glyceraldehyde-3-phosphate dehydrogenase with secondary and tertiary RNA structural elements of the hepatitis A virus 3′ translated and non-translated regions. J. Gen. Virol. 84, 403–414.

Glaser, P., Gross, R., 1995. Rapid plasmenylethanolamine-selective fusion of membrane bilayers catalyzed by an isoform of glyceraldehyde-3-phosphate dehydrogenase: discrimination between glycolytic and fusogenic roles of individual isoforms. Biochemistry 34, 12194–12203.

Huang, Z.-M., Zang, W.-Q., Yen, T., 1996. Cellular proteins that bind to the hepatitis B virus post-transcriptional regulatory element. Virology 217, 573–581.

Ikeda, Y., Yamaji, R., Irie, K., Kioka, N., Murakami, A., 2012. Glyceraldehyde-3-phosphate dehydrogenase regulates cyclooxygenase-2 expression by targeting mRNA stability. Arch. Biochem. Biophys. 528, 141–147.

Karpel, R., Burchard, A., 1981. A basic isozyme of yeast glyceraldehyde-3-phosphate dehydrogenase with nucleic acid helix-destabilizing activity. Biochim. Biophys. Acta 654, 256–267.

Kishimoto, N., Onitsuka, A., Kido, K., et al., 2012. Glyceraldehyde-3-phosphate dehydrogenase negatively regulates human immunodeficiency virus type 1 infection. Retrovirology 9, 107.

Kondo, S., Kubota, S., Mukudai, Y., et al., 2011. Binding of glyceraldehyde-3-phosphate dehydrogenase to the cis-acting element of structure-anchored repression in ccn2 mRNA. Biochem. Biophys. Res. Commun. 405, 382–387.

Kusov, Y., Weitz, M., Dollenmeier, G., Gauss-Müller, V., Siegl, G., 1996. RNA-protein interactions at the 3′ end of the hepatitis A virus RNA. J. Virol. 70, 1890–1897.

Lin, T., Allen, R., 1986. Isolation and characterization of a 37,000-dalton protein associated with the erythrocyte membrane. J. Biol. Chem. 261, 4594–4599.

Lin, S.-S., Chang, S., Wang, Y.-H., Sun, C.-Y., Chang, M.-F., 2000. Specific interaction between the hepatitis delta virus RNA and glyceraldehyde-3-phosphate dehydrogenase: an enhancement on ribozyme catalysis. Virology 271, 46–57.

McGowan, K., Pekala, P., 1996. Dehydrogenase binding to the 3-untranslated region of GLUT1 mRNA. Biochem. Biophys. Res. Commun. 221, 42–45.

Mezquita, J., Pau, M., Mezquita, C., 1998. Several novel transcripts of glyceraldehyde-3-phosphate dehydrogenase expressed in adult chicken testis. J. Cell. Biochem. 71, 127–139.

Millet, P., Vachharajani, V., McPhail, L., Yoza, B., McCall, C., 2016. GAPDH binding to TNF-α mRNA contributes to posttranscriptional repression in monocytes: a novel mechanism of communication between inflammation and metabolism. J. Immunol. 196, 2541–2551.

Nagy, E., Rigby, W., 1995. Glyceraldehyde-3-phosphate dehydrogenase selectively binds AU-rich RNA in the NAD$^+$-binding region (Rossmann fold). J. Biol. Chem. 270, 2755–2763.

Nakagawa, T., Hirano, Y., Inomata, A., et al., 2003. Participation of a fusogenic protein, glyceraldehyde-3-phosphate dehydrogenase, in nuclear membrane assembly. J. Biol. Chem. 278, 20395–20404.

Petrik, J., Parker, H., Alexander, G., 1999. Human hepatic glyceraldehyde-3-phosphate dehydrogenase binds to the poly(U) tract of the 3′ non-coding region of hepatitis C virus genomic RNA. J. Gen. Virol. 80, 3109–3113.

Rodríguez-Pascual, F., Rendondo-Horcajo, M., Magán, N., et al., 2008. Glyceraldehyde-3-phosphate dehydrogenase regulates endothelin-1 expression by a novel, redox-sensitive mechanism involving mRNA stability. Mol. Cell. Biol. 28, 7139–7155.

Ryazanov, A., 1985. Glyceraldehyde-3-phosphate dehydrogenase is one of the three major RNA-binding proteins of rabbit reticulocytes. FEBS Lett. 192, 131–134.

Ryzlak, M., Pietruszko, R., 1988. Heterogeneity of glyceraldehyde-3-phosphate dehydrogenase from human brain. Biochim. Biophys. Acta 954, 309–324.

Schultz, D., Hardin, C., Lemon, S., 1996. Specific interaction of glyceraldehyde-3-phosphate dehydrogenase with the 5′-nontranslated RNA of hepatitis A virus. J. Biol. Chem. 14134–14142.

Singh, R., Green, M., 1993. Sequence-specific binding of transfer RNA by glyceraldehyde-3-phosphate dehydrogenase. Science 259, 365–368.

Sioud, M., Jespersen, L., 1996. Enhancement of hammerhead ribozyme catalysis by glyceraldehyde-3-phosphate dehydrogenase. J. Mol. Biol. 257, 775–789.

Sirover, M., 1999. New insights into an old protein: the functional diversity of mammalian glyceraldehyde-3-phosphate dehydrogenase. Biochim. Biophys. Acta 1432, 159–184.

Sirover, M., 2011. On the functional diversity of glyceraldehyde-3-phosphate dehydrogenase: biochemical mechanisms and regulatory control. Biochem Biophys Acta 1810, 741–751.

Sirover, M., 2014. Structural analysis of glyceraldehyde-3-phosphate dehydrogenase functional diversity. Int. J. Biochem. Cell Biol. 57, 20–26.

Soukri, A., Valverde, F., Hafid, N., Elkebbaj, M., 1995. Characterization of muscle glyceraldehyde-3- phosphate dehydrogenase isoforms from euthermic and induced hibernating *Jaculus orientalis*. Biochim. Biophys. Acta 1243, 161–168.

Susor, W., Kochman, M., Rutter, W., 1973. Structure and determination of FDF aldolase and the fine resolution of some glycolytic enzymes by isoelectric focusing. Ann. N.Y. Acad. Sci. 209, 328–344.

Yi, M., Schultz, D., Lemon, S., 2000. Functional significance of the interaction of hepatitis A virus RNA with glyceraldehyde-3-phosphate dehydrogenase (GAPDH): opposing effects of GAPDH and polypyrimidine tract binding protein on internal ribosome entry site function. J. Virol. 74, 6459–6468.

Zang, W.-Q., Fieno, A., Grant, R., Yen, T., 1998. Identification of glyceraldehyde-3-phosphate dehydrogenase as a cellular protein that binds to the hepatitis B virus posttranscriptional regulatory element. Virology 248, 46–52.

Zeng, T., Dong, Z.-F., Liu, S.-J., et al., 2014. A novel variant in the 3′UTR of human SCN1A gene from a patient with Dravet syndrome decreases mRNA stability mediated by GAPDH's binding. Hum. Genet. 133, 801–811.

Zhou, Y., Yi, X., Stofffer, J., et al., 2008. The multifunctional protein glyceraldehyde-3-phosphate dehydrogenase is both regulated and controls colony-stimulating factor-1 messenger RNA stability in ovarian cancer. Mol. Cancer Res. 6, 1375–1384.

## FURTHER READING

Evguenieva-Hackenberg, E., Schiltz, E., Klug, G., 2002. Dehydrogenases from all three domains of life cleave RNA. J. Biol. Chem. 277, 46145–46150.

White, M., Khan, M., Deredge, D., et al., 2015. A dimer interface mutation in glyceraldehyde-3- phosphate dehydrogenase regulates its binding to AU-rich RNA. J. Biol. Chem. 290, 1770–1785.

# Chapter 4

# The Role of Moonlighting GAPDH in Membrane Structure and Function: Membrane Fusion and Iron Metabolism

*Save the best for last*

Philip Galdston, Wendy Waldman, Jon Lind

The complexity of membrane function is reflected by the intricacy of its structure. Its dynamic nature is illustrated by its role in macromolecular import/export and in receptor-mediated cell signaling. Further, its significance is indicated by its obvious function in the maintenance not only of cell integrity but also that of intracellular structures. Each function is defined by specific protein–protein and protein–lipid interactions. In that regard, early studies indicated not only the structural association of GAPDH with cell membranes but also provide the initial reports on the role of specific GAPDH isozymes in that association.

That being said, recent studies indicate two major functional roles for moonlighting GAPDH in membrane structure and activity. These include its role in membrane fusion and in iron metabolism. The former relates to the required participation of GAPDH in nuclear membrane fusion as one of the penultimate steps in cell division. The latter relates to the role of cell membrane GAPDH as a $Fe^{++}$-transferrin receptor-binding protein—the intracellular movement of the $Fe^{++}$-transferrin–GAPDH complex to the endosome followed by the release of the bound $Fe^{++}$ for distribution. It also relates to the role of membrane-bound GAPDH in the acquisition of extracellular $Fe^{++}$ through its secretion, its binding of exogenous $Fe^{++}$, and the subsequent intercellular transport of the bound $Fe^{++}$. Intriguingly, the role of GAPDH in membrane fusion and iron metabolism appears to involve specific GAPDH isozymes. Further, as GAPDH contains a phosphatidylserine (PS) binding site, the role of that GAPDH domain will be considered. Through this discussion, it is hoped that the reader will gain an understanding with respect to the physiological significance of GAPDH membrane function. It should be noted that membrane GAPDH fulfills an important role in infection and immunity, which is discussed in Chapter 13.

Glyceraldehyde-3-Phosphate Dehydrogenase (GAPDH). http://dx.doi.org/10.1016/B978-0-12-809852-3.00004-2

## 1. INITIAL IDENTIFICATION OF GAPDH–MEMBRANE ASSOCIATION

As discussed in Chapters 1–3, early studies functioned as the trailblazers for the investigations, which followed defining the functional diversity of moonlighting GAPDH. Similarly, early studies on the role of GAPDH in membrane structure and function marked the path through which its membrane-related moonlighting functions were identified.

### 1.1 GAPDH Binding to Erythrocyte Membranes

As indicated in Table 4.1, a series of studies were performed indicating the selective binding of GAPDH to the erythrocyte membrane. These studies were facilitated by the preparation of what were termed "erythrocyte ghosts." The latter are prepared by hemolysis through which hemoglobin and cytoplasmic material are removed. This leaves behind membrane and cytoskeletal structures that, intriguingly, exhibit "normal" morphology (Dodge et al., 1963). This permits the examination of those proteins that provide the foundation for membrane structure, which defines its function. As such, this early technological advance has much in common with the previously described development of DNA-cellulose columns (Chapter 2), which permitted the identification of DNA-binding proteins, i.e., it provided the means through which relevant questions of the time could be posed and answered.

Analysis of erythrocyte ghosts identified a major membrane-bound protein, termed Protein K, which was selectively extracted with EDTA (Tanner and Gray, 1971). Intriguingly, as defined by urea gel, it appeared to be homogeneous. SDS gel analysis indicated an $M_r = 32$ and $33\,kDa$. Tryptic peptide mapping and N-terminal sequencing suggested its resemblance to pig muscle GAPDH. This was confirmed by enzymatic assay in which Protein K exhibited comparable catalysis as that observed for human erythrocyte GAPDH (48 units/mg protein vs. 50 units/mg protein), respectively. That analysis of enzyme activity suggested that Protein K/GAPDH comprises 4%–5% of the total membrane protein.

In a subsequent study, the specificity of GAPDH binding was examined using an experimental paradigm, which included several different ghost types as well as GAPDH and cytochrome c as candidate-binding proteins (Kant and Steck, 1973). The former included what were termed standard, depleted, resealed, and inside-out ghosts. The latter two proteins were used to provide a membrane-binding protein and a negative control.

Quantitation of catalysis indicated that 60%–70% of total erythrocyte GAPDH activity was present in the ghost preparation. That being said, use of the various ghost types demonstrated that GAPDH activity was present in sealed inside-out vesicles but not in sealed right side-out. This was taken to suggest an asymmetry in GAPDH localization such that the protein was bound to

**TABLE 4.1 GAPDH Binding to Erythrocyte Membranes[a]**

| Cell Source/Structure | Experimental Paradigm | Experimental Result | "Unique" Finding | References |
|---|---|---|---|---|
| Human erythrocyte "ghosts" | Purification of membrane-bound protein K | Protein K identified as GAPDH | Protein K exhibits GAPDH activity | Tanner and Gray (1971) |
| Human erythrocyte "ghosts" | Characterization of GAPDH versus cytochrome c membrane-binding capacity | Specificity of GAPDH binding; localization to inner plasma membrane | Glyceraldehyde-3-phosphate, NAD+ affect membrane binding | Kant and Steck (1973) |
| Human erythrocyte "ghosts" | Removal of endogenous GAPDH→GAPDH readdition | Identification of two affinity-binding sites | Specificity of binding-yeast GAPDH does not bind | McDaniel et al. (1974) |
| Erythroleukemic cells | In vitro mRNA translation; immunoprecipitation of ($^{35}$S)-methionine radiolabeled protein | Immunoprecipitation of ($^{35}$S)-p37 protein | Translocation from outer membrane to inner membrane during erythrocyte maturation | Allen and Hoover (1983) |
| Leukemic cell lines, erythrocytes | In vitro mRNA translation; immunoprecipitation of ($^{35}$S)-methionine radiolabeled protein | p37 translation protein is p37 membrane protein | No protein processing of p37 | Allen and Hoover (1985) |
| Leukemic cell lines, erythrocytes | Protein purification; two-dimensional gel electrophoresis | Purification of a 37-kDa membrane-associated protein | Characterization of 37-kDa protein isozymes | Lin and Allen (1986) |
| Erythrocytes, erythroid progenitor cells | Protein purification; restriction mapping of cloned cDNA | Identification of 37 kDa protein as GAPDH | Comparative restriction mapping and cDNA sequencing of p37 protein and GAPDH | Allen et al. (1987) |

[a]Chronological order.

the inner but not the outer membrane, i.e., it is located on the cytoplasmic side. Cytochrome c, which is not found on isolated erythrocyte membranes bound to both sides, which was considered the result expected for nonspecific ligand binding. That conclusion was supported by the ease with which low ionic strength conditions released membrane-bound cytochrome c, while GAPDH remained membrane associated. In contrast, over 90% of membrane-bound GAPDH was released by incubation with both glyceraldehyde-3-phosphate (G3P) and NAD$^+$. That being said, negligible membrane-bound GAPDH was released by incubation with G3P by itself, and only 20% was released by incubation with NAD$^+$ by itself. It was suggested that these metabolites could compete for the membrane-binding site. Alternatively, reduced membrane binding could result from conformational changes results from G3P and NAD$^+$ binding.

The nature and specificity of GAPDH binding was examined by the removal/readdition of GAPDH from the erythrocyte ghost preparations (McDaniel et al., 1974). The experimental paradigm was to remove bound GAPDH by incubation with NaCl then, to determine the kinetics of GAPDH–membrane binding as a function of their reassociation. A salient feature of these studies was the use of three different GAPDH species: human erythrocyte, rabbit muscle, and yeast.

Using the depletion/readdition model, GAPDH readily reassociated with depleted ghost erythrocytes, thereby establishing that GAPDH dissociation was reversible. Further, kinetic analysis revealed a biphasic pattern of reassociation. It was noted that this biphasic uptake was indicative of two different binding sites, one with high affinity and one with low affinity. This pattern was observed for both human erythrocyte GAPDH and for rabbit muscle GAPDH. Accordingly, even though there was a 4% difference in amino acids, that distinction did not affect binding. In contrast, little binding was observed with the yeast enzyme. It was noted that the latter differed by 8% in total amino acid residues. Accordingly, it would seem that the additional 4% difference between human erythrocyte GAPDH and yeast GAPDH is meaningful functionally.

Using a different experimental paradigm, a major p37 membrane protein was identified (Allen and Hoover, 1983). In this investigation RNA from the erythroid cell line K562 was isolated, translated in vitro using ($^{35}$S)-methionine to radiolabel-synthesized proteins. Those proteins were then immunoprecipitated using an antiserum prepared against erythrocyte ghosts. SDS-page identified a number of radiolabeled proteins among them a p37 protein.

Of interest, preincubation of the antiserum with B-lymphoid or T-lymphoid cells removed the anti-p37 antibodies. From that finding, it was suggested that the p37 protein was localized to the outer membrane. However, preincubation with intact erythrocytes did not remove the anti-p37 antibodies nor did preincubation with erythrocyte ghosts, which had been washed with 0.5 M NaCl. Accordingly, it was suggested that, in both cell populations, the p37 protein was present on the inner membrane.

In a further study, the membrane localization of the p37 protein was examined as a function of erythrocyte maturation (Allen and Hoover, 1985). The experimental paradigm was to use the in vitro translation/protein biosynthesis protocol described above or to radiolabel cell proteins with ($^{35}$S)-methionine. Subsequent immunoprecipitation, SDS-PAGE, and autoradiographic analysis were used to identify the p37 protein.

The p37 protein was identified by immunoprecipitation of the in vitro synthesized proteins or of K562 membrane proteins using an antierythrocyte antiserum. Preabsorption with K562 cells eliminated detection of the p37 protein. The p37 translated protein (p37T) was then compared to the p37 membrane protein (p37M). Protease digestion followed by one-dimensional electrophoresis indicated their identity, suggesting that no protein processing occurred during erythrocyte maturation. Further, in common with the previous study, the p37 protein was localized on the outside of the K562 cell membrane and on the cytoplasmic side of the mature red blood cell.

The p37 protein was then purified and characterized by immunological and biochemical criteria (Lin and Allen, 1986). The two p37 proteins were immunologically related as defined by antibody competition and by preabsorption experiments. In addition, each yielded identical peptide maps on *Staphylococcus aureus* V8 protease digestion. Further, in both erythroleukemic cells and in erythrocytes, two-dimensional electrophoresis of immunoprecipitated polypeptides revealed a series of isozymes with a pH range of 7.0–7.8.

In a subsequent study, the p37 protein purified from erythrocyte and K562 membranes was identified as GAPDH (Allen et al., 1987). Peptide mapping studies were performed not only with each of those samples but also with commercially available GAPDH. Identical V8 peptide maps were obtained in each sample. A p37 cDNA clone was isolated, which, in the in vitro translation/protein synthesis protocol, resulted in the synthesis of the p37 protein. Further, restriction mapping of the cDNA produced a virtually identical map to that for GAPDH. Sequence analysis of the p37 cDNA was identical to that of GAPDH. In toto, these four studies demonstrate not only that the p37 protein was indeed GAPDH but also that its membrane localization was variable as a function of its differentiated state.

## 1.2 GAPDH Binding to Band 3—The Major Erythrocyte Membrane Protein

As indicated in Table 4.2, a series of studies were performed indicating the specific interaction between GAPDH and the Band 3 protein. The significance of these investigations became evident on the later identification of the Band 3 protein as Anion Exchange-1 (AE-1) involved in membrane-mediated exchange of chloride and bicarbonate anion transport (rev. in Alper, 2009).

An early indication of GAPDH–Band 3 (AE-1) binding was reported during an initial Band 3 purification and characterization study (Yu and Steck, 1975). The purified Band 3 protein exhibited an $M_r$ c.90 kDa, was a glycoprotein, and was dimeric in nature, even in the presence of Triton X-100. That being said,

**TABLE 4.2 Functional Significance of Membrane-Bound GAPDH: Band 3 (AE1)–GAPDH Interactions[a]**

| Cell Source/ Structure | Experimental Paradigm | Experimental Result | "Unique" Finding | References |
|---|---|---|---|---|
| Erythrocyte membrane | Band 3 protein purification | Band 3 purified and characterized | GAPDH binds to Band 3 | Yu and Steck (1975) |
| Erythrocytes, ghosts | Quantitation of GAPDH activity | Ghost structure affects activity | Ternary complex (Band 3–GAPDH–NADH) formed | Kliman and Steck (1980) |
| Erythrocytes | Quantitation of GAPDH activity when bound to Band 3 | GAPDH activity inhibited by ghost Band 3-concentration dependent | 23 kDa N-terminal band 3 fragment inhibits GAPDH activity concentration dependent | Tsai et al. (1982) |
| Erythrocytes | Phosphorylation/ dephosphorylation of Band 3 | Band 3 phosphorylation inhibits GAPDH binding | Binding site located on inner (cytoplasmic) Band 3 domain | Low et al. (1987) |
| Erythrocytes | Indirect immunofluorescence using human GAPDH antibody | GAPDH localized to membrane; hemoglobin to cytosol | Antibody did not react with rabbit or yeast GAPDH | Rogalski et al. (1989) |
| Erythrocytes, kidney | Antibody colocalization analysis | GAPDH binds to rat erythrocyte and kidney Band 3 | Alternative GAPDH-binding site on kidney Band 3 | Ercolani et al. (1992) |
| Erythrocytes, ghosts | Effect of FP[b] binding | FP induces Band 3[Pc], membrane GAPDH activity | Increase in GAPDH activity independent of Band 3[P] | Omodeo-Salè et al. (2007) |
| Kidney | Yeast 2-hybrid screen using kidney cDNA library | GAPDH cDNA identified as binding partner to AE1 and AE2 | GAPDH silencing inhibits Band 3 basolateral residency | Su et al. (2011) |

[a]Chronological order.
[b]Ferriprotoporphyrin IX (Heme).
[c]Band 3 phosphorylation.

even though the focus of the investigation was Band 3, it was noted that it was necessary during the purification procedure to separate it from membrane protein Band 6, which had been identified as GAPDH. Further, it was noted that as a dimer, in the presence of Triton X-100, the Band 3 protein retained its ability to bind to GAPDH.

In a further investigation the role of GAPDH–Band 3 binding with respect to GAPDH enzyme kinetics (Kliman and Steck, 1980) was studied. The experimental paradigm utilized was a filter binding retention protocol using erythrocyte ghosts to determine catalysis as a function of membrane GAPDH release. The ghosts were prepared in two forms as previously described, i.e., "normal" and "inside-out" ghosts. Initial quantitation indicated that the ghosts retained c.70% of cell GAPDH. Further, no difference was observed in GAPDH release from either of the ghost forms. This is of particular importance given the localization of GAPDH on the inner membrane.

As expected, GAPDH release was dependent on ionic strength. Kinetic analysis indicated a bimolecular–molecular reaction, permitting quantitation of association and dissociation rate constants. Those studies demonstrated that $k'+1$ decreased 27-fold and the $k'+1$ increased 1200-fold as NaCl was increased from 0 to 150 mM NaCl. Similarly, the effect of NADH on the kinetics of GAPDH release was examined. Its effect was hyperbolic for both rate constants. It was suggested that NADH altered GAPDH–Band 3 binding through its interaction at the GAPDH-binding site. However, it was also noted that, as both kinetic constants are still measureable, it was possible that a ternary membrane–enzyme–nucleotide complex could be formed.

Subsequently, the reverse question was posed, i.e., what would be the effect of erythrocyte membrane addition on GAPDH catalysis? In these studies, human erythrocyte or rabbit muscle GAPDH were used (Tsai et al., 1982). The addition of erythrocyte ghosts devoid of GAPDH inhibited enzyme activity in a dose-dependent manner. The $K_i$ of $1.8 \times 10^{-8}$ M was comparable to the previously observed $K_d$ of binding. Band 3 inhibition of GAPDH activity was competitive with respect to $NAD^+$ and noncompetitive with respect to G3P. Similar results were obtained with both the human erythrocyte and rabbit muscle enzyme.

Previous studies indicated aldolase bound to a 23 kDa N-terminal Band 3 peptide (Murthy et al., 1981). Addition of the 23-kDa fragment inhibited enzyme catalysis in a concentration-dependent manner. In agreement with the effect of the entire Band 3 protein, binding was competitive with $NAD^+$ and noncompetitive with G3P. Further, peptide fragments from the 23-kDa Band 3 protein fragment inhibited also GAPDH activity, albeit at higher concentrations.

Cumulatively, the studies described above present compelling evidence for the membrane association of GAPDH with the Band 3 protein. That being said, it may be axiomatic that such studies need to be correlated in intact cells in vivo. Further, ideally, there should be a specificity of such in vivo binding. In intact human erythrocytes, a double immunolabeling protocol was used to examine the intracellular localization of GAPDH and of hemoglobin (Rogalski et al., 1989).

The experimental paradigm was to use a biotinylated anti-GAPDH antibody and an antihemoglobin antibody modified with dichlorotriazinylaminofluorescein.

These studies demonstrated the cytoplasmic localization of hemoglobin and the membrane localization of GAPDH. Control studies indicated that the order of antibody addition or their simultaneous application did not alter the subcellular distribution of either protein. Further, similar findings were observed using whole, fixed, and Triton X-100–permeabilized cells. A second protocol using fixed but non–Triton X-100–permeabilized cells yielded similar findings. In toto, this series of studies provide the requisite documentation of the membrane association of moonlighting GAPDH with Band 3.

A common theme throughout GAPDH studies is the role of GAPDH posttranslational modification as it pertains to its moonlighting functions. Similarly, Band 3 may be phosphorylated on tyr[8], which is located in its N-terminal domain. The effect of that phosphorylation on GAPDH binding was examined (Low et al., 1987). The experimental paradigm utilized an immobilized Band 3 fragment, cdb3, which contained the phosphorylation site. This permitted binding analysis using either phosphorylated or nonphosphorylated cdb3 as the immobilized probe. This study demonstrated that the incorporation of 1.4 mol of phosphate/mol of immobilized cdb3 inhibited GAPDH binding by 70% as compared with the unmodified cdb3 control.

In a further study, the effect of ferriprotoporphyrin (FP IX, a.k.a. heme) on both Band 3 phosphorylation and GAPDH activity was examined (Omodeo-Salè et al., 2007). The experimental paradigm was to determine phosphorylation by immunoblot analysis and GAPDH activity by in vitro assay. Intact erythrocytes and isolated membranes were examined.

In these studies, exposure of erythrocytes to FP resulted in a time-dependent increase in Band 3 phosphorylation. Analysis using staurosporine, a tyrosine kinase inhibitor, reduced FP-mediated Band 3 phosphorylation. Using erythrocyte ghosts, FP stimulated Band 3 phosphorylation. Determination of GAPDH activity indicated that FP exposure stimulated GAPDH activity. Intriguingly, incubation with the tyrosine kinase inhibitor staurosporine, which inhibited Band 3 phosphorylation, had no effect on the ability of FP to stimulate GAPDH activity. As such, it would appear that FP has two effects, which appear to be independent of each other, i.e., Band 3 phosphorylation and stimulation of GAPDH catalytic activity.

One constant in moonlighting GAPDH studies is its transcription from a single gene in somatic cells; the production of a single mRNA; no alternative transcripts have been identified; and a single polypeptide of 37 kDa is produced. In contrast, differences exist in the rodent erythrocyte and kidney isoforms. In particular, the latter is shorter than the former. This size difference may arise from a second promoter in the Band 3/AE1 gene such that the kidney protein lacks the first 1–79 amino acid residues (Kudrycki and Shull, 1989).

The functional significance of this erythrocyte/kidney distinction was examined in relation to the binding and colocalization of GAPDH (Ercolani et al., 1992).

These studies used two antibodies, which recognized residues 214–228 and 918–929 of erythrocyte Band 3/AE1 and one antibody directed against residues 256–267 in the GAPDH catalytic site. A singular characteristic of the latter is that it does not recognize human GAPDH. Initial analysis using ghost membranes indicated that the anti-AE1 918–929 recognized AE1 and the anti-GAPDH 256–267 antibody-detected GAPDH. This demonstrated their specificity. Further, the latter did not recognize purified human GAPDH except at concentrations ≥1 μg.

Colocalization studies were performed in both kidney and in erythrocytes. In the kidney, GAPDH was localized in the basolateral surface of collecting duct cells. Similarly, Band 3/AE1 was observed at that location. Erythrocyte analysis demonstrated membrane localization for each protein as well. The specificity of kidney GAPDH and Band 3/AE1 colocalization was demonstrated by determining that membrane-bound GLUT1 transporter did not colocalize with the Band 3/AE1 protein. In toto, these studies strongly suggest that GAPDH and the Band 3/AE1 proteins colocalize in the kidney. However, as discussed above the latter does not contain the established N-terminal GAPDH-binding site. Accordingly, it was suggested that another binding site is present, which is recognized by GAPDH.

The functional significance of kidney GAPDH–Band 3/AE1 binding was examined recently using a series of molecular analyses as experimental paradigms (Su et al., 2011). Yeast two-hybrid analysis utilized two different constructs as bait to screen a human kidney cDNA library; AE1C-WT, which contained the entire AE1 sequence and AE1C-Δ11, which lacked the last 11C-terminal residues. Two clones were obtained, which contained the entire GAPDH coding sequence. Mating tests indicated that they interacted with AE1C-WT but not the AE1C-Δ11 construct. It was suggested that this indicated that those 11 residues contained the GAPDH-binding site. The latter may be distinct from the two previously identified N-terminal Band 3-binding sites (Chu and Low, 2006).

Coimmunoprecipitation with α-AE1 antiserum resulted in the identification of GAPDH as an AE1-binding partner in both human and rat kidney. Intriguingly, using a α-AE2 antibody resulted in the coimmunoprecipitation of GAPDH as well. Of note, sequence comparison indicated similar motifs in the C-termini. Comparable results were obtained using a GST-tagged AE1C-WT fusion protein, i.e., the fusion protein–bound GAPDH. ELISA determination indicated the binding of AE1C-WT with GAPDH in a concentration-dependent manner. Other studies indicated that phosphorylation of a C-terminal tyrosine residue ($Y^{904}$) did not affect GAPDH binding.

As such, these studies demonstrate in an elegant manner, the binding of kidney GAPDH to the AE1 isoform expressed in that organ. That being said, they do not indicate the significance of those interactions. Accordingly, siRNA GAPDH knockdown studies were performed. In control cells expressing eGFP-tagged WT kidney AE1 (kAE1) exhibited normal basolateral residency as previously reported. In contrast, in siRNA GAPDH-treated cells, minimal AE1 was now located in the surface membrane fraction. Control studies established the specificity of this effect, i.e., there was normal expression of AE1, and depletion of GAPDH did not affect the membrane localization or aberrations in cytoskeletal structure of a number of other proteins. In toto, this molecular analysis establishes firmly the significant role of moonlighting GAPDH in kidney structure and function.

## 2. PHYSIOLOGICAL SIGNIFICANCE OF MEMBRANE-BOUND GAPDH-I: FUSOGENIC ACTIVITY OF MOONLIGHTING GAPDH

It may be axiomatic to state that the identification of a new structural GAPDH characteristic presupposes a new function for this moonlighting protein. Accordingly, the studies described above for membrane-bound GAPDH would be the precursors for the functional studies to follow.

In this instance, as the structural studies described above were progressing, parallel investigations sought to determine the functional ramifications of membrane-bound GAPDH. Initially, the interaction of GAPDH with artificial membrane structures was probed using ($^{14}$C)-radiolabeled rabbit muscle GAPDH with a monolayer composed of phosphatidic acid or on a mixed layer of phosphatidic acid and phosphatidylcholine (Wooster and Wrigglesworth, 1976).

These studies demonstrated that the radiolabeled protein bound to the phosphatidic acid monolayer with first-order kinetics, which was sensitive to both changes in temperature and to pH. Of interest, adsorption of GAPDH to mixed monolayers was dependent on the composition of the monolayer, i.e., low binding was observed at low phosphatidic acid concentrations, while no binding was observed to a monolayer composed solely of phosphatidylcholine. Increased ionic strength promoted GAPDH binding to the mixed polymer while a change in pH from 7.6 to 6.0 diminished binding. Finally, the addition of 1 mM $Ca^{2+}$ diminished GAPDH binding to a phosphatidic acid monolayer. In toto, these basic studies not only established the binding of GAPDH to defined lipid structures but also indicated the parameters, which influenced that protein–lipid interaction.

The observation that $Ca^{2+}$ could stimulate the fusion of acidic phospholipids (Liao and Prestegard, 1979) suggested the possibility that other cellular constituents, especially proteins, might also exhibit comparable fusogenic properties (Morero et al., 1985). The experimental paradigm was to determine the fusion of a mixture of phosphatidylcholine and phosphatidic acid vesicles by a resonance energy transfer assay between a fluorescence donor (CA9C) and an acceptor (NBD-E). This activity could be monitored by changes in emission spectra.

Using this protocol, a series of proteins were tested for their fusogenic activity. $Ca^{2+}$-induced fusion was used as a positive control. The strongest activity was observed with GAPDH. Control studies eliminated the possibility of

$Ca^{2+}$ contamination; gel filtration analysis revealed a pattern very similar to that observed for $Ca^{2+}$; electron microscopy confirmed phospholipid vesicular fusion; and, lastly, heat treatment of GAPDH (5 min at 100°C) resulted in a loss of fusogenic ability. In toto, these were the first studies to indicate the fusogenic activity of moonlighting GAPDH using model membrane structures.

In a further study, the fusogenic properties of rabbit muscle GAPDH were characterized (Lopez Vinals et al., 1987). Again the experimental paradigm was quantitation of resonance energy transfer between the CA9C fluorescent probe donor and the N-NBD-PE acceptor using a vesicle with a phosphatidylcholin e:phosphatidic acid ratio of 9:1. These studies demonstrated that the GAPDH fusogenic activity was pH dependent with a maximum at pH 5, although considerable activity was observed at neutral pH (c.75% of that observed at pH 5). Kinetic analysis indicated that the reaction was rapid, complete at < 30 s at pH 5 but taking 4 min at pH 7.4. Similarly, the reaction was essentially complete at concentrations of GAPDH ≤ 60 nM at either pH 5 or at pH 7.4. As expected, GAPDH fusogenic activity was proportional to increases in temperature up to 37°C and was dependent on vesicle charge with maximal activity observed with negatively charged vesicles although reduced activity was observed with neutral or positively charged species.

Subsequently, using different parameters, the fusogenic activity of GAPDH was examined and yielded some unexpected, intriguing findings (Glaser and Gross, 1994, 1995). The initial experimental paradigm for these studies was the utilization of stop–flow kinetics coupled with quantitation of increases in fluorescent intensity. Calcium-induced fusion rates were determined using PS liposomes and mixtures containing phosphatidylcholine and ethanolamine glycerophospholipids in varying ratios. Through these studies it was possible to define the stereochemical basis through which plasmenylethanolamine could facilitate membrane fusion (Glaser and Gross, 1994).

Based on those findings it was postulated that cells may contain proteins with fusogenic properties based on the stereochemical parameters described in that initial study. To test that hypothesis, the fusion of membranes comprised of phosphatidylcholine, plasmenylethanolamine, **PS**, and cholesterol (27%, 27%, 6%, and 40%) was used as the experimental paradigm. The reason for highlighting PS will be discussed later.

Using this protocol, a cytosolic fusion activity was detected in rabbit brain homogenates. Its protein identification was suggested by its sensitivity to trypsin digestion and by its heat lability. In addition, the fusion activity was calcium independent. Further, purification by conventional column chromatography revealed a predominant 38-kDa protein. N-terminal sequencing of the protein indicated that it was GAPDH.

That being said, using GTP-agarose affinity column chromatography, GAPDH activity, which was recovered in the eluent did not possess significant membrane fusion activity. In contrast, the 38-kDa GAPDH protein, which was retained, exhibited fusogenic activity but reduced GAPDH catalytic activity.

Each species was recognized by a polyclonal anti-GAPDH antibody. Two-dimensional electrophoresis identified two GAPDH species, one with a pI = 7–8 and another with a pI = 8–9. The latter contained the fusogenic activity, while the former exhibited catalytic activity. Considering the acidic nature of the lipid vesicles, the basic nature of the fusogenic GAPDH species is intriguing.

Control studies indicated that fusogenic activity was inhibited by G3P and by a series of anti-GAPDH monoclonal antibodies. It was not inhibited by koningic acid, which covalently binds to the GAPDH active site cysteine. In toto, these studies suggest that a specific, basic GAPDH isoform is responsible for its fusogenic activity. As will be noted in various chapters of this book, GAPDH isoforms fulfill a number of its diverse moonlighting activities.

## 2.1 Identification of Tubulin as an Inhibitor of GAPDH Fusogenic Activity

As discussed above, investigation of GAPDH as a fusogenic protein was hampered initially by the presence of an inhibitor, which was present in crude cell extracts (Glaser and Gross, 1995). In particular, only by further purification could this new function of moonlighting GAPDH be pursued.

Accordingly, the nature of that inhibitory activity was examined (Glaser et al., 2002). Its purification indicated that it was a cytosolic protein as defined by heat lability and by trypsin sensitivity. SDS-PAGE analysis indicated it was a protein with an $M_r = 55$ kDa. Edman degradation of a 15-amino acid sequence demonstrated its complete identity to human and mouse α-tubulin. Immunoblot analysis with an antitubulin antibody was used as a further confirmation of its identity.

Mechanistic analysis indicated that tubulin inhibition was effected by the formation on a GAPDH–tubulin protein complex. This conclusion was based on affinity chromatography studies, demonstrating that the GAPDH fusogenic isoform was adsorbed to a tubulin-Sepharose affinity resin; determination of the protein complex stoichiometry by mixing experiments using different molar ratios of the two proteins; and by kinetic analysis using the stopped-flow system initially developed to analysis membrane fusion (Glaser and Gross, 1994). In toto, this study indicated a role for tubulin in the modulation of GAPDH fusogenic activity.

## 2.2 Identification of a GAPDH Phosphatidylserine-Binding Site

Although, as described above, considerable evidence indicated the fusogenic role of GAPDH, wherein the specific protein–lipid interactions, which underlay that function, were unknown. That being said, an intriguing immunological approach was used, which not only identified a unique PS binding site within the GAPDH protein but also defined the mechanisms through which GAPDH mediated membrane fusion (Kaneda et al., 1997).

An antibody (αPSD-2) was raised against a peptide (FNFRLKAGQKIRFGC), which had been identified as a binding site based on studies examining PS binding to protein kinase C and to PS decarboxylase. This antibody bound both to the band 6/GAPDH protein in the erythrocyte membrane and to rabbit muscle GAPDH isozymes following two-dimensional electrophoresis. ELISA analysis indicated that phospholipid vesicles containing PS competed with αPSD-2 for GAPDH binding. No competition was observed with vesicles containing phosphatidylcholine and/or phosphatidylethanolamine. Other studies indicated that GAPDH bound to PS-immobilized membranes. Further, incubation with the peptide inhibited GAPDH fusogenic activity.

Protein mapping studies were then performed to identify the GAPDH PS binding site. Cyanogen bromide cleavage of GAPDH yielded a number of peptides, which were resolved by reverse phase HPLC then examined for immunoreactivity with the αPSD-2 antibody. An immunoreactive 6.5-kDa peptide was subjected to Edman degradation, and a single sequence identified, which mapped to GAPDH amino acids 45–103. Further, protein analysis localized the PS binding site to amino acids 70–94. To verify that identification, an antibody was raised against that sequence. It inhibited membrane fusion in a concentration-dependent manner. Further, that sequence by itself also inhibited membrane fusion, again in a concentration-dependent manner. Accordingly, these studies suggest that PS is the membrane component, which interacts with GAPDH, and that this protein–lipid interaction provides the foundation for the fusogenic activity of this moonlighting protein. This may be of special importance as sequence analysis demonstrated that GAPDH lacks a typical membrane localization domain.

## 2.3  Fusogenic GAPDH and Insulin Exocytosis

As described earlier, structural studies on moonlighting GAPDH are usually the harbinger of investigations, which reveal its new functional activities. In this regard, the fusogenic GAPDH isoform previously identified in rat brain (Glaser and Gross, 1995) was tested for its ability to facilitate insulin exocytosis (Han et al., 1998). The latter requires membrane fusion as one of the penultimate steps in insulin secretion.

The experimental paradigm of these studies was to examine whether the brain GAPDH isoform could catalyze the fusion of islet secretory granules with the corresponding plasma membrane. These studies demonstrated that addition of the partially purified isoform enhanced secretory granule–membrane fusion in a concentration-dependent manner. Pretreatment of the partially purified factor with trypsin abolished fusogenic activity. In contrast, trypsin treatment of the secretory granule or membrane preparations did not affect their fusion. Pretreatment with a GAPDH monoclonal antibody, which binds the fusogenic GAPDH isoform–inhibited fusogenic activity. Further, fusion by the GAPDH isoform was $Ca^{2+}$ independent. As such, these studies are in accord with a previous report that mutations in GAPDH can affect exocytosis in CHO cells (Robbins et al., 1995).

## 2.4 Fusogenic GAPDH and Neutrophil Degranulation

In common with the studies described above, membrane fusion is required as part of the inflammatory response. Accordingly, a search was conducted for neutrophilic proteins, which may be involved in membrane fusion (Hessler et al., 1998). The experimental paradigm in these studies was to examine liposome fusion using fluorescence analysis similar to that described in this chapter. A change in the protocols involved the use of $Ca^{2+}$ in the fusion assay.

The use of this assay as a detection mechanism resulted in the purification of a 36-kDa protein from neutrophils, which was identified as GAPDH by enzymatic analysis. The latter activity coincided with fusogenic activity until the final preparatory step. In addition, although commercial chicken GAPDH promoted $Ca^{2+}$ fusion, human erythrocyte GAPDH did not. Accordingly, although these results are not as definitive as desired, they do suggest the possibility that the fusogenic GAPDH isoform previously identified (Glaser and Gross, 1995) is also present in neutrophils.

## 2.5 Role of GAPDH in Nuclear Membrane Fusion

As described in Chapter 1, one of the earliest demonstrations of GAPDH functional diversity was the dynamic changes in subcellular localization during cell proliferation. The latter is characterized by many complex events one of which is reformation of the nuclear membrane in daughter cells.

As related in Section 2.1, above, a GAPDH PS binding site was identified (Kaneda et al., 1997). Subsequently, the role of that binding site in nuclear envelope fusion was examined (Nakagawa et al., 2003). The experimental paradigm used was fluorescence microscopy using *Xenopus* egg extracts as well as cytosol and membrane preparations. A clinical aspect of these studies was the use of a serum, K199, collected from an autoimmune disease patient. It was stated that sera obtained from patients with this clinical phenotype produce autoantibodies against nuclei or the nuclear envelope.

Initial studies demonstrated the selective inhibition of *Xenopus* nuclear envelope fusion by the K199 autoimmune serum. Analysis of *Xenopus* cytosol and membrane indicated that the inhibitory factor was present in the former. Immunoblot analyses indicated that an antigen recognized by the K199 autoimmune serum was GAPDH. Further, studies demonstrated that commercially available GAPDH antibody not only inhibited nuclear membrane assembly but also recognized a c.39 kDa *Xenopus* egg protein. An antibody directed against the PS GAPDH binding site or the peptide itself also inhibited nuclear membrane assembly. Confocal microscopy indicated that both the antibody and the peptide inhibited the membrane fusion step of nuclear assembly. A model for this GAPDH membrane function is illustrated in Fig. 4.1D.

# 3. PHYSIOLOGICAL SIGNIFICANCE OF MEMBRANE-BOUND GAPDH-II: THE FUNCTIONAL DIVERSITY OF GAPDH IN IRON UPTAKE, TRANSPORT, AND SEQUESTRATION

In common with the studies described in Section 2, the role of GAPDH in iron metabolism highlights the temporal sequence through which new moonlighting GAPDH structures and activities are identified.

In a rigorous investigation, the transferrin receptor function of membrane-bound GAPDH was discovered (Raje et al., 2007). In particular, through the use of five separate protocols, it was established that not only could GAPDH bind transferrin at the membrane surface but also that membrane GAPDH transported transferrin internally to the early endosome.

Initially, transmission and scanning electron microscopy were used to demonstrate the presence of membrane GAPDH in murine macrophages and human monocytes. Of significance, these studies not only established that GAPDH was localized on the cell membrane surface but also that it was enzymatically active. Further, it was demonstrated that the extent of macrophage surface GAPDH was dependent on the presence of iron in the culture media, i.e., surface GAPDH was increased in iron-depleted cells.

The formation of a GAPDH–transferrin complex was established using three different protocols. In an in vitro analysis using dot blot and ELISA capture assays, transferrin bound to rabbit muscle GAPDH as detected by an antitransferrin antibody. In the reverse analysis, GAPDH was captured by bound transferrin in the dot blot assay. Subsequently, confocal microscopy was used to colocalize GAPDH and transferrin using double immunofluorescence staining protocols. Finally, coimmunoprecipitation studies were performed using membrane preparations from intact cells. GAPDH was coimmunoprecipitated by antitransferrin antibodies, and transferrin was coprecipitated by an anti-GAPDH antibody.

The functional significance of this interaction was demonstrated using confocal microscopy to demonstrate the internalization of the transferrin–membrane GAPDH complex to the early endosome. The latter may be considered as an "end point" for iron transport, where it can then be dispersed to recipient cell molecules. Immunoprecipitation analysis was performed to demonstrate the presence of membrane-derived GAPDH in the endosome. In these studies, cells were surface biotinylated prior to transferrin absorption. After 5 min at 37°C to allow for internalization, the early endosome fraction was purified and immunoprecipitation performed with an anti-GAPDH antibody. That study demonstrated the presence of biotinylated GAPDH in the early endosome fraction, indicating its membrane→endosome translocation as a function of transferrin binding.

In a subsequent investigation (Kumar et al., 2012), the role of membrane-bound GAPDH was examined in CHO-TRVb cells, which do not contain either transferrin receptor 1 (TfR1) or transferrin receptor 2 (TfR2). Following GAPDH silencing, surface GAPDH and transferrin levels were reduced, and there was a

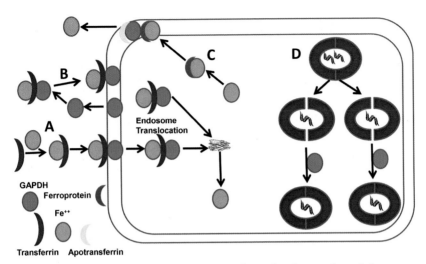

**FIGURE 4.1** Role of GAPDH in iron metabolism and nuclear membrane fusion.

30% decrease in transferrin mediated $^{55}$Fe uptake. Inhibition studies were used as a mechanistic analysis of GAPDH–transferrin binding and internalization. As exposure to agents that affect cholesterol sequestration (nystatin and filipin) diminished transferrin internalization in CHO-TRVb cells, it was suggested that GAPDH–transferrin uptake may be mediated by lipid rafts. Using an inhibitor of macropinocytosis (wortmannin), there was a 70% reduction in transferrin uptake. Further, inhibition of clathrin-mediated endocytosis reduced transferrin uptake by 20%–40%, depending on the inhibitor utilized. The latter finding is consistent with the previously reported role of GAPDH in endocytosis (Robbins et al., 1995). In toto, these studies not only demonstrate the role of GAPDH as a transferrin receptor but also indicate that several pathways may be utilized for the internalization and transport of the GAPDH–transferrin receptor–ligand complex. A model for this new GAPDH membrane function is illustrated in Fig. 4.1A.

In a third study, a "hidden" property of membrane-bound GAPDH was uncovered, i.e., GAPDH could be secreted into the extracellular milieu to acquire transferrin and iron (Sheokand et al., 2013). In this investigation, the presence of GAPDH in the media was determined in 22 different cell lines as a function of iron depletion. There was a significant variance in cellular response with fold changes ranging from 0.49 (HeLa cells) to 5.76 and 6.05 in K562 lymphoblasts and J774 macrophage cells, respectively. It may be of interest that the greatest increases were observed in cells whose function is intrinsic to the immune response (see Chapter 13 for the role of GAPDH in infection and immunity). Control studies indicated that there was no change in cell viability as defined by the release of lactate dehydrogenase in the media.

The functional significance of GAPDH secretion may be indicated by the uptake of Fe-transferrin in J774 and CHO-TRVb cells incubated with media from iron-depleted cell cultures. Further, addition of purified GAPDH to a variety of cells increased transferrin uptake. The latter was both time dependent and a function of the amount of purified GAPDH. Intriguingly, the uptake of GAPDH in both cell samples was saturable, indicating either an active transport mechanism or a protein–protein GAPDH complex responsible for its uptake. Coimmunoprecipitation studies indicated the formation of a GAPDH–uPAR protein complex. The latter, urokinase plasminogen activator receptor, is a membrane-bound protein. Its binding to GAPDH is discussed in detail in Chapter 13. Inhibitor analysis demonstrated a sensitivity to lipid raft and macropinocytosis inhibitors but not to those implicated as disruptors of clathrin-mediated pathways. The latter finding is in contrast to that described above (Kumar et al., 2012). In toto, the results of this third study demonstrate again the dynamic role of moonlighting GAPDH, i.e., it is not a static, passive intracellular protein. Instead, in this instance, it is used by the cell to respond to a significant nutritional deficiency. A model for this GAPDH membrane function is illustrated in Fig. 4.1B.

## 3.1 Role of Membrane-Bound GAPDH in Iron Export

The detailed studies described above demonstrated a role for membrane-bound GAPDH in iron import. Surprisingly, a study also suggests that it may also fulfill an important function in iron export (Sheokand et al., 2014). In these studies, the experimental paradigm was to incubate cells in media containing $100\,\mu M$ $FeCl_3$ or to increase macrophage iron concentration by erythrophagocytosis (coculturing with lysed erythrocytes). The change in surface GAPDH ranged from 0.77 in rat enterocytes to 3.38 and 4.0 in J774 cells and in rat peritoneal cells, respectively.

Analysis of apotransferrin, an iron-chelating protein, indicated its increase as well in macrophages (2.17-fold), in J774 cells (2.77-fold) in cells treated with $FeCl_3$ and 5.50-fold increase in macrophages cocultured with lysed erythrocytes. ELISA studies using bound GAPDH and purified apotransferrin demonstrated the in vitro formation of a GAPDH–apotransferrin complex. Confocal microscopy in iron-treated J774 cells demonstrated the colocalization of both proteins on the cell surface. Further, a similar study demonstrated the colocalization of ferroportin, an iron exporter, with both apotransferrin and GAPDH. Coimmunoprecipitation analysis demonstrated the physical interaction of GAPDH with ferroportin. The physiological significance of these GAPDH interactions was determined in iron-treated cells. The inclusion of apotransferrin in the media resulted in increased iron export as defined both by chromogenic iron assay and by determination of $^{55}Fe$ radiolabel in the media. Silencing of GAPDH eliminated apotransferrin facilitated iron export. A model for this new membrane GAPDH function is illustrated in Fig. 4.1C.

In an intriguing analysis, characterization of membrane-bound GAPDH as a function of iron overload versus iron depletion revealed a novel finding. First, two-dimensional gel electrophoresis demonstrated the presence of more basic GAPDH isoforms in the former as compared to the latter. Second, consistent with that finding, there was a defined difference in posttranslational modifications in membrane GAPDH from iron loaded versus iron-depleted cells. These included differences in oxidation, dimethylation, acetylation, nitrosylation, and phosphorylation. Each posttranslational modification was higher in membrane GAPDH from iron depleted cells. In toto, these cumulative findings suggest an alternative GAPDH-mediated pathway in which it facilitates iron export. That being said, these studies suggest that different GAPDH isoforms function in iron import and in iron export, respectively. As such, these findings indicate again the role of distinct moonlighting GAPDH isozymes in its diverse activities.

## 3.2 Membrane-Bound GAPDH as a Lactoferrin Receptor Protein

As with transferrin, lactoferrin also functions in iron metabolism. Accordingly, the possibility that membrane-bound GAPDH may bind to lactoferrin was examined (Rawat et al., 2012). The experimental paradigm was similar to that used previously, i.e., studies were performed in the mouse macrophage cell line J774, the human monocytic cell line THP-1, and in CHO-TrVb cells.

Initially, biotinylated lactoferrin binding to immobilized GAPDH demonstrated the physical interaction of these two proteins in vitro. In a complementary study, membrane coimmunoprecipitation analysis using lactoferrin pull down demonstrated their physical association in vivo. Confocal microscopy demonstrated their colocalization in intact cells. Finally, subcellular localization studies indicated the presence of both proteins in the early endosome fraction, demonstrating the intracellular movement of the lactoferrin–GAPDH protein complex from the membrane to that intracellular locale. As these studies extend this new moonlighting GAPDH activity to another iron-related cellular protein, they, along with the role of GAPDH in iron import, indicate a new cellular function, which requires the active modulation of this moonlighting protein. This is in accord with a recent analysis of iron-related proteins, which demonstrated an increase in both GAPDH protein and mRNA in macrophages exposed to low levels of ferric ammonium citrate (Polati et al., 2012).

## 4. SUMMARY

Analysis of membrane GAPDH may be considered as a prime example of the temporal interrelationship between the structural analysis and the functional determination of its moonlighting activities. First, it was defined structurally as a membrane-bound protein. Second, the structural formation of a Band 3/AE1–GAPDH membrane complex was reported, which was an a priori requirement for Band 3/AE1 localization within the membrane. Third, fusogenic GAPDH activity was defined in a model membrane system followed by reports of this new function in vivo. Fourth, membrane GAPDH activity was essential for iron metabolism including its acquisition, transport into the cell as well as its export from the cell. In toto, the temporal sequence of membrane GAPDH structure and function analyses is characteristic of most moonlighting GAPDH studies.

# REFERENCES

Allen, R., Hoover, B., 1983. Reorientation of membrane polypeptides during erythrocyte maturation. Blood 61, 803–806.

Allen, R., Hoover, B., 1985. Characterization of the processed form of a ubiquitous protein displaying variable membrane orientation in erythroid cells. Blood 65, 1048–1055.

Allen, R., Trach, K., Hoch, J., 1987. Identification of the 37-kDa protein displaying a variable interaction with the erythroid cell membrane as glyceraldehyde-3-phosphate dehydrogenase. J. Biol. Chem. 262, 649–653.

Alper, S., 2009. Molecular physiology and genetics of Na$^+$-independent SLC4 anion exchangers. J. Exp. Biol. 212, 1672–1683.

Chu, H., Low, P., 2006. Mapping of glycolytic enzyme-binding sites on human erythrocyte band 3. Biochem. J. 400, 143–151.

Dodge, J., Mitchell, C., Hanahan, D., 1963. The preparation and chemical characteristics of hemoglobin-free ghosts of human erythrocytes. Arch. Biochem. Biophys. 100, 119–130.

Ercolani, L., Brown, D., Sturt-Tilley, A., Alper, S., 1992. Colocalization of GAPDH and band 3 (AE1) in rat erythrocytes and kidney intercalated membranes. Am. J. Physiol. 262, F892–F896.

Glaser, P., Gross, R., 1994. Plasmenylethanolamine facilitates rapid membrane fusion: a stopped-flow kinetic investigation correlating the propensity of a major plasma membrane constituent to adopt an H$_{II}$ phase with its ability to promote membrane fusion. Biochemistry 33, 5805–5812.

Glaser, P., Gross, R., 1995. Rapid plasmenylethanolamine-selective fusion of membrane bilayers catalyzed by an isoform of glyceraldehyde-3-phosphate dehydrogenase: discrimination between glycolytic and fusogenic roles of individual isoforms. Biochemistry 34, 12193–12203.

Glaser, P., Han, X., Gross, R., 2002. Tubulin is the endogenous inhibitor of the glyceraldehyde-3-phosphate dehydrogenase isoform that catalyzes membrane fusion: implications for the coordinated regulation of glycolysis and membrane fusion. Proc. Natl. Acad. Sci. U.S.A. 99, 14104–14109.

Han, X., Ramanadham, S., Turk, J., Gross, R., 1998. Reconstitution of membrane fusion between pancreatic islet secretory granules and plasma membrane: catalysis by a protein constituent recognized by monoclonal antibodies directed against glyceraldehyde-3-phosphate dehydrogenase. Biochim. Biochem. Acta 1414, 95–107.

Hessler, R., Blackwood, R., Brock, T., et al., 1998. Identification of glyceraldehyde-3-phosphate dehydrogenase as a Ca$^{2+}$-dependent fusogen in human neutrophil cytosol. J. Leuk. Biol. 63, 331–336.

Kaneda, M., Takeuchi, K., Inoue, K., Umeda, M., 1997. Localization of the phosphatidylserine binding site of glyceraldehyde-3-phosphate dehydrogenase responsible for membrane fusion. J. Biochem. 122, 1233–1249.

Kant, J., Steck, T., 1973. Specificity in the association of glyceraldehyde-3-phosphate dehydrogenase with isolated human erythrocyte membranes. J. Biol. Chem. 248, 8457–84464.

Kliman, H., Steck, T., 1980. Association of glyceraldehyde-3-phosphate dehydrogenase with the human red cell membrane: a kinetic analysis. J. Biol. Chem. 255, 6314–6321.

Kudrycki, K., Shull, G., 1989. Primary structure of the rat kidney band 3 anion exchange protein deduced from a cDNA. J. Biol. Chem. 264, 8185–8192.

Kumar, S., Sheokand, N., Mhadeshwar, M., Raje, C., Raje, M., 2012. Characterization of glyceraldehyde-3-phosphate dehydrogenase as a novel transferrin receptor. Int. J. Biochem. Cell Biol. 44l, 189–199.

Liao, M.-J., Prestegard, J., 1979. Fusion of phosphatidic acid-phosphatidylcholine mixed lipid vesicles. Biochim. Biophys. Acta 550, 157–173.

Lin, T., Allen, R., 1986. Isolation and characterization of a 37,000-Dalton protein associated with the erythrocyte membrane. J. Biol. Chem. 261, 4594–4599.

Lopez Vinals, A., Farias, R., Morero, R., 1987. Characterization of the fusogenic properties of glyceraldehyde-3-phosphate dehydrogenase: fusion of phospholipid vesicles. Biochem. Biophys. Res. Commun. 143, 403–409.

Low, P., Allen, D., Zioncheck, T., et al., 1987. Tyrosine phosphorylation of Band 3 inhibits peripheral protein binding. J. Biol. Chem. 262, 4592–4596.

McDaniel, C., Kirtley, M., Tanner, M., 1974. The interaction of glyceraldehyde-3-phosphate dehydrogenase with human erythrocyte membranes. J. Biol. Chem. 249, 6478–6485.

Morero, R., López Viñals, A., Bloj, B., Farías, 1985. Fusion of phospholipids induced by muscle glyceraldehyde-3-phosphate dehydrogenase in the absence of calcium. Biochemistry 24, 1904–1909.

Murthy, S., Liu, N., Kaul, T., Köhler, R., Steck, 1981. The aldolase-binding site of the human erythrocyte membrane is at the $NH_2$ terminus of Band 3. J. Biol. Chem. 256, 11203–11208.

Nakagawa, T., Hirano, Y., Inomata, A., et al., 2003. Participation of a fusogenic protein, glyceraldehyde-3-phosphate dehydrogenase, in nuclear membrane assembly. J. Biol. Chem. 278, 20395–20404.

Omodeo-Salè, F., Corgelezzi, L., Riva, E., Vanzulli, E., Taramelli, D., 2007. Modulation of glyceraldehyde-3-phosphate dehydrogenase activity and try-phosphorylation of Band-3 in human erythrocytes treated with ferriprotoporphyrin IX. Biochem. Pharm. 74, 1383–1389.

Polati, R., Castagna, A., Bossi, A., et al., 2012. Murine macrophages response to iron. J. Proteomics 76, 10–27.

Raje, C., Kumar, S., Harle, A., Nanda, J., Rage, M., 2007. The macrophage cell surface glyceraldehyde-3-phosphate dehydrogenase is a novel transferrin receptor. J. Biol. Chem. 282, 3252–3261.

Rawat, P., Kumar, S., Sheokand, N., Raje, C., Raje, M., 2012. The multifunctional glycoprotein glyceraldehyde-3-phosphate dehydrogenase (GAPDH) is a novel macrophage lactoferrin receptor. Biochem. Cell Biol. 90, 3290338.

Robbins, A., Ward, R., Oliver, C., 1995. A mutation in glyceraldehyde-3-phosphate dehydrogenase alters endocytosis in CHO cells. J. Cell Biol. 130, 1093–1104.

Rogalski, A., Steck, T., Waseem, A., 1989. Association of glyceraldehyde-3-phosphate dehydrogenase with the plasma membrane of the intact human red blood cell. J. Biol. Chem. 264, 6438–6446.

Sheokand, N., Kumar, S., Malhotra, H., et al., 2013. Secreted glyceraldehyde-3-phosphate dehydrogenase is a multifunctional autocrine transferrin receptor for cellular iron acquisition. Biochim. Biophys. Acta 1830, 3816–3827.

Sheokand, N., Malhotra, H., Kumar, S., et al., 2014. Moonlighting cell-surface GAPDH recruits apotransferrin to effect iron egress from mammalian cells. J. Cell Sci. 127, 4279–4291.

Su, Y., Blake-Palmer, K., Fry, A., et al., 2011. Glyceraldehyde-3-phosphate dehydrogenase is required for band 3 (anion exchanger 1) membrane residency in the mammalian kidney. Am. J. Physiol. Ren. Physiol. 300, F157–F166.

Tanner, M., Gray, 1971. The isolation and functional identification of a protein from the human erythrocyte ghost. Biochem. J. 125, 1109–1117.

Tsai, I., Murthy, S., Steck, T., 1982. Effect of red cell membrane binding on the catalytic activity of glyceraldehyde-3-phosphate dehydrogenase. J. Biol. Chem. 257, 1438–1442.

Wooster, M., Wrigglesworth, J., 1976. Adsorption of glyceraldehyde-3-phosphate dehydrogenase on condensed monolayers of phospholipid. Biochem. J. 153, 93–100.

Yu, J., Steck, T., 1975. Isolation and characterization of Band 3, the predominant polypeptide of the human erythrocyte membrane. J. Biol. Chem. 250, 9170–9175.

Chapter 5

# The Role of Moonlighting GAPDH in Intracellular Membrane Trafficking: Tubulin Regulation and Modulation of Cytoskeletal Structure

*Follow the Yellow Brick Road*

The Munchkins to Judy Garland as Dorothy in the Wizard of Oz.

As with the complex processes through which cells import macromolecules from the external environment, the mechanisms through which they export macromolecules or provide for their intracellular transport are equally complicated. In particular, the latter require highly ordered protein and membrane structures as well as a "highway" on which such macromolecules can be transported. The endoplasmic reticulum (ER) and the Golgi apparatus may represent the former, while tubulin microtubules may represent the latter. Movement between the ER and the Golgi may be termed as intracellular membrane trafficking, which is not a "one-way" street. ER to Golgi transport is termed anterograde, while Golgi to ER transport is termed retrograde.

Recent evidence suggests not only that GAPDH fulfills a major role in intracellular membrane trafficking but also that it may be the means through which the tubulin microtubule "highway," i.e., the yellow brick road, is constructed. As with other moonlighting GAPDH functions, early structural studies defined the binding of GAPDH to tubulin as well as its ability to effect tubulin structure, the conversion of soluble tubulin to insoluble microtubules, termed microtubule bundling. Later functional studies defined its physical association with Rab2 vesicular tubule structures, the requirement for its sequential phosphorylation by two separate kinases followed by its modulation of tubulin structure altered cytoskeletal structure and cell morphology. In toto, these studies may provide evidence indicating the role of moonlighting GAPDH in another critical cell pathway.

## 1. INITIAL IDENTIFICATION OF GAPDH–TUBULIN BINDING

Similar to the discussions in Chapters 1–4, the early studies described in Table 5.1 provided the structural basis for the subsequent functional investigations that identified and characterized this new moonlighting GAPDH activity.

As with many GAPDH studies, the initial investigation focused on another question, the identification of tubulin-binding proteins (Kumagai and Sakai, 1983). The experimental paradigm was conventional in nature, i.e., the use of microtubule-Sepharose 4B affinity chromatography to isolate those proteins that were bound to that matrix. An unconventional aspect of this approach was to use 2 mM ATP as the eluting agent (the significance of this will be discussed below).

Using this purification scheme, a 35-kDa protein was purified to homogeneity and its effect on tubulin polymerization was determined. An enhancement in turbidity ($\Delta A_{350}$) was observed as a function of increasing amounts of the protein in the reaction mixture. The digestion of the 35-kDa protein by trypsin diminished turbidity, while the addition of a trypsin inhibitor reversed that effect. Cosedimentation analysis demonstrated the physical association of the 35-kDa protein with tubulin. Electron microscopic analysis revealed the presence of "thick bundles of microtubules" on addition of the 35-kDa protein (Fig. 5.1,IB and C). One dimensional peptide mapping identified the 35-kDa protein as GAPDH. Affinity analysis of rabbit muscle GAPDH demonstrated its adsorption to tubulin Sepharose 4B as well as its elution by ATP.

In a subsequent study, the nucleotide dependence of GAPDH-induced microtubule formation was examined (Huitorel and Pantaloni, 1985). The experimental paradigm was to examine the effect of GTP, ATP, and adenosine 5'-[β,γ-imido]biphosphate (Ado$PP$[NH]$P$) on bundling activity as defined by turbidity changes. The latter analogue is nonhydrolyzable. In each instance, a decrease in $\Delta A_{350}$ was observed. This finding demonstrated that ATP hydrolysis was not required for microtubule bundling. Further, a model was presented in which tetrameric GAPDH in the absence of ATP induced microtubule bundling, while ATP-induced GAPDH tetrameric dissociation to a dimeric form was responsible for the observed decrease in bundling activity.

Stoichiometric analysis of GAPDH–tubulin binding was monitored by using coprecipitation as a means to define the physical association of GAPDH and microtubules (Durrieu et al., 1987a). As expected, binding was proportional to GAPDH with half saturation observed at 70 µg/mL of enzyme, a concentration of 0.5 µM, which was comparable to the physiological concentration of GAPDH (estimated at between 1 and 5 µM). Similarly, in vitro tubulin concentrations utilized were within its physiological concentration (estimated at 10–25 µM). Accordingly, an important result of this study was to validate the biological relevance of in vitro GAPDH–tubulin studies.

As maximal GAPDH-binding capacity was calculated as 0.1 mole GAPDH bound per mole of assembled tubulin, another significant finding was the

*Continued*

**TABLE 5.1** Identification of GAPDH–Tubulin Binding[a]

| Cell Source/ Organism | Experimental Paradigm | Experimental Finding | "Unique" Finding | Reference |
|---|---|---|---|---|
| Porcine brain | Tubulin affinity chromatography | 35-kDa tubulin-binding protein is GAPDH | Microtubule bundling | Kumagai and Sakai (1983) |
| Porcine brain, rabbit muscle erythrocyte ghost | Tubulin affinity chromatography | ATP inhibition of GAPDH-induced tubulin polymerization | ATP hydrolysis not required for inhibition | Huitorel and Pantaloni (1985) |
| Rabbit muscle GAPDH | Microtubule: GAPDH cosedimentation | Quantitation of GAPDH-binding sites | GAPDH–tubulin binding at physiological protein concentrations | Durrieu et al. (1987a) |
| Rabbit muscle GAPDH | Microtubule: GAPDH cosedimentation | Microtubule inhibition of GAPDH activity | GAPDH monomer binds to microtubules | Durrieu et al. (1987b) |
| Rabbit muscle GAPDH | Agarose gel electrophoresis | GAPDH cathode migration, tubulin anode migration | Complex migration, GAPDH dependent | Karkhoff-Schweizer and Knull (1987) |
| Human colon cancer cell line | Affinity chromatography | Bundling activity comparable to rabbit GAPDH | Bundling at 1:10 GAPDH:tubulin molar ratio | Launay et al. (1989) |
| Rabbit muscle GAPDH | Centrifugation, anisotropy | Tubulin binding increases GAPDH anisotropy | GAPDH, tubulin concentrations at physiological levels | Walsh et al. (1989) |

**TABLE 5.1** Identification of GAPDH–Tubulin Binding[a]—cont'd

| Cell Source/ Organism | Experimental Paradigm | Experimental Finding | "Unique" Finding | Reference |
|---|---|---|---|---|
| *Trypanosomatid* | Radioiodination, sedimentation, precipitation | GAPDH-g[b] binds microtubules | GAPDH-g identified in cytoskeleton | Kambadur et al. (1990) |
| Rabbit muscle GAPDH | In vitro analysis, coprecipitation | Effect of ionic strength on tubulin binding | No effect of MAPS[c] on tubulin binding | Somers et al. (1990) |
| Rabbit muscle | Copelleting–Sepharose matrix | Similar effect of ionic strength compounds | Immobilization blocks soluble GAPDH binding | Muronetz et al. (1994) |
| Rabbit muscle | Electron microscopy | Microtubule cross-linking | Multiplicity of structures | Volker et al. (1995) |
| Rabbit muscle | Peptide cleavage | Mapping of binding sites | Site on tubulin identified | Volker and Knull (1997) |

[a]*Chronological order.*
[b]*Glycosomal GAPDH.*
[c]*Microtubule-associated proteins.*

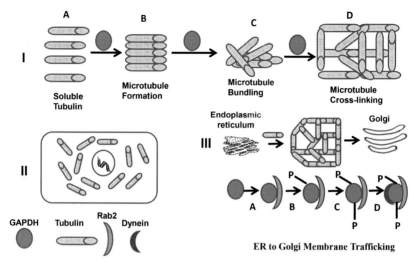

**FIGURE 5.1** Role of GAPDH–tubulin binding in intramembrane trafficking. *ER*, endoplasmic reticulum.

observation that there were a limited number of GAPDH-binding sites within the tubulin molecule. As discussed later, that implied a specific GAPDH-binding site. These studies also demonstrated the specificity of GAPDH binding. Using a crude rat liver soluble fraction incubated with rat brain microtubules, the only protein that coprecipitated with the microtubule sample was a 35-kDa protein subsequently identified as GAPDH by both peptide mapping and immunoblot analysis. Finally, binding of GAPDH was found to impair microtubule dissociation. In toto, these studies not only defined the stoichiometry of GAPDH–tubulin binding but also demonstrated their physiological relevance.

Further analysis demonstrated that the binding of GAPDH to tubulin resulted in an inhibition of enzyme activity (Durrieu et al., 1987b). Analysis of precipitated and soluble GAPDH indicated a recovery of approximately 85% of the added enzyme activity with 17% of the initial activity contained in the precipitate. The latter contained 40% of the protein suggesting that GAPDH binding to tubulin diminished its catalytic activity. Increasing the amount of tubulin in the in vitro assay resulted in a further decrease in enzyme activity.

To begin to consider the mechanism that underlay that finding, tetrameric GAPDH was dissociated into the monomeric form by incubation with 15 mM ATP for 14 days. Intriguingly, whereas incubation of tetrameric GAPDH resulted in its partial association with tubulin (as defined by coprecipitation), incubation of the monomeric form resulted in its complete association with tubulin in the precipitate. It was noted that these findings were in accord with the model previously proposed that tetrameric GAPDH binds to tubulin, while its dissociation into dimers resulted in its dissociation from the complex in an ATP-dependent manner (Huitorel and Pantaloni, 1985).

In the next study, an agarose gel electrophoresis protocol was utilized to examine GAPDH–tubulin binding (Karkhoff-Schweizer and Knull, 1987).[1] It was observed that, by themselves, commercial rabbit muscle GAPDH migrated toward the cathode, while tubulin migrated toward the anode. In contrast, 1:10 mixture of GAPDH and tubulin resulted in their comigration, i.e., only a single protein band was identified by Coomassie brilliant blue staining. However, that band did not migrate to either the GAPDH or the tubulin position. Instead, it was observed toward the anode but in a lower position from that observed with tubulin alone.

Increasing the concentration of GAPDH progressively in the reaction mixture (ultimately to a 1: 1 ratio) resulted in a sequential change in the migration pattern. In particular, the position of the complex continued to move toward the cathode. At the final 1:1 concentration, it appeared that the complex migrated to a position approximately equidistant from both tubulin and GAPDH. This suggests that there was an equivalence of charges between both proteins.

One of the tenets of moonlighting GAPDH activity is the generality of its new structural identification as well as that of its new functional activity. So, the observation that GAPDH binds microtubules in a colon cancer line is a further testament to that GAPDH property (Launay et al., 1989). Affinity chromatography was performed using taxol-stabilized microtubules. SDS-PAGE analysis identified a 35-kDa protein following its adsorption and elution from the matrix. Gel permeation studies indicated an $M_r = 140\,kDa$. Immunoblot analysis confirmed GAPDH as the 35-kDa protein. Enzyme assay demonstrated that the 35-kDa protein exhibited GAPDH activity.

Changes in $A_{350}$ indicated that the 35-kDa protein enhanced tubulin turbidity comparable to that observed for rabbit muscle GAPDH. Centrifugation studies demonstrated its physical association with tubulin. Immunoblot analysis identified the 35-kDa protein as GAPDH. Stoichiometric analysis revealed the binding of one GAPDH tetramer per two tubulin dimers. Electron microscopic studies demonstrated that incubation with GAPDH resulted in the formation of microtubule bundles (Fig. 5.1,IC). Thus, these studies extend the ability of moonlighting GAPDH to another cell type to function in tubulin polymerization.

In a subsequent study, centrifugation and fluorescence anisotropy were used as the experimental paradigms to examine the interaction of GAPDH with tubulin (Walsh et al., 1989).[1] As before, commercial rabbit muscle GAPDH and purified bovine brain tubulin were utilized. As expected, significant GAPDH activity was recovered in the pellet (60% of activity). This was considerably higher than previous estimates. Noticeably, addition of polyethylene glycol (to hydrate tubulin) did not affect GAPDH activity. Studies using FITC-labeled GAPDH demonstrated a tubulin concentration–dependent increase in anisotropy, which saturated at 5 µM tubulin. Again, this range of

---

1. Other glycolytic enzymes were examined as well. The discussion is limited to GAPDH.

concentrations is within that observed in vivo, indicating the physiological relevance of these in vitro studies.

## 1.1 GAPDH as a *Trypanosomatid* Cytoskeletal Tubulin-Binding Protein

*Trypanosomatids* are infectious protozoans that induce human pathologies (most notably sleeping sickness). Three proteins, COP-33, COP-41, and COP-61, were isolated from the *trypanosomatid* cytoskeleton. The latter is composed of the plasma membrane and associated microtubules. COP-41 was identified as GAPDH. Further, *trypanosomatids* contain an intracellular glycosome in which glycolytic enzymes are located.

A recent study compared the microtubule-binding properties of the COP-33, COP-61, and GAPDH-g and GAPDH-m (Kambadur et al., 1990). For the purposes of this discussion, only those findings with respect to GAPDH will be considered (GAPDH-g is a glycosomal protein but is also isolated from the cytoskeleton; GAPDH-m is muscle GAPDH). The native $M_r$ of GAPDH-g was 165,000 or 172,000 as determined by gel filtration and by equilibrium sedimentation, respectively. Each was tetrameric. Immunoassay indicated that there was a 1:3 ratio of GAPDH in the cytoskeleton as compared to that detected in the glycosome.

GAPDH-g binding to tubulin was quantitated using radioiodinated ligand pelleted by centrifugation. These studies demonstrated that GAPDH-g and GAPDH-m bound both to microtubules and to tubulin in a concentration-dependent manner. Analysis of cooperativity was considered equivocal for GAPDH-g. GAPDH-m exhibited cooperativity on when high levels of detergent were present. In toto, these studies, along with those in colon cancer cells, demonstrate the generality of GAPDH–tubulin binding.

Subsequent studies used an in vitro analysis to examine the effects of multiple factors on GAPDH-induced microtubule bundling (Somers et al., 1990). In this investigation, cosedimentation, determination of soluble and precipitate GAPDH activity as well as electron microscopic analysis were used as experimental paradigms. As expected, the binding of GAPDH to microtubules was GAPDH dependent, achieving saturation at 20 μM GAPDH. However, the enzyme concentration curves varied considerably as a function of ionic strength, i.e., at low ionic strength, 0.03 μM, a biphasic profile was observed along with an approximately twofold increase of GAPDH bound as compared to that observed at high ionic strength (0.1 μM). It was suggested that this ionic strength effect may be due to its effect on GAPDH binding to high-affinity versus low-affinity binding sites.

The combination of changes in ionic strength and the addition of 1 mM p[NH]ppA also were examined. The partial inhibition observed with p[NH]ppA was independent of changes in ionic strength. As incubation of the inhibitor with GAPDH by itself did not change its sedimentation as a tetramer, it was postulated that the inhibitor effect was not due to a tetrameric to monomeric dissociation. Instead a change in tetrameric GAPDH conformation was postulated as the basis for the partial inhibition observed. This observation would seem to contradict the models proposed earlier for that change in oligomeric structure as the basis for inhibition of GAPDH–tubulin binding (Huitorel and Pantaloni, 1985). Finally, the effect of microtubule-associated proteins was examined. No effect was observed. This observation is not in accord with that observed previously in both porcine and in colon cancer cells (Kumagai and Sakai, 1983; Launay et al., 1989). However, the latter two results are referred to as "unpublished findings" or "data not shown." Accordingly, the basis for this discrepancy is unknown.

An important question in in vitro studies of this kind is whether or not the experimental paradigm utilized may influence not only the experimental findings but also the conclusions derived from those results. In a subsequent study, a series of protocols were used to define the binding constants and the stoichiometries of GAPDH–tubulin bindings (Muronetz et al., 1994). In particular, each protein was coupled to active Sepharose so that the binding of the other could be determined.

The addition of soluble tubulin diminished GAPDH activity in a concentration manner with complete elimination of catalysis at c. 25 nM tubulin. Stoichiometric quantitation indicated that a tubulin dimer:GAPDH ratio of 1:2 was required for complete activity inhibition. NAD$^+$ analysis indicated a concentration dependent on soluble tubulin inhibition of GAPDH catalysis. At 0.1 mM there was no effect. Increasing NAD$^+$ concentration decreased tubulin inhibition. As a similar effect was detected with 80 mM KCl, the effect appears to be due to ionic strength and not in its capacity as a GAPDH-binding molecule.

Using the Sepharose protocol, it was determined that bound tubulin inhibited GAPDH activity to a comparable extent as that observed with soluble tubulin. Stoichiometric analysis indicated a 0.91 GAPDH:tubulin dimer ratio for GAPDH saturation, which was comparable to that observed in the soluble tubulin reaction. This suggested that bound tubulin contained the same structural parameters that defined GAPDH binding. Intriguingly, reversing the two proteins (Sepharose-bound GAPDH vs. free tubulin), resulted in the lack of a tubulin inhibition of GAPDH activity even at a 7 μM tubulin concentration. This suggested that the GAPDH tubulin binding was blocked as a function of GAPDH immobilization on the Sepharose matrix.

Although the previously described studies demonstrate microtubule bundling, the ability of microtubules to form extended structures remained unknown. Analogous to enzyme studies, it is necessary to define structure from the primary sequence to the quaternary conformation to understand function. With respect to GAPDH–tubulin interactions, the latter were examined using transmission electron microscopy (Volker et al., 1995).[1]

As the experimental paradigm, these studies quantitated structure as a function of tubulin and of GAPDH concentrations. The formation of GAPDH-induced cross-linked microtubular structures was dependent on both (Fig. 5.1,ID). Of note, the concentrations of each utilized represent those detected in vivo, thereby demonstrating the physiological relevance of the studies. In addition, the protocols used permitted the detection of what were termed "large-scale multiple bundling complexes." Several models were proposed, which traced complex formation as a function of tubulin concentration. Most notably, in each model, the interaction of GAPDH resulted in a significant increase in network complexity.

As described in the previous study, it was proposed that tubulin must contain multiple binding sites for GAPDH interaction. In a subsequent report, the investigations that defined that site were described (Volker and Knull, 1997). In this study, a chemical procedure using a mixture of DMSO/HCl/HBr was used to cleave tubulin at tryptophan residues. Subsequent DEAE-HPLC and reversed phase column chromatography were used for peptide purification. Their binding to GAPDH, lactate dehydrogenase muscle (LDH$_m$), and pyruvate kinase (PK) was determined by affinity chromatography and by enzyme activity. A α-43 peptide (α-mer) displayed the greatest affinity. It was localized to the α C-terminal 409–451 domain.

In inhibition assays, it displayed the greatest levels of binding inhibition of GAPDH to tubulin and also inhibited GADPH enzyme activity by itself in a concentration-dependent manner. With respect to the latter, incubation with 5 µM of the peptide inhibited catalysis by 75%. In contrast, an α-30-mer of the peptide inhibited catalysis by c. 50%. The ability of the peptide to inhibit GAPDH activity was diminished by NADH. It was noted that the GAPDH-binding site of the Band 3 protein was located in its acidic N-terminal region and that the C-terminal domain of tubulin is similarly acidic, suggesting a common configuration of both.

In summary, in this section, exhaustive in vitro studies are presented, which describe the structural interactions between GAPDH and tubulin. Notably, they report that this interaction occurs at GAPDH and tubulin physiological concentrations; there is a pronounced effect of ATP on microtubule bundling; the dissociation of the GAPDH–tubulin complex may or may not result in changes in GAPDH oligomeric structure; and, lastly, that GAPDH forms actively large cross-linked microtubule structures. Evidence for the functional significance of this protein–protein interaction is described below.

## 2.  ROLE OF GAPDH–TUBULIN INTERACTIONS IN MEMBRANE TRAFFICKING

Initial studies indicated the role of Rab2 GTPase with respect to intracellular transport from the ER to the Golgi apparatus (Tisdale et al., 1992). In particular, mutational analysis demonstrated that alterations in Rab2 structure inhibited transport as compared to that observed for the wild-type (wt) protein (Tisdale and Balch, 1996). Further study indicated that Rab2 was involved in the formation of pre-Golgi intermediates, termed vesicular tubular clusters (VTCs, Tisdale, 1999). Rab2 is responsible for the recruitment of the β-coat protein (β-COP) and requires protein kinase Cι/λ (PKCι/λ) during the process of vesicle construction (Tisdale and Jackson, 1998; Tisdale, 2000, respectively). In toto, VTCs contained Rab2, Coatomer proteins, PKCι/λ, and the p53/gp58 recycling proteins.

As described above, Rab2 mediated the formation of vesicular tubular complexes involved in membrane trafficking. Since evidence was presented that GAPDH was not only structurally a membrane-bound protein but also that it exhibited discrete moonlighting membrane activities (Chapter 4), a series of studies were performed to determine whether it may also be a VTC protein with defined functions as well. As indicated in Table 5.2, those studies defined it as a structural VTC protein, its functional role in microtubule bundling as the "yellow brick road" of ER to Golgi transit and the intrinsic requirement for its sequential, dual posttranslational phosphorylation to accomplish the construction of that intracellular highway.

## 2.1 Structural Determination of GAPDH Binding Within the Vesicular Tubular Cluster

Initially, the physical association of GAPDH within the VTC was determined and the resultant functional consequences were examined (Tisdale, 2001). The experimental paradigms utilized were density gradient cosedimentation coupled with immunoblot analysis to determine structure. ER to Golgi transport of a temperature-sensitive vesicular stomatitis virus glycoprotein (VSV-G) was used to assess function.

To assess binding, microsomes were incubated with recombinant Rab2 to stimulate vesicle formation. Subsequent to isolation, the membrane fraction was analyzed by equilibrium density centrifugation. The sedimentation of β-COP as a representative VTC protein and that of GAPDH were determined by immunoblot analysis. Cosedimentation of both was observed providing the initial indication that Rab2 may induce VTC GAPDH binding. The effect of Rab2 on GAPDH binding was dose dependent as increased levels of bound GAPDH were observed as a function of Rab2 concentration. Further, as defined by the use of a 13-mer, the Rab2 GAPDH-binding site was localized to its N-terminus.

Inhibition of ER to Golgi transport of VSV-G by an anti-GAPDH antibody was used as the protocol to examine the functional consequences of GAPDH as a VTC protein. Normally, ER to Golgi VSV-G transport occurs within 20 min at the permissive temperature. Addition of 2 μg of the GAPDH antibody inhibited transport by 50%, while the addition of 10–15 μg eliminated its movement. Further, while immunofluorescence demonstrates the localization of VSV-G in the Golgi at the permissive temperature, in GAPDH

**TABLE 5.2 Role of GAPDH in Membrane Trafficking[a]**

| Cell Source Organism Recombinant Constructs | Experimental Paradigm | Experimental Finding | "Unique" Finding | Reference |
|---|---|---|---|---|
| Rat kidney fibroblasts | Density gradient, centrifugation, immunoblot | Rab2 recruits GAPDH to vesicular tubular clusters (VTCs) | GAPDH antibodies inhibit ER to Golgi transport | Tisdale (2001) |
| Rabbit muscle GAPDH, rat kidney fibroblasts | PKCι/λ kinase assay, microtubule bundling | GAPDH phosphorylated, GAPDH$^{serP}$ present in VTCs, required for β-tubulin binding | GAPDH required for microtubule bundling and for microtubule organization in vivo | Tisdale (2002) |
| Recombinant constructs, HeLa cells | Immunoprecipitation, transfection, microsomal binding | Rab2 binds to PKCι/λ regulatory domain | Rab2: PCKι/λ binding inhibits GAPDH phosphorylation | Tisdale (2003) |
| Recombinant constructs, HeLa, BL21(DE3)pLysS cells | Mammalian two-hybrid assay, immunoprecipitation | Rab2 contains a GAPDH-binding site | GAPDH glycolytic activity not required for ER to Golgi transit | Tisdale et al. (2004) |
| Normal rat kidney cells, recombinant constructs | In vitro kinase assay, Rab2-mediated microsome binding | Rab2 mediates Src microsome binding, Src tyrosine phosphorylates PKCι/λ | Tyr$^P$ required for PKCι/λ membrane binding, Src binds GAPDH | Tisdale and Artalejo (2006) |
| Normal rat kidney, HeLa cells | Construction of GAPDH mutants (GAPDH$^{Y41F}$) | Src tyrosine phosphorylates GAPDH$^{41}$ | GAPDH$^{41tyrP}$ required for ER to Golgi transport | Tisdale and Artalejo (2007) |
| HeLa cells, recombinant constructs | Membrane, microtubule-binding assays, immunofluorescence | Rab2 and PKCι/λ colocalize with motor protein dynein | GAPDH antibody blocks cynein binding | Tisdale et al. (2009) |

[a]*Chronological order.*

antibody–treated cells, it was localized in small structures that surrounded the nucleus. Further analysis demonstrated that the GAPDH antibody did not inhibit Rab2-stimulated vesicle budding but did inhibit transport in a reconstituted ER to Golgi assay. As such, this was the first indication of a functional consequence of VTC-bound GAPDH.

Structure–function studies also examined the role of GAPDH catalysis on this new moonlighting activity (Tisdale et al., 2004). As noted, GAPDH contains an active site cysteine that is required for catalysis. As will be discussed in other chapters (see Chapter 8), modification of that amino acid reducing catalytic activity is a not uncommon feature of moonlighting GAPDH function.

In these studies, GAPDH[cys149] was mutated (C149G), which abolished enzyme activity. In spite of that change in protein structure, Rab2 facilitated the incorporation of GAPDH[C149G] into VTCs in a dose-dependent manner. Similarly, using the previously described kinase assay, the mutant GAPDH protein was phosphorylated by PKCɩλ to a similar level observed for the wt protein. Further using the VSV-G model, the effect of the mutant on membrane trafficking was negligible as compared to the wt protein. In toto, these findings demonstrate that GAPDH-mediated ER to Golgi transport is independent of its enzyme activity. These findings suggest that the characteristics of this mutant are comparable to the fusogenic GAPDH identified previously (Glaser and Gross, 1995). Finally, structural analyses of wt GAPDH and of the mutant GAPDH indicate that both are tetrameric when incorporated into the VTC. This suggests that their binding to microtubules forming the intracellular "yellow brick road" is not accompanied by their oligomeric dissociation into dimers or monomers.

## 2.2 The Role of Kinase-Mediated Phosphorylation in GAPDH–Vesicular Tubular Cluster Structure and Function

By definition, kinase reactions, as a mechanism of cell singling, provide a means to initiate or to terminate the activity of biochemical pathways through this now classical posttranslational protein modification. As will be discussed below, membrane trafficking is no exception, providing a prototypic model for kinase regulation of cell function.

Initially, VTC–GAPDH structure and function were analyzed (Tisdale, 2002). In particular, with respect to the former, its potential phosphorylation, along with that of β-COP and Rab2, by PKCɩ/λ was examined. The rationale was that, since this kinase was present in the VTC, along with each protein, it may be that not only could one or more of the VTC proteins be phosphorylated but also that this posttranslational modification was necessary for membrane trafficking.

In an in vitro kinase reaction, no phosphorylation was observed using either β-COP or Rab2 as substrates. The former was observed using either rat liver cytosol or immunoprecipitated protein, thereby demonstrating the lack of a contaminant phosphatase. The latter was observed using wt Rab2 or several mutant constructs, which were either constitutively active or which exhibited different affinities for GTP.

In contrast, rabbit muscle GAPDH or purified rat liver GAPDH were suitable substrates for PKCι/λ. Further analysis indicated the physical association of GAPDH with PKCι/λ. Intriguingly, PKCι/λ is a phosphatidylserine (PS)-binding protein, as is GAPDH (Kaneda et al., 1997). The PKCι/λ PS-binding site is located in its regulatory domain. Accordingly, GAPDH binding was examined using two PKCι/λ constructs, in which one contained the regulatory domain and the other contained the catalytic domain. Nitrocellulose slot blot analysis demonstrated that GAPDH bound only to the regulatory domain.

In a complementary study, a microsomal assay was performed, which contained both [$\gamma$-$^{32}$P] ATP and Rab2 (to promote VTC formation). Following its isolation by centrifugation, radiolabel in the VTC was determined. No radiolabel was detected in the control reaction. In contrast, radiolabel was detected in VTC isolated from assays that contained Rab2. The radiolabeled protein was identified as GAPDH. Using phosphoserine and phosphotyrosine antibodies in the immunoblot suggested that the modified GAPDH amino acid was serine. Further, using competition with a mutant PKCι/λ construct defective in kinase activity, no radiolabeled GAPDH was detected. Cumulatively, these two studies suggest that GAPDH is serine phosphorylated by PKCι/λ when it is present in the VTC. Further, the observation that both GAPDH and PKCι/λ contain PS-binding sites suggests the possibility of a defined intra-VTC structural alignment between these two proteins.

Subsequently, the role of Rab2 in regulation of VTC-associated PKCι/λ activity was examined (Tisdale, 2003). Coimmunoprecipitation using an anti-Rab2 antibody demonstrated the physical association of both proteins in HeLa cells. Their binding was confirmed in vivo using a mammalian two-hybrid assay with CAT expression as the experimental paradigm. The latter was increased approximately 15-fold in cotransfected cells. Similarly, in a GST pull-down study, immobilized GST-PKCι/λ-bound Rab2.

Recombinant constructs were prepared to determine binding sites. As the Rab2 amino terminus contained a GAPDH-binding site (Tisdale, 2001), Rab2 deletion mutants were tested for PKCι/λ binding. Mutants missing residues 1–8 or 1–14 exhibited reduced binding, while no association was observed with Rab2N'Δ19 (which lacked the first 19 amino acids). In contrast, Rab2 constructs encoding sequences 2–7, 2–9, and 2–19 exhibited increased PKCι/λ binding. As no association was observed with Rab2-containing residues 20–211, it is reasonable to conclude that the Rab2 amino terminus contained the PKCι/λ-binding site. Reversing the protocol using PKCι/λ constructs, Rab2 bound to the PKCι/λ regulatory domain.

The physiological significance of the Rab2 amino terminus was demonstrated by the use of recombinant Rab2 or the Rab2N'Δ19 construct in the microsomal assay. Whereas addition of the recombinant Rab2 protein resulted in PKCι/λ incorporation into the VTC, no such effect was seen with Rab2N'Δ19. Further, the physical association of GAPDH with the VTC exhibited a similar response, i.e., association was observed with the wt protein but not the mutant.

The functional significance of the Rab2–PKCι/λ interaction was tested in an in vitro kinase assay using recombinant PKCι/λ, which quantitated its phosphorylation of GAPDH. Addition of Rab2 resulted in the virtual elimination of GAPDH$^{serP}$. Similarly, addition of the Rab2 13-mer diminished GAPDH phosphorylation. In contrast, no inhibitory effect was observed using Rab2N'Δ19. Therefore, it was postulated that Rab2 binding may modulate PKCι/λ activity.

### 2.2.1 Role of Src in Protein Kinase Cι/λ Structure and Function

As indicated above, initial studies indicated the role of PKCι/λ in the formation of GAPDH$^{serP}$. Given its role in membrane trafficking, the regulation of PKCι/λ activity may also be of interest, i.e., other analyses had indicated an Src–PKCι/λ association (Wooten et al., 2001). For that reason, the role of Src modification of PKCι/λ on its structure and function was determined (Tisdale and Artalejo, 2006).

Using the standard microsomal assay, Rab2 increased membrane Src binding in a dose-dependent manner. Addition of an anti-Rab2 antibody blocked Src binding. Immunoprecipitation analysis indicated not only the association of Src with Rab2 but also demonstrated the presence of GAPDH and PKCι/λ in the immune complex. Intriguingly, using a pull-down assay, Src interacted with PKCι/λ and with GAPDH but not with Rab2. It was suggested that Src interacts indirectly with Rab2 due to its PKCι/λ- and GAPDH-binding capacity.

The ability of Src to phosphorylate PKCι/λ was demonstrated in an in vitro kinase assay. Analysis of PKCι/λ in the VTC structure indicated that it, also, was tyrosine phosphorylated. Further, addition of an Src inhibitor to the standard microsome assay diminished both PKCι/λ phosphorylation and its presence in the VTC. In addition, as PKCι/λ is required for β-COP recruitment, it was noted that there was a coordinate decrease in its presence in the VTC. In toto, these findings indicate a further complexity in the role of protein phosphorylation in VTC formation.

### 2.2.2 Role of Src in Vesicular Tubular Cluster GAPDH Structure and Function

The observations that Rab2 formed a protein "quartet" composed of itself, Src, GADPH, PKCι/λ; Src-induced formation of PKCι/λ$^{tryP}$; and that PKCι/λ$^{tryP}$ was required for VTC formation suggested the possibility that Src-mediated phosphorylation of other VTC proteins may also be a requirement for membrane trafficking.

As previous studies had indicated modification of GAPDH$^{tyr41}$ as a potential candidate (Rush et al., 2005), a GAPDH mutant (GAPDH$^{Y41F}$) was constructed to consider the importance of Src-mediated phosphorylation of GAPDH$^{try41}$ in membrane trafficking (Tisdale and Artalejo, 2007). Incubation in vitro of Src with both GAPDH$^{try41}$ and GAPDH$^{Y41F}$ demonstrated the phosphorylation of the former but not the latter. Subsequently, both were tested for Rab2-mediated membrane binding. Each was membrane incorporated by Rab2. Intriguingly, kinetic analysis suggested a specific temporal sequence such that Src is bound first and GAPDH$^{tyr41}$ second. An in vitro VSV-G transport assay was used to define the role of Src-mediated GADPH$^{try41}$ and GAPDH$^{Y41F}$ on membrane trafficking. The latter was detected when wt GAPDH was added to the assay. In contrast, incubation with GAPDH$^{Y41F}$ diminished trafficking by c. 67%.

The mechanism through which GAPDH$^{Y41F}$ affects membrane trafficking was defined by a nitrocellulose-binding assay in which GAPDH$^{tyr41}$ or GAPDH$^{Y41F}$ was bound. The blot was overlaid with Rab2 or with PKCι/λ. Binding of either protein was determined by antibody–second antibody chemiluminescence. Both GAPDH species bound Rab2, while only GAPDH$^{try41}$ bound to PKCι/λ. As a further confirmation, coimmunoprecipitation analysis demonstrated recognition of GAPDH$^{try41}$ but not GAPDH$^{Y41F}$ by PKCι/λ. In toto, these findings suggest that membrane trafficking requires two sequential GAPDH phosphorylations (Fig. 5.1,IIIB and C). The first is by Src at GAPDH$^{tyr41}$; the second is a serine modification to produce GAPDH$^{serP}$. The location of that modification is unknown at the present time. In toto, these studies highlight both the complexity of GAPDH structure and function in VTCs and that of kinase function in the regulation of membrane trafficking.

## 2.3 The Role of Vesicular Tubular Cluster–Associated GAPDH in Tubulin Structure and Function

As discussed earlier in this chapter, detailed structural studies defined GAPDH–tubulin interactions. Accordingly, the role of GAPDH$^{serP}$ in relation to VTC–tubulin binding was examined (Tisdale, 2002).[2] Following the standard microsomal assay, differential centrifugation was used to separate microsomal-associated microtubules. As defined by immunoblot analysis, this protocol established that Rab2 increased membrane-associated GAPDH and membrane-associated microtubules (detected as β-tubulin) in a coordinate manner. The latter finding may be of particular interest as previous studies demonstrated that GAPDH bound to α-tubulin and did not bind to β-tubulin (Volker and Knull, 1997).

To assess the role of GAPDH in β-tubulin binding, anti-GAPDH antibody was added to the standard microsomal assay. As with the previously observed decrease in GAPDH binding, there was a decrease in the binding of β-tubulin. Further, using two different GAPDH antibody concentrations,

---

2. The studies identifying the dual phosphorylation were performed later.

the reduction in β-tubulin binding was proportional to the amount of GAPDH antibody added. Inhibition of PKCι/λ in the assay decreased both GAPDH$^{serP}$ and microtubule membrane association. Lastly, dephosphorylation of GAPDH$^{serP}$ with alkaline phosphatase resulted in a significant reduction of membrane-associated microtubules (detected as diminished β-tubulin by immunoblot analysis).

To define the physiological significance of these findings, immunofluorescence was used to examine the intracellular microtubule distribution and structure as a function of Rab2-mediated VTC formation. In the absence of Rab2, microtubules exhibited a cytosolic localization with some concentration near the nucleus (illustrated in Fig. 5.1,II). In contrast, in cells transfected with activated Rab2, an extensive cross-linked, filamentous microtubule structure was detected (Fig. 5.1,ID). Parallel examination of GAPDH indicated that, in the absence of Rab2, GAPDH exhibited its classical pattern of cytoplasmic localization (Cool and Sirover, 1989; Sirover, 1997). However, in the presence of activated Rab2, GAPDH now was localized in the microtubule filamentous structures as defined by costaining. Transfection of the anti-GAPDH antibody eliminated the filamentous microtubule structure. Cumulatively, these findings demonstrate the role of GAPDH$^{serP}$ as a microtubule bundling protein associated with membrane trafficking. This is illustrated in Fig. 5.1,IIIA and B. Further, these studies validate the physiological relevance of the in vitro GAPDH–tubulin structural studies described in Section 1.

The studies as described above demonstrated in an elegant manner the formation of VTCs and the construction of the microtubule network on which transport can occur. That being said, there is one component missing, i.e., the component that is capable of moving cargo from the ER to the Golgi. In a subsequent study, that component was identified as the motor protein dynein (Tisdale et al., 2009).

Sucrose gradient centrifugation was used to determine whether Rab2 bound to microtubules. In this instance the microtubules were localized in the pellet and Rab2 in the supernatant. In contrast, incubation of the microtubules with GAPDH or with PCKι/λ resulted in their cosedimentation with microtubules. However, when the Rab2–microtubule binding assay was performed in the presence of GAPDH and PCKι/λ, Rab2 now cosedimented with the microtubules. It was suggested that Rab2 binding was indirect, using GAPDH and PCKι/λ as either adapter or scaffolding proteins. Colocalization of Rab2 and tubulin as defined by indirect immunofluorescence into VTCs was observed in HeLa cells transfected with Rab2 cDNA. Further, these two proteins colocalized with the p53/p58 recycling protein.

The requirements for Rab2–tubulin association were defined using a membrane-binding assay. As expected, both PCKι/λ and GAPDH were required. Substitution of the recombinant mutant protein, Rab2N'Δ19 (which did not contain the N-terminus Rab2-binding site), eliminated tubulin association. In addition, the membrane association of both the motor proteins, kinesin and dynein, were examined by probing the immunoblot with the respective antibodies. This analysis revealed not only the detection of dynein but also that its membrane association was Rab2 dose dependent. Addition of the Rab2N'Δ19 mutant failed to recruit dynein into the membrane fraction. Addition of the GAPDH antibody blocked both PCKι/λ and dynein membrane binding.

Characterization of dynein as a membrane component demonstrated that the kinase activity of PCKι/λ was required, suggesting that GAPDH$^{serP}$ was necessary for dynein binding. Indirect immunofluorescence demonstrated the colocalization of dynein and PCKι/λ in vivo as well as its codistribution with Rab2. Accordingly, these studies demonstrate that VTC-mediated membrane trafficking is accomplished through the use of dynein as a motor protein through the GAPDH constructed microtubule highway. A model is presented in Fig. 5.1,IIIA–D.

## 3. SUMMARY

The investigations described in this chapter highlight a novel function of moonlighting GAPDH, i.e., its interactions with tubulin and the consequences thereof. That being said, there is a distinctive difference in the structural as compared to the functional characteristics of GAPDH–tubulin binding. The former are straightforward, clear, and distinct. Although of elegant design, they are "simple" in nature as illustrated in Fig. 5.1,IA–D. They describe GAPDH-mediated formation of microtubules (Fig. 5.1,IB); GAPDH-mediated microtubule bundling (Fig. 5.1,IC); and GAPDH-mediated formation of complex cross-linked microtubule networks (Fig. 5.1,D). In contrast, the functional studies are highly complex, describing not only a sequential series of protein–protein interactions but also the role of kinase function in the regulation of VTC formation, membrane trafficking, and GAPDH-mediated formation of the highway on which membrane trafficking may occur (Fig. 5.1,II and IIIA–D). As such, each makes a substantial contribution to our understanding of moonlighting GAPDH structure and function.

## REFERENCES

Cool, B., Sirover, M., 1989. Immunocytochemical localization of the base excision repair enzyme uracil DNA glycosylase in quiescent and proliferating normal human cells. Cancer Res. 49, 3029–3036.

Durrieu, C., Bernier-Valentin, F., Rousset, B., 1987a. Binding of glyceraldehyde-3-phosphate dehydrogenase to microtubules. Mol. Cell. Biochem. 74, 55–65.

Durrieu, C., Bernier-Valentin, F., Rousset, B., 1987b. Microtubules bind glyceraldehyde-3-phosphate dehydrogenase and modulate its enzyme activity and quaternary structure. Arch. Biochem. Biophys. 252, 32–40.

Glaser, P., Gross, R., 1995. Rapid plasmenylethanolamine-selective fusion of membrane bilayers catalyzed by an isoform of glyceraldehyde-3-phosphate dehydrogenase: Discrimination between glycolytic and fusogenic roles of individual isoforms. Biochemistry 24, 12193–12203.

Huitorel, P., Pantaloni, D., 1985. Bundling of microtubules by glyceraldehyde-3-phosphate dehydrogenase and its modulation by ATP. Eur. J. Biochem. 150, 265–269.

Kambadur, R., Lewis, M., Chang, S., Flavin, M., 1990. Characterization of putative cytoskeletal proteins from a *trypanosomatid* and their comparative binding to microtubules and soluble tubulin. J. Biol. Chem. 265, 20959–20965.

Kaneda, M., Takeuchi, K., Inoue, K., Umeda, M., 1997. Localization of the phosphatidylserine binding site of glyceraldehyde-3-phosphate dehydrogenase responsible for membrane fusion. J. Biochem. 122, 1233–1249.

Karkhoff-Schweizer, R., Knull, H., 1987. Demonstration of tubulin-glycolytic enzyme interactions using a novel electrophoretic approach. Biochem. Biophys. Res. Commun. 146, 827–831.

Kumagai, H., Sakai, H., 1983. A porcine brain protein (35K protein) which bundles microtubules and its identification as glyceraldehyde-3-phosphate dehydrogenase. J. Biochem. 93, 1259–1269.

Launay, J., Jellali, A., Vanier, M., 1989. Glyceraldehyde-3-phosphate dehydrogenase is a microtubule binding protein in a human colon tumor cell line. Biochim. Biophys. Acta 996, 103–109.

Muronetz, V., Wang, Z.-X., Keith, T., Knull, H., Srivastava, D., 1994. Binding constants and stoichiometries of glyceraldehyde-3-phosphate dehydrogenase-tubulin complexes. Arch. Biochim. Biophys. 313, 253–260.

Rush, J., Moritz, A., Lee, K., et al., 2005. Immunoaffinity profiling of tyrosine phosphorylation in cancer cells. Nat. Biotechnol. 23, 94–101.

Sirover, M., 1997. Role of the glycolytic protein, glyceraldehyde-3-phosphate dehydrogenase, in normal cell function and in cell pathology. J. Cell. Biochem. 66, 133–140.

Somers, M., Engelborghs, Y., Baert, J., 1990. Analysis of the binding of glyceraldehyde-3-phosphate dehydrogenase to microtubules, the mechanism of bundle formation and the linkage effect. Eur. J. Biochem. 193, 437–444.

Tisdale, E., 1999. A Rab2 mutant with impaired GTPase activity stimulates vesicle formation from pre- Golgi intermediates. Mol. Biol. Cell. 10, 1837–1849.

Tisdale, E., 2000. Rab2 requires PKCι/λ to recruit β-COP for vesicle formation. Traffic 1, 702–712.

Tisdale, E., 2001. Glyceraldehyde-3-phosphate dehydrogenase is required for vesicular transport in the early secretory pathway. J. Biol. Chem. 276, 2480–2486.

Tisdale, E., 2002. Is phosphorylated by protein kinase Cι/λ and plays a role in microtubule dynamics in the early secretory pathway. J. Biol. Chem. 277, 3334–3341.

Tisdale, E., 2003. Rab2 interacts with atypical protein kinase C (aPKC) ι/λ and inhibits a PKCι/λ-dependent glyceraldehyde-3-phosphate dehydrogenase phosphorylation. J. Biol. Chem. 278, 52524–52530.

Tisdale, E., Artalejo, C., 2006. Src-dependent aprotein kinase Cι/λ (aPKCι/λ) tyrosine phosphorylation is required for aPKCι/λ association with Rab2 and glyceraldehyde-3-phosphate dehydrogenase on pre-Golgi intermediates. J. Biol. Chem. 281, 8436–8442.

Tisdale, E., Artalejo, C., 2007. A GAPDH mutant defective in Src-dependent tyrosine phosphorylation impedes Rab2-mediated events. Traffic 8, 733–741.

Tisdale, E., Azizi, F., Artalejo, C., 2009. Rab2 utilizes glyceraldehyde-3-phosphate dehydrogenase and protein kinase Cι to associate with microtubules and to recruit dynein. J. Biol. Chem. 284, 5876–5884.

Tisdale, E., Balch, W., 1996. Rab2 is essential for the maturation of pre-Golgi intermediates. J. Biol. Chem. 271, 29372–29379.

Tisdale, E., Bourne, J., Khosravi-Far, R., Der, C., Balch, W., 1992. GTP-binding mutants of Rab1 and Rab2 are potent inhibitors of vesicular transport from the endoplasmic reticulum to the Golgi complex. J. Cell. Biol. 119, 749–761.

Tisdale, E., Jackson, M., 1998. Rab2 protein enhances coatomer recruitment to pre-Golgi intermediates. J. Biol. Chem. 273, 17269–17277.

Tisdale, E., Kelly, C., Artalejo, C., 2004. Glyceraldehyde-3-phosphate dehydrogenase interacts with Rab2 and plays an essential role in endoplasmic reticulum to Golgi transport exclusive of its glycolytic activity. J. Biol. Chem. 279, 54046–54052.

Volker, K., Reinitz, C., Knull, H., 1995. Glycolytic enzymes and assembly of microtubule networks. Comp. Biochem. Physiol. 112B, 503–514.

Volker, K., Knull, H., 1997. A glycolytic enzyme binding domain on tubulin. Arch. Biochem. Biophys. 338, 237–243.

Walsh, J., Keith, T., Knull, H., 1989. Glycolytic enzyme interactions with tubulin and microtubules. Biochim. Biophys. Acta 999, 64–70.

Wooten, M., Vandenplas, M., Seibenhener, M., Geetha, T., Diaz-Meco, M., 2001. Neuropeptide-induced androgen independence in prostate cancer cells: roles of nonreceptor tyrosine kinases Etk/Bmx, Src and focal adhesion kinase. Mol. Cell. Biol. 21, 8414–8427.

Chapter 6

# Moonlighting GAPDH and the Maintenance of DNA Integrity: Preservation of Genetic Information and Cancer Facilitation

*Everything has its advantages and disadvantages*

A common idiom

The conservation of genetic information in cellular DNA is an intrinsic requirement for cell function and for cell viability. Accordingly, DNA modification or chromosomal alteration may present existential threats to both cells and to organisms. As such, DNA repair pathways and protocols that ensure chromosomal stability represent advantageous mechanisms through which cells recognize and remove DNA lesions following which resynthesis may restore the original, correct nucleotide sequences. In terms of organization, such enzymatic pathways are comparable to other types of cell metabolic systems with respect to their complexity and with respect to structural interrelationships. That being said, DNA repair pathways may be uniquely disadvantageous, given their function in cancer cells, they may serve to mitigate the efficacy of radio- or chemotherapeutic regimens designed to cure the respective malignancy.

Recent evidence indicates that GAPDH fulfills important functions with respect to the preservation of cellular genetic information and with respect to the maintenance of chromosomal structure. These moonlighting activities involve the recognition of DNA damage, the functioning of DNA repair pathways, the structural organization of DNA damage, and repair protein complexes, the structural integrity of telomeres, and, perhaps surprisingly, the repair of DNA repair proteins. By definition, each of these separate functions requires GAPDH nuclear localization. As with the description of moonlighting GAPDH membrane activities, these studies redefine the conventional dogma of GAPDH as a cytosolic housekeeping protein of little interest.

Glyceraldehyde-3-Phosphate Dehydrogenase (GAPDH). http://dx.doi.org/10.1016/B978-0-12-809852-3.00006-6

## 1. GAPDH AS A DNA DAMAGE RECOGNITION AND DNA REPAIR PROTEIN

As indicated in Table 6.1, a series of studies identified GAPDH as a required protein for the recognition and binding of multiprotein complexes to modified DNA. In the initial study, the recognition of mismatched DNA bases was examined using a mercaptopurine modified guanine residue (G$^s$): T or a G$^s$: C base pair within a 34 oligonucleotide duplex DNA structure (Krynetski et al., 2001). Six different human lymphoblastic leukemia cell lines were used. As expected, using an electrophoretic mobility shift assay (EMSA), the mismatched G$^s$ duplexes were recognized in five of the cell lines, which contained the mismatch repair (MMR) proteins. In contrast, those proteins were absent in the sixth cell line.

Analysis of that cell line identified a different protein complex, which bound the DNA duplex containing the dG$^s$-T mismatch. Proteins within that complex exhibited $M_r$'s between 30 and 200 kDa. A 37 kDa protein was identified as GAPDH by N-terminal sequencing. Addition of an anti-GAPDH monoclonal antibody dissociated the complex while RT-PCR indicated that there was no mutation within the GAPDH gene in that sixth cell line. Cellular analysis demonstrated that mercaptopurine treatment resulted in the cytoplasmic to nuclear translocation of GAPDH.

In a further study, the components of this nuclear multiprotein complex were identified (Krynetski et al., 2003). Apart from GAPDH, following EMSA, SDS-PAGE indicated that the complex was composed of four to five major proteins. Mass spectroscopy identified these proteins as heat shock cognate protein 70 (HSC70), ERp60, high mobility group protein 1 (HMGB1), and high mobility group protein 2 (HMGB2). Incubation with an antibody specific to the HMGB1 protein resulted in a supershift in the EMSA analysis. Further, competition analysis with a 4–5 fold excess of a 34-mer oligo DNA duplex containing only natural bases did not result in dissociation of the DNA–protein complex, indicating its high affinity for the mismatched base sequence. Additionally, EMSA analysis indicated the formation of the DNA–protein complex when cytosine arabinoside or 5-fluorouracil was substituted for G$^s$ in the mismatched DNA duplex. Each may produce a DNA mismatch. Lastly, coimmunoprecipitation analysis was performed in untreated cells using an anti-HMGB1 antibody. This study demonstrated the physical association of HMGB1 with three components of the complex (HMGB2, ERp60, and HSC70). GAPDH was not detected as a complex component. Although, at first this may seem surprising, given the recruitment of cytosolic GAPDH to the nucleus following mercaptopurine exposure (Krynetski et al., 2001), it may be suggested that GAPDH association activates the complex to bind to mismatched DNA. As related in Chapter 1, further studies demonstrated the mechanisms through which nuclear GAPDH may be exported to the cytoplasm subsequent to mercaptopurine exposure (Brown et al., 2004). The role of GAPDH and this DNA damage recognition complex in DNA repair is illustrated in Fig. 6.1.

**TABLE 6.1 GAPDH and the Maintenance of DNA Integrity**

| Cell Source/Organism | Experimental Paradigm | Experimental Finding | "Unique" Finding | References |
| --- | --- | --- | --- | --- |
| Human acute lymphoblastic leukemia cells (MMR+ and −)[a] | Recognition and binding to mismatched DNA duplexes | Protein complex in MMS-cells, which binds to mismatched DNA duplexes | GAPDH is a component of the multiprotein complex | Krynetski et al. (2001, 2003) and Brown et al. (2004) |
| Human colon carcinoma, lung carcinoma cells | Recognition and binding to S23906-1[b] alkylated DNA | GAPDH–HMGB1 protein complex binds to S23906-1 alkylated DNA | GAPDH binds to alkylated DNA; HMGB1[c] binds to GAPDH | Lenglet et al. (2013) and Savreux-Lenglet et al. (2015) |
| Escherichia coli | DNA repair in GAPDH mutant cells | E. coli GAPDH- cells hypersensitive to DNA damage | Accumulation of abasic sites, increased spontaneous mutation frequency | Ferreira et al. (2015) |
| Human fibroblasts, colon carcinoma cells | Quantitation of APE1 structure and activity | GAPDH repairs APE1, restoring activity | APE1 sensitivity to redox status, inactivation of GAPDH reduces APE1 activity | Azam et al. (2008) and Karmahapatra et al. (2014) |
| E. coli | Affinity analysis of GAPDH cross-linked proteins | Binding of GAPDH to Gph[d] | Bleomycin exposure increases GAPDH–Gph binding | Ferreira et al. (2013) |
| Human placenta, human fibroblasts, Hela cells, pea chloroplasts | Isolation and characterization of GAPDH protein and gene | GAPDH monomer exhibits uracil DNA glycosylase (UDG) activity | Coordinate reduction of nuclear GAPDH and UDG activity during apoptosis | Meyer-Siegler et al. (1991), Baxi and Vishwanatha (1995), Saunders et al. (1997) and Wang et al. (1999) |
| Human lung carcinoma, mouse lung epithelial; Trypanosoma cruzi | Analysis of telomere binding proteins | GAPDH binds to and stabilizes telomeric DNA | GAPDH–telomere DNA binding in a protozoan parasite | Sundararaj et al. (2004), Demarse et al. (2009) and Pariona-Lianos et al. (2015) |

[a]MMR, mismatch repair.
[b]S23906-1, cis-1,2-diacetoxy-1,2-dihydro-benzo[b]acronycine
[c]High mobility group protein B1.
[d]Phosphoglycolate phosphatase.

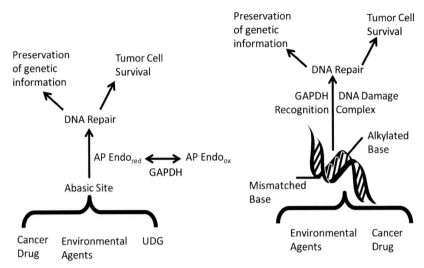

**FIGURE 6.1** Role of GAPDH in the maintenance of DNA integrity and cancer facilitation. *UDG*, uracil DNA glycosylase.

## 1.1 Recognition of S23906-1 Alkylated DNA

One of the hallmarks of moonlighting GAPDH function is the generality of its identified activities. So, recent evidence indicates that the multiprotein complex discussed above recognizes and binds to other types of DNA lesions (Lenglet et al., 2013; Savreux-Lenglet et al., 2015). In these studies, *cis*-1,2-diacetoxy-1,2-dihydro-benzo[*b*]acronycine (S23906) alkylated DNA was used in affinity chromatography to determine whether binding proteins could be isolated using HT-29 human colon carcinoma nuclear extracts. Two protein bands were isolated following NaCl elution. Tryptic digestion identified the two proteins as GAPDH and HMGB1. EMSA analysis revealed an intriguing finding, i.e., purified GAPDH but not purified HMGB1 bound to S23906-alkylated DNA.

As described, nuclear extracts were used to define GAPDH binding. To determine whether exposure of cells to S23906 resulted in a GAPDH cytoplasmic→nuclear translocation, subcellular fractionation studies were performed following S23906 exposure. These studies demonstrated that there was no change in either the cytoplasmic or soluble nuclear GAPDH levels as defined by both immunoblot analysis and by confocal microscopy. In contrast, there was a twofold and a fourfold increase in chromatin-associated GAPDH in HT-29 and A549 cells, respectively. Further chromatin analysis demonstrated an increase in chromatin-associated HMGB1 as well.

GAPDH silencing was used to determine its influence on S23906 cytotoxicity in the A549 (p53-wt) and in the HT-29 (p53 mutant) cells. Cell survival was low in the A549 cells after si-GAPDH exposure (48–72h). In contrast, in the p53 mutant HT-29 cells, there was only a slight change in cell viability at the

48–72 h interval. However, diminution of GAPDH protein in the HT-29 cells resulted in an increase in S23906-induced cytotoxicity. In toto, these findings establish the role of GAPDH in the recognition of S23906 DNA adducts, the chromatin recruitment of both GAPDH and HMGB1 in response to S23906 DNA modification, and the role of GAPDH in cell survival in p53 mutant HT-29 cells. Further, the observation that only GAPDH binds to S23906 modified DNA suggests that GAPDH binding is required for the binding of HMGB1, either to chromatin itself or to GAPDH.

## 1.2 Role of GAPDH in Cell-Mediated DNA Repair

As noted in previous chapters, for the most part, structural studies of moonlighting GAPDH functions precede those studies that establish new GAPDH activities. As the studies described above indicated structural determinants of moonlighting GAPDH in the recognition and binding to DNA lesions, a subsequent study indicated the functional consequences of those structural interactions (Ferreira et al., 2015).

In these studies, now classical mutational analyses were used to determine the role of GAPDH in DNA repair. In particular, a GAPDH-deficient cell strain, MC4100 $\Delta gapA$ was constructed and demonstrated to be deficient in GAPDH enzymatic activity and in GAPDH protein deficiency as defined by immunoblot analysis. To assess DNA repair capacity, cell sensitivity to bleomycin (BM) or methyl methanesulfonate (MMS) was determined. Each is well known as DNA-damaging agents. These studies indicated that the GAPDH mutant exhibited increased sensitivity to both agents. This effect was time dependent and was considered significant ($P < .05$). Further, there was a much greater sensitivity to BM as compared to MMS. This difference was replicated in GAPDH silenced wt cells. Morphological analysis indicated distinct differences in cell structure of the $\Delta gapA$ mutant as compared with its wt parent following exposure to either BM or MMS.

Subsequently, the level of abasic sites [apurinic/apyrimdinic (AP)] was examined in both the mutant and in wt cells as a function of exposure. The rationale was that abasic sites can occur subsequent to damage by either agent. The experimental paradigm was a genomic degradation assay, which used alkaline pH to abasic sites that are susceptible. In this study, both mutant and wt cells were treated with MMS then analyzed either immediately after exposure or after a 3 h recovery interval. Immediate analysis revealed the presence of degraded DNA in both cell types. In contrast, following the 3 h recovery, wt DNA was characterized by high-molecular weight species while analysis of mutant DNA revealed not only reduced high-molecular weight species but also the continued presence of lower molecular weight species. It was stated that these findings were in accord with the inability of the mutant to repair DNA lesions.

The physiological significance of these findings was examined by determining spontaneous mutation frequencies in both the mutant and wt cells. These studies revealed a fivefold increase in the $\Delta gapA$ cells. Quantitation of abasic sites indicated a 50% increase in the $\Delta gapA$ cells as compared to wt. Cumulatively, these studies demonstrate that the structural role of GAPDH in the recognition of damaged DNA described above may have functional consequences.

## 1.3 GAPDH and the Repair of Abasic DNA

The studies described above suggest that the formation of abasic sites and their repair may be a focal point for moonlighting GAPDH DNA function. As such, the "viability" of cellular AP endonucleases would be an a priori requirement, i.e., their inactivation would result in the failure to repair abasic sites resulting in higher mutation rates.

In that regard, a recent report describes a novel moonlighting GAPDH function, which repairs inactivated AP endonuclease 1 (APE1), restoring its activity (Azam et al., 2008) As with many GAPDH studies, the intent was not to study GAPDH but to probe the structure and function of mammalian AP endonucleases. During that study, AP enzyme activity copurified with a 37 kDa protein, which was identified as GAPDH.

This interaction was probed using His-tagged APE1 affinity chromatography. A mixing experiment was performed, which contained the His-APE1-tagged affinity beads plus recombinant GST-GAPDH. That experiment demonstrated that the latter bound to the former as detected by immunoblot analysis using an anti-GST monoclonal antibody. Significantly, three mutant GAPDH constructs, GAPDH$^{C152G}$, GAPDH$^{C156G}$ , and GAPDH$^{C247G}$ also bound to APE1. As indicated previously, GAPDH$^{C152}$ is the active site cysteine. Its alteration abolishes GAPDH glycolytic activity, suggesting that GAPDH–APE1 binding is independent of its glycolytic function. Control experiments demonstrated that the glycolytic enzymes pyruvate dehydrogenase and 6-phosphofructokinase did not bind to APE1. Further, coimmunoprecipitation with an anti-APE1 antibody using colon cancer cell (HCT116) extracts followed by immunoblot analysis with an anti-GAPDH antibody demonstrated the in vivo formation of the APE1–GAPDH complex.

Although these studies indicated the formation of an APE1–GAPDH complex, the significance of that interaction was unknown. As previous studies indicated that oxidative stress could inactivate APE1, the ability of GAPDH to recover APE1 activity was examined. Incubation of APE1, which exhibited no activity with increased GAPDH concentrations restored enzyme activity. As APE1 contains both oxidized and reduced forms, this study suggests that GAPDH can convert inactive oxidized APE1 to the active reduced form. This stimulatory effect was not observed with either GAPDH$^{C152G}$ or with GAPDH$^{C156G}$, indicating that, although the mutants could bind to APE1, each cysteine residue was required for this new GAPDH function.

To specifically determine whether GAPDH could reactivate oxidized APE1, the latter was exposed to $H_2O_2$, a chemical oxidant. Following exposure, no

APE1 activity could be detected. In contrast, when the $APE1_{oxd}$ was incubated for 5 min with increasing GAPDH concentrations, catalytic activity was restored. Control experiment indicated the specificity of this effect, i.e., addition of N-acetylcysteine, glutathione, dithiothreitol (DTT), or purified GST did not reactivate APE1. Further, the addition of diverse metal ions, including $Mg^{++}$, failed to restore activity. The latter is significant as it precluded the possibility that GAPDH acted as a metal donor.

The in vitro studies described above suggest that alterations in GAPDH levels in vivo should affect DNA repair capacity. GAPDH knock down cells were hypersensitive to either BM or to MMS. Determination of AP sites in GAPDH silenced cells as compared with controls demonstrated a 1.5-fold increase in the former as compared with the latter. Further analysis revealed a decrease in AP endonuclease activity, consistent with the increase in genomic AP sites in GAPDH silenced cells. In toto, as illustrated in Fig. 6.1, these studies reveal a new moonlighting GAPDH function, i.e., the repair of a DNA repair protein.

As discussed in other chapters, regrettably, we gain insight into fundamental cell functions due to their deficiency in human genetic disorders. In this instance, Wilson's disease, a syndrome in which copper is aberrantly regulated, exhibits chronic inflammation and presents with an increased probability of hepatocellular cancer. An animal model was developed, the Long–Evans Cinnamon (LEC) rat, which has a mutation in the copper-transporting *Atp7b* gene. Using this model, APE1 activity was measured in LEC liver as a function of liver pathology in comparison to a normal control (Karmahapatra et al., 2014). As indicated in Fig. 6.1 of that study, the "advantage" of this mutant is that it exhibits a defined temporal sequence of hepatic pathology, i.e., prehepatitis (0–15 weeks), acute hepatitis (15–40 weeks), chronic hepatitis (30–80 weeks) culminating in hepatocellular carcinoma (75–100 weeks).

Within this temporal sequence, there was a noticeable increase in AP sites as a function of liver pathology, especially in the acute hepatitis and early chronic hepatitis phase (up to 30%–40% as compared to controls). APE1 activity was measured using a 50-mer oligonucleotide containing a single AP-site analogue in both control animals and in copper-overloaded LEC animals. Perhaps surprisingly, no change in APE1 activity was observed in the acute hepatitis stage (8–40 weeks). As defined by immunoblot analysis, no change in APE1 expression was detected. However, it was noted that those quantitations were performed in the presence of DTT, a reducing agent. In the absence of DTT, a 20%–30% decrease in APE1 activity was observed in the LEC animals although no change in protein expression was detected.

Analysis of GAPDH activity indicated that it was decreased by 25%–50% during the acute hepatitis phase (16–18 weeks). No change in protein expression was detected. Analysis of glutathione levels indicated a 60%–90% increase in oxidized glutathione. Cumulatively, from these studies, it was postulated that there was a correlation between the increased levels of AP sites and diminution of APE1 activity on the one hand and the reduced GAPDH activity on the other hand, i.e., as GAPDH activity was diminished, its ability to reactivate $APE1_{ox}$ to the reduced form decreased. This conclusion would be in accord not only with the studies described above but also results in this investigation demonstrating that two reducing agents, DTT and $\alpha$-lipoic acid, could restore APE1 activity in LEC rat liver extracts.

A further indication of abasic sites as the focus of moonlighting GAPDH in DNA damage and repair, a detailed analysis of GAPDH protein interactions identified phosphoglycolate phosphatase (Gph) as a GAPDH-binding protein (Ferreira et al., 2013). The latter enzyme functions to hydrolyze 2-phosphoglycolate formed during the abasic site repair process.

In this study, the experimental paradigm was to use formaldehyde crosslinking in vivo to identify GAPDH-V5–binding proteins followed by immunoblot analyses using either anti-GAPDH or anti-V5 antibodies. As expected, a wide variety of proteins were so identified given the functional diversity of GAPDH. That being said, the identification of Gph was notable, given the studies indicating the importance of GAPDH-mediated abasic DNA repair.

The copurification of GAPDH with Gph was examined in cells expressing a GST-Gph protein, which would be selected by a glutathione-sepharose 4B resin. That analysis demonstrated not only the retention of the GST-Gph protein on the resin but also its coelution with GAPDH as defined by immunoblot analysis of the eluate. A subsequent, and intriguing, study examined whether Gph bound to a specific GAPDH isoform. Two-dimensional gel electrophoresis identified multiple GAPDH species with pI between 5 and 8. Only a species with a pI c. is 6.5 bound to Gph. As such, these results are similar to those indicating that the membrane fusogenic properties of GAPDH are restricted to a specific isoform (Chapter 4; Glaser and Gross, 1995).

The physiological significance of the GAPDH–Gph interaction was examined in cells exposed to BM. A Gph mutant strain containing a pGEX-Gph plasmid was incubated with $BM + FeSO_4$ (to increase the number of DNA strand breaks). GAPDH–Gph binding was assessed by glutathione-sepharose 4B resin retention followed by immunoblot analysis of the eluate. These studies indicated a time-dependent increase in the formation of the GAPDH–Gph complex, suggesting a functional role for this structural protein–protein complex.

## 1.4 GAPDH and the Enzymatic Generation of Abasic Sites

As described earlier, abasic sites may be produced by a variety of DNA-damaging agents. Such sites may arise also from the action of a series of enzymes, termed DNA glycosylases. The latter functions to remove altered DNA bases, thereby producing an abasic site. Recent evidence from a number of independent laboratories suggests that GAPDH may function as a uracil DNA glycosylase (UDG; Meyer-Siegler et al., 1991; Baxi and Vishwanatha, 1995; Saunders et. al., 1997; Wang et al., 1999). Uracil is formed in DNA primarily from deamination of preexisting cytosine residues creating a G–U mismatch.

As with many GAPDH studies, the intent of the first study was to examine DNA repair processes, both in vivo and in vitro. For that purpose, a series of antihuman placental monoclonal antibodies were prepared against partially purified UDG (Arenaz and Sirover, 1983). Each antibody inhibited UDG activity. Further, in an in vitro biosynthesis study using Poly (A⁺) mRNA, an antiplacental antibody (40.10.09) recognized a newly synthesized 37 kDa protein as defined by immunoprecipitation of ($^{35}$S)-methionine radiolabeled protein followed by SDS-PAGE (Vollberg et al., 1987a).

Subsequently, the antibodies were used to immunoscreen a λgt11 cDNA library. A cDNA was isolated and sequenced (Vollberg et al., 1989; Meyer-Siegler et al., 1991). Surprisingly, comparison with known sequences identified the unknown cDNA as GAPDH. A β-galactosidase fusion protein was isolated and exhibited UDG activity. That activity was inhibited by incubation with either the antiplacental antibodies or by a β-galactosidase antibody.

In a second study, an Ap$_4$A-binding protein was purified from HeLa cells to homogeneity and identified as GAPDH by amino acid sequence analysis (Baxi and Vishwanatha, 1995). Further investigation demonstrated not only that it exhibited UDG activity but also that its activity was comparable to that of *Escherichia coli* UDG (0.229 units vs. 0.565 units, respectively, using 1.5 µg of HeLa cell Ap$_4$A-binding protein vs. 0.25 units of *E. coli* UDG).

In a third study, the UDG activity of pea chloroplast GAPDH was examined (Wang et al., 1999). This investigation used both GAPDH isolated from pea chloroplasts as well as the recombinant subunit B enzyme. Immunological analysis demonstrated that an antichloroplast dehydrogenase antibody inhibited recombinant subunit B GAPDH glycolytic activity and UDG activity by >than 90%. In contrast, incubation with an anti-*E. coli* UDG antibody failed to inhibit significantly the recombinant subunit B UDG activity with >90% of the latter remaining following antibody incubation.

In a fourth study, the physiological significance of GAPDH UDG activity was indicated (Saunders et al., 1997). As described in detail in Chapter 8, not only does moonlighting GAPDH fulfill a significant role in apoptosis but also its S-nitrosylation at its active site cysteine is required for that new activity. The latter resulted in the reduction of GAPDH glycolytic activity. Temporal analysis of nuclear GAPDH during apoptosis demonstrated the simultaneous reduction of both glycolytic activity and UDG catalysis. In toto, these studies indicate that GAPDH possesses a UDG activity, that this new function may not be restricted to animal cells, and that reduction of UDG catalysis may be a characteristic of programmed cell death.

## 1.5 GAPDH and the Cellular Phenotype of Bloom's Syndrome

As indicated previously, the pathology of human disease provides an unfortunate experimental paradigm for the investigation of cell processes. Bloom's syndrome is an autosomal recessive human genetic disorder, and Bloom's syndrome individuals are cancer prone. The molecular defect in Bloom's syndrome was identified as a defective DNA helicase (Karrow et al., 1997). Phenotypically, Bloom's syndrome cells exhibit increased spontaneous mutation frequencies and are hypersensitive to DNA-damaging agents. Cellular analysis indicated that Bloom's syndrome cells displayed an alteration in the cell cycle regulation of DNA excision repair pathways (Gupta and Sirover, 1980, 1984).

Although the mechanism that underlies the alteration of DNA repair regulation in Bloom's syndrome cells may be unknown (Sirover et al., 1990), immunological and biochemical analysis of its GAPDH revealed unique changes in the Bloom's syndrome protein. As previously indicated, a series of antihuman placental GAPDH antibodies were prepared (Arenaz and Sirover, 1983). Immunological analysis indicated that two of the antibodies, 37.04.12 and 42.08.07, recognized the native Bloom's syndrome protein as defined by ELISA. In contrast, a third antibody, 40.10.09, failed to recognize the native protein yet recognized it by immunoblot analysis (Vollberg et al., 1987b, Seal et al., 1988). It was suggested that a conformational change was responsible for the lack of recognition of the native Bloom's syndrome protein as the immunoblot results indicated the presence of the primary amino acid sequence, which defined the 40.10.09 binding site. Purification of the Bloom's syndrome GAPDH protein indicated structural and catalytic differences as compared with the normal human protein (Seal et al., 1991). These included changes in pI and in thermal stability. As GAPDH is implicated in DNA repair and as it is structurally altered in Bloom's syndrome, it was hypothesized that the latter may underlie some of the alterations, which occur in the former.

## 2. GAPDH AND THE MAINTENANCE OF TELOMERE STRUCTURE

As with the conservation of genomic DNA sequences, the preservation of teleomeric structure is necessary for the integrity of chromosomal organization. The latter may be an a priori stem cell requirement. Conversely, telomere shortening may be a basis for cell aging, i.e., the continual loss of genomic structure results in the phenotypic changes, which result in cell senescence. Finally, as with DNA repair, such mechanisms in cancer cells may ensure their survival.

An initial study indicated the role of moonlighting GAPDH in the maintenance of telomere structure. Human lung carcinoma cells (A549) were used as the cell source (Sundararaj et al., 2004). In an analysis of the sphingolipid ceramide's effect on telomere structure, a nuclear telomere–protein complex was identified in the control sample, i.e., cells grown in the absence of ceramide The formation of this complex was reduced in ceramide-treated cells. UV cross-linking studies subsequent to gel shift assay indicated that the protein exhibited an $M_r$ between 36 and 40 kDa.

Affinity chromatography using the teleomeric sequence $(TTAGGG)_3$ resulted in the elution of one major abundant protein. Protein sequence analysis identified it as GAPDH. Immunoblot analysis with an anti-GAPDH antibody confirmed the identification. Further, commercially available GAPDH bound a single-strand (ss) telomeric DNA probe as defined by UV cross-linking. In a dose-dependent manner, $NAD^+$ reduced telomere binding of purified GAPDH and reduced the formation of the telomere–GAPDH complex in nuclear extracts. In interphase nuclei, dual labeling immunofluorescent analysis in situ indicated that GAPDH was localized with telomeric DNA.

Analysis of the basis not only for GAPDH–telomere binding but also for the effect of ceramide on that protein–nucleic acid interaction yielded intriguing findings. First, in the absence of ceramide, the cell cycle dependency of GAPDH nuclear localization was determined (as described in Chapter 1). Cells were blocked at the $G_1/S$ border by a thymidine block, and, subsequent to release, GAPDH localization was determined by immunofluorescence. In S phase, GAPDH was localized in the nucleus followed by a perinuclear then a cytosolic localization in $G_0/G_1$. Examination of the intracellular localization of a transfected $3'$-His tagged recombinant GAPDH demonstrated a similar cell cycle–dependent intracellular localization. In contrast, a different pattern of GAPDH subcellular distribution was observed in ceramide-treated cells, i.e., GAPDH nuclear localization was inhibited in ceramide-treated cells. In particular, GAPDH distribution appeared to be similar to that observed in untreated cells in $G_0/G_1$. Accordingly, it was suggested that one mechanism through which ceramide affected telomere–GAPDH binding was through the control of GAPDH nuclear localization.

Analysis of nuclear and cytoplasmic GAPDH demonstrated differences both in their structure and in their function. Nuclear GAPDH was basic in nature with a pI between 8.3 and 8.7 while cytoplasmic (and commercial GAPDH) exhibited a pI of 7.0–7.5. Again, this is in accord with respect to the differences observed in membrane isozymes with different functional characteristics (Chapter 4; Glaser and Gross, 1995).

The physiological significance of nuclear GAPDH–telomere binding was examined by both GAPDH silencing and by GAPDH overexpression. Downregulation of nuclear GAPDH expression resulted in a GAPDH siRNA concentration–dependent decrease in telomere restriction fragment length (120 and 500 bp decreases in cells treated with 200 and 400 nm GAPDH siRNA, respectively). In contrast, overexpression of GAPDH (His-tagged recombinant GAPDH) abrogated the ceramide-mediated telomere shortening effects of the cancer chemotherapeutic agents gemcitabine (GEM) and doxorubicin (DOX). Similarly, overexpression of the recombinant GAPDH construct reduced the telomere-shortening effects of $C_6$-ceramide analog. In toto, this initial study identified the role of GAPDH in the maintenance of telomere integrity, indicated its role in ceramide-mediated telomere shortening, and its effect on the efficacy of cancer treatment. As such, these findings are in accord with the duality of GAPDH function in the maintenance of DNA and chromatin integrity.

Subsequent analysis characterized the GADPH–telomeric DNA interaction (Demarse et al., 2009). The experimental paradigm was the utilization of an EMSA protocol through which the differential binding of recombinant human GAPDH to specific telomeric oligomeric constructs could be examined. The basic protocol was established using a triplet repeat of human telomeric DNA ($[^{32}P]$ss-5′(TTAGGG)$_3$-3′). GAPDH dose-dependent studies indicated a $K_d$ of $70 \pm 5$ nM (n = 3). The use of ss and double-stranded oligomers indicated a preference for ss telomeric DNA. The specific nucleotide sequence recognized by human GAPDH in ss telomeric DNA was determined through the use of a series of oligomers which contained single base changes in the hexameric sequence. Those studies demonstrated that GAPDH recognized the GGT triplet sequence.

One of the basic questions that underlies moonlighting GAPDH function is the oligomeric structure of GAPDH responsible for the specific activity. With respect to its binding to telomeric DNA, EMSA indicated complete binding of the DNA probe at 50 nM GAPDH. It was stated that this would be consistent with the binding of one oligonucleotide per dimer or two oligonucleotides per tetramer. Native PAGE was utilized to examine the GAPDH oligomeric form contained within that complex. This analysis indicated the formation of a high-molecular weight complex with an $M_r$ between 242 and 480 kDa. Given that molecular mass, it was concluded that the GAPDH–telomeric DNA complex contained only the GAPDH tetrameric form.

Mutational analysis was then utilized to determine GAPDH structural parameters, which defined its binding to telomeric DNA. These focused on the GAPDH catalytic site, i.e., GAPDH$^{C149A}$ and GAPDH$^{D32A}$, which altered the active site cysteine and aspartic$^{32}$, which is required for NAD$^+$ binding. As defined by EMSA, each mutant failed to bind to telomeric DNA, indicating that both the catalytic and NAD$^+$ sites were necessary for formation of the GAPDH–telomere DNA complex. Analysis of binding in vivo using FLAG-tagged constructs that, as defined by ChIP assay, neither mutant bound to telomeric DNA in transfected A549 cells. Immunoblot analysis determined equivalent expression of each construct indicating that reduced synthesis could not account for the reduction in binding capacity. Further, there was no difference in nuclear localization, as defined by immunofluorescence and confocal microscopy, in GEM/DOX–treated cells.

The physiological significance of these two binding sites was assessed in vivo using protection of transfected cells to the telomere-shortening ability of the cancer chemotherapeutic agents. As expected in control studies (GEM/DOX exposure in vector-transfected cells), DNA was observed as lower molecular weight species. Further, as expected, treatment of wt GAPDH transfected cells protected against telomere shortening. Expression of either mutant did not significantly protect against telomeric shortening. Accordingly, these studies provide a rigorous structure–function analysis of moonlighting GAPDH activity in the maintenance of DNA integrity as it relates to telomeric structure.

Another of the basic questions, which underlies moonlighting GAPDH function, is the generality of the new GAPDH activity. As will be described in virtually all of the chapters, these multiple GAPDH functions are observed not only in diverse cells and organisms but also, in many cases, across the phylogenetic scale. In that regard, the role of GAPDH in telomere structure and function is no exception. GAPDH–telomere DNA binding was detected in both cancerous and noncancerous cell lines (Demarse et al., 2009).

Evidence indicates also that GAPDH binds to telomeric DNA as well in the protozoan pathogen *Trypanosoma cruzi* (Pariona-Lianos et al., 2015). The initial experimental paradigm was to examine the binding of a recombinant *T. cruzi* GAPDH protein to ss or double-stranded oligonucleotides containing (TTAGGG)$_6$ telomeric repeats. As defined by EMSA, the recombinant protein bound to the ss oligonucleotide but not to the double-stranded oligomer. These results were confirmed using an anti-GAPDH antibody. Incubation with the antibody resulted in a supershift using the ss oligonucleotide. Further, addition of NAD$^+$-inhibited GAPDH–oligomer binding in a concentration-dependent manner. The latter is in accord with previous studies indicating that the GAPDH telomeric binding site is located within the GAPDH NAD$^+$ domain (Demarse et al., 2009).

Subsequent studies examined whether GAPDH bound in vivo to *T. cruzi* telomeric DNA. Initial studies used detergent precipitation followed by DNase digestion to determine bound proteins. Immunoblot analysis demonstrated that GAPDH was present in the insoluble precipitate. Given the severity of the detergent treatment, it may be suggested that GAPDH bound strongly to *T. cruzi*. This is in accord with previous studies demonstrating the tight association of GAPDH to DNA during programmed cell death (Chapter 8; Sawa et al., 1997). Chromatin immunoprecipitation confirmed GAPDH binding to DNA while a colocalization analysis (FISH/IF assay) demonstrated the binding of GAPDH to *T. cruzi* telomeric DNA.

Further investigation indicated that the cellular NAD$^+$/NADH balance could affect GAPDH–telomeric DNA binding. Noteworthy, the *T. cruzi* life cycle includes both an insect-living replicative epimastigote form and mammalian-living nonreplicative trypomastigote form. The latter exhibits greater NAD$^+$ levels than NADH. Binding studies indicated that GAPDH from this *T. cruzi* form did not form a GAPDH–telomeric DNA complex. Further, treatment of epimastigote cells with exogenous NAD$^+$ reduced GAPDH–telomere association. These findings are in accord with the in vitro NAD$^+$ competition analysis. In toto, these studies demonstrate this new moonlighting GAPDH activity in a protozoan parasite, which is in accord with studies demonstrating the role of moonlighting GAPDH in infection and immunity (Chapter 13). Further, the role of GAPDH–telomere binding in tumorigenesis is considered in Chapter 11 (Nicholls et al., 2012).

## 3. SUMMARY

The investigations described in this chapter highlight not only a novel function of moonlighting GAPDH but also the duality of this new activity. With respect to the former, GAPDH is involved deeply in the maintenance of DNA integrity in normal cells. In particular, as illustrated in Fig. 6.1, its role in the formation and repair of abasic sites as well as its role in a DNA damage recognition protein complex may be of singular importance. The same may be said for its telomere binding especially with respect to stem cell DNA integrity. In contrast, with respect to the latter, both mechanisms may be utilized by tumor cells to mitigate the effects of radiation therapy or of chemotherapeutic protocols. In toto, these nuclear activities of moonlighting GAPDH form another part of the matrix, which comprises its functional diversity.

## REFERENCES

Arenaz, P., Sirover, M., 1983. Isolation and characterization of monoclonal antibodies directed against the DNA repair enzyme uracil DNA glycosylase from human placenta. Proc. Natl. Acad. Sci. U.S.A. 80, 5822–5826.

Azam, S., Jouvet, N., Jilani, A., et al., 2008. Human glyceraldehyde-3-phosphate dehydrogenase plays a direct role in reactivating oxidized forms of the DNA repair enzyme APE1. J. Biol. Chem. 283, 30632–30641.

Baxi, M., Vishwanatha, J., 1995. Uracil DNA-glycosylase/glyceraldehyde-3-phosphate dehydrogenase is an Ap$_4$A binding protein. Biochemistry 34, 9700–9707.

Brown, V., Krynetski, E., Krynetskaia, N., et al., 2004. A novel CRM1-mediated nuclear export signal governs nuclear accumulation of glyceraldehyde-3-phosphate dehydrogenase following genotoxic stress. J. Biol. Chem. 279, 5984–5992.

Demarse, N., Ponnusamy, S., Spicer, E., et al., 2009. Direct binding of glyceraldehyde-3-phosphate dehydrogenase to teleomeric DNA protects telomeres against chemotherapy-induced rapid degradation. J. Mol. Biol. 394, 789–803.

Ferreira, E., Gimenez, Aguilera, L., et al., 2013. Protein interaction studies point to new functions for *Escherichia coli* glyceraldehyde-3-phosphate dehydrogenase. Res. Microbiol. 164, 145–154.

Ferreira, E., Gimenez, R., Canas, M., et al., 2015. Glyceraldehyde-3-phosphate dehydrogenase is required for efficient repair of cytotoxic DNA lesions in *Escherichia coli*. Int. J. Biochem. Cell Biol. 60, 202–212.

Glaser, P., Gross, R., 1995. Rapid plasmenylethanolamine-selective fusion of membrane bilayers catalyzed by an isoform of glyceraldehyde 3 phosphate dehydrogenase: discrimination between glycolytic and fusogenic roles of individual isoforms. Biochemistry 34, 12193–12203.

Gupta, P., Sirover, M., 1980. Sequential stimulation of DNA repair and DNA replication in normal human cells. Mutat. Res. 72, 273–284.

Gupta, P., Sirover, M., 1984. Altered temporal expression of DNA repair in hypermutable Bloom's syndrome cells. Proc. Natl. Acad. Sci. U.S.A. 81, 757–761.

Karmahapatra, S., Saha, T., Adhikari, S., Woodrick, J., Roy, R., 2014. Redox regulation of apurinc/apyrimidinic endonuclease 1 activity in Long-Evans Cinnamon rats during spontaneous hepatitis. Mol. Cell. Biochem. 388, 185–193.

Karrow, J., Chakraverty, R., Hickson, I., 1997. The Bloom's syndrome gene product is a 3′-5′ DNA helicase. J. Biol. Chem. 272, 30611–30614.

Krynetski, E., Krynetskaia, N., Bianchi, M., Evans, W., 2003. A nuclear protein complex containing high mobility group proteins B1 and B2, heat shock cognate protein 70, ERp60, and glyceraldehyde-3-phosphate dehydrogenase is involved in the cytotoxic response to DNA modified by incorporation of anticancer nucleoside analogues. Cancer Res. 63, 100–106.

Krynetski, E., Krynetskaia, N., Gallo, A., Murti, K., Evans, W., 2001. A novel protein complex distinct from mismatch repair binds thioguanylated DNA. Mol. Pharm. 59, 367–374.

Lenglet, G., Depauw, S., Mendy, D., David-Cordonnier, M.-H., 2013. Protein recognition of the S23906-1-DNA adduct by nuclear proteins: direct involvement of glyceraldehyde-3-phosphate dehydrogenase (GAPDH). Biochem. J. 452, 147–159.

Meyer-Siegler, K., Mauro, D., Seal, et al., 1991. A human nuclear uracil DNA glycosylase is the 37-kDa subunit of glyceraldehyde-3-phosphate dehydrogenase. Proc. Natl. Acad. Sci. U.S.A. 88, 8460–8464.

Nicholls, C., Pinto, A.R., Li, H., Ling, L., Wang, L., Simpson, R., Liu, J.-P., 2012. Glyceraldehyde-3-phosphate dehydrogenase (GAPDH) induces cancer cell senescence by interacting with telomerase RNA component. Proc. Natl. Acad. Sci. U.S.A. 109, 13308–13313.

Pariona-Lianos, R., Pavani, R., Reis, M., et al., 2015. Glyceraldehyde-3-phosphate dehydrogenase-telomere association correlates with redox status in *Trypanosoma cruzi*. PLoS One. http://dx.doi.org/10.1371/journal.pone.0120896.

Saunders, P., Chalecka-Franaszek, E., Chuang, D.-M., 1997. Subcellular distribution of glyceraldehyde-3-phosphate dehydrogenase in cerebellar granule cells undergoing cytosine arabinoside-induced apoptosis. J. Neurochem. 69, 1820–1828.

Savreux-Lenglet, G., Depauw, S., David-Cordonnier, M.-H., 2015. Protein recognition in drug-induced DNA alkylation: when the moonlight protein GAPDH meets S23906-1/DNA minor groove adducts. Int. J. Mol. Sci. 16, 26555–26581.

Sawa, A., Khan, A., Hester, L., Snyder, S., 1997. Glyceraldehyde-3-phosphate dehydrogenase: nuclear translocation participates in neuronal and nonneural cell death. Proc. Natl. Acad. Sci. U.S.A. 94, 11669–11674.

Seal, G., Brech, K., Karp, S., Cool, B., Sirover, M., 1988. Immunological lesions in human uracil DNA glycosylase: association with Bloom's syndrome. Proc. Natl. Acad. Sci. U.S.A. 83, 2339–2343.

Seal, G., Tallarida, R., Sirover, M., 1991. Purification and properties of the uracil glycosylase from Bloom's syndrome. Biochem. Biophys. Acta 1097, 299–308.

Sirover, M., Seal, G., Vollberg, T., 1990. DNA repair and the molecular mechanisms of Bloom's syndrome. Crit. Rev. Oncog. 2, 19–33.

Sundararaj, K., Wood, R., Ponnusamy, S., et al., 2004. Rapid shortening of telomere length in response to ceramide involves the inhibition of telomere binding activity of nuclear glyceraldehyde-3-phosphate dehydrogenase. J. Biol. Chem. 279, 6152–6162.

Vollberg, T., Cool, B., Sirover, M., 1987a. Biosynthesis of the human base excision repair enzyme uracil DNA glycosylase. Cancer Res. 47, 123–128.

Vollberg, T., Seal, G., Sirover, M., 1987b. Monoclonal antibodies detect conformational abnormality of uracil DNA glycosylase in Bloom's syndrome cells. Carcinogenesis 8, 1725–1729.

Vollberg, T., Siegler, K., Cool, B., Sirover, M., 1989. Isolation and characterization of the human uracil DNA glycosylase gene. Proc. Natl. Acad. Sci. U.S.A. 86, 8693–8697.

Wang, X., Sirover, M., Anderson, L., 1999. Pea chloroplast glyceraldehyde-3- phosphatedehydrogenase has uracil glycosylase activity. Arch. Biochem. Biophys. 367, 348–353.

## FURTHER READING

Loecken, E., Guengerich, F., 2008. Reactions with glyceraldehyde-3-phosphate dehydrogenase sulfhydryl groups with bis-electrophiles produce DNA-protein cross-links but not mutations. Chem. Res. Toxicol. 21, 453–458.

Chapter 7

# Moonlighting GAPDH in Neuronal Structure and Function: Regulation of Ion Channels and Signal Transduction

*All's well that ends well*

William Shakespeare

The structure and function of neuronal cells and tissues is, perhaps, by far, the *sine qua non* of existence and, in humans, of consciousness itself. Indeed, the development of higher organisms may be commensurate with the organization and complexity of their neural structures. Further, at its most basic and at its most complex, neuronal function and the regulation thereof may be dependent on those proteins and membranes of which it is comprised. The synapse is constructed in such a manner not only to perform the basic task of information transfer but also contains within it the means through which an organism may regulate synaptic activity. The postsynaptic density (PSD) itself contains a distinct structure that defines its specific function. Further, perturbations in neuronal macromolecules may provide the bases for nervous system pathologies, which are present in afflicted individuals.

Recent evidence indicates that moonlighting GAPDH fulfills important roles with respect to both the organization of neuronal structure and to basic mechanisms, which underlie neuronal function. These include its role in synaptic function as a nucleotide transporter, a determinant of synaptic glutamate accumulation, its enzyme activity as a phosphorylase, which is an a priori requirement for $GABA_AR$ function, the regulation of neuronal $Na^+$ channels through its effect on gene expression and its role in $K^+$ ion channels. By definition, each of these disparate functions requires a specific, separate subcellular localization. Intriguingly, GAPDH may fulfill a role in neuronal inhibition by its regulation of $GABA_AR$ function and, in apparent contradiction, be involved in neuronal excitation by its regulation of synaptic glutamate levels. Further, as an extracellular protein, it may not only

Glyceraldehyde-3-Phosphate Dehydrogenase (GAPDH). http://dx.doi.org/10.1016/B978-0-12-809852-3.00007-8

regulate neurite outgrowth but also may affect microglial cells. Accordingly, this final chapter on the role of moonlighting GAPDH in normal cell functions illustrates both its requirement for, and the diversity of, its multiple activities.

## 1. GAPDH AS A COMPONENT OF NEURONAL STRUCTURE

As indicated in Table 7.1, a series of studies identified GAPDH as a common element in a variety of neuronal structures. That being said, the initial study did not identify GAPDH as a neuronal protein per se. Instead, it described a 34 kDa cholinergic synaptic protein isolated from *Torpedo marmorata* that functioned as an ATP translocase (Stadler and Fenwick, 1983). The experimental paradigm was to use the binding of [$^3$H] atractyloside, an ATP translocase inhibitor, as a probe to identify the unknown protein. Initial analysis demonstrated the binding of [$^3$H] atractyloside to synaptic vesicles as defined by sucrose density step gradient centrifugation. The purified ligand–binding material was analyzed by gel filtration, which identified it as a binding protein with an $M_r = 34$ kDa. Isoelectric focusing characterized the protein as basic in nature. Comparative analysis indicated that it exhibited a biochemical similarity to the mitochondrial ADP/ATP carrier protein.

Subsequently, a second investigation focused on the identification of the 34 kDa cholinergic synaptic vesicle transporter protein (Schläfer et al., 1994). It was purified from mammalian brain or *T. marmorata* electric organ using radiolabeling with [$\gamma$-$^{32}$P]azido ATP as a detection mechanism. Conventional purification protocols included differential centrifugation, Sephacryl gel filtration, and affinity chromatography (AMP-Sepharose). Sequence analysis of the 34 kDa protein following limited proteolysis identified it as GAPDH. Antibodies prepared against the purified 34 kDa protein recognized both the 34 kDa protein in synaptic vesicles as well as commercially available porcine muscle GAPDH. Labeling the 34 kDa protein with [$^{14}$C] N-ethylmaleimide followed by two-dimensional analysis indicated the formation of a triplet protein with a pI range between 7 and 8. Significant GAPDH enzyme activity was detected both in synaptic vesicle preparations and in the 34 kDa protein preparation isolated by AMP-Sepharose affinity chromatography. Thus, this second study identified not only the presence of GAPDH in synaptic vesicles but also defined a new neuronal moonlighting GAPDH function.

As is typical of moonlighting GAPDH studies, another investigation was undertaken to examine the effect of calmodulin on synaptic membrane proteins (Leung et al., 1987). The rationale was that calmodulin may regulate neuronal membrane functions. Accordingly, the experimental paradigm was to examine the effects of trifluoperazine (TFP), a modulator of calmodulin reactions, on synaptic plasma membrane proteins.

**TABLE 7.1** Identification of GAPDH in Neuronal Structures[a]

| Neuronal Structure | Experimental Paradigm | Experimental Finding | "Unique" Finding | References |
|---|---|---|---|---|
| *Torpedo marmorata*; cerebral cortex synaptic vesicle | Purification of a cholinergic vesicular nucleotide transporter | Nucleotide transporter identified as a basic 34 kDa protein | Nucleotide transporter identified as GAPDH | Stadler and Fenwick (1983) and Schläfer et al. (1994) |
| Cerebral cortex synaptic plasma membrane (SPM) | Treatment with the calmodulin modulator Trifluoperazine (TFP) | GAPDH tightly associated with synaptic plasma membrane | Bound GAPDH identified as a basic isoform (pI unspecified) | Leung et al. (1987) |
| Cerebral cortex synaptosome Postsynaptic density (PSD) | Immunological, biochemical analysis of neuronal GAPDH | GAPDH distribution: PSD > SPM | GAPDH: F-actin physical association in synaptosomes, PSD | Rogalski-Wilk and Cohen (1997)[b] |
| PSD | Immunological, biochemical analysis of PSD GAPDH | Glycolytic production of ATP; GAPDH phosphorylation | GAPDH function regulated by nitric oxide modification | Wu et al. (1997)[b] |
| Synaptic vesicle | Characterization of endogenous ATP production/glutamate uptake | GAPDH: PGK ATP producing protein complex | GAPDH modulates glutamate neurotransmission | Ikemoto et al. (2003) |

[a]*Chronological Order.*
[b]*Alphabetical Order.*

Two-dimensional electrophoretic analysis indicated that TFP treatment released a number of SPM bound proteins as defined by their presence in the soluble fraction, while others remained in the particulate fraction. Further, TFP treatment activated several proteins as defined by enzyme assay. The salient finding related to this discussion was that a 34 kDa protein remained bound to the synaptic plasma membrane. It was noted that this protein was basic in nature although the pI was not determined. Comparison of the 34 kDa protein with rabbit muscle GAPDH yielded identical V8 protease digestion patterns. Comparison of a rat brain cDNA clone to that of a rat liver GAPDH clone verified the identity of the 34 kDa protein. In toto, this initial study established not only that GAPDH was an SPM protein but also that it bound tightly to the membrane even when other glycolytic proteins were released by TFP treatment.

As evidenced by studies discussed in earlier chapters, the physical association of GAPDH with other proteins may be a basic requirement for its moonlighting activities. In a further study, immunochemical and biochemical protocols were used to probe the structural relationship between GAPDH and F-Actin in both cerebral cortex synaptosomes and in the PSD (Rogalski-Wilk and Cohen, 1997).

Immunofluorescent analysis using an affinity-purified GAPDH antibody indicated the presence of GAPDH in synaptosomes isolated from the porcine cerebral cortex. No immunofluorescence was observed using a preimmune control. Immunoblot analysis was performed in synaptosomal, synaptosomal plasma membrane (SPM), and PSD preparations using both anti-GAPDH and anti-actin antibodies. Densitometric quantitation revealed that PSDs have a 272% and a 173% increase in GAPDH and actin immunoreactivity, respectively, as compared to that observed for each protein in SPMs. Examination of GAPDH glycolytic activity demonstrated a comparable increase in enzyme catalysis in the PSD fraction. Intriguingly, determination of phosphoglycerate kinase (PGK) indicated its PSD coenrichment as well.

To determine the physical association of GAPDH with actin, a phalloidin shift sedimentation analysis was performed. Phalloidin, an actin-binding toxin, prevents actin depolymerization. Accordingly, in phalloidin-treated samples, F-actin and its binding proteins would sediment at an increased sucrose density as compared to control samples. Using this protocol, as detected by immunoblot analysis using a 15%–40% sucrose density gradient, both F-actin and GAPDH cosedimented in at a higher sucrose density in the presence of phalloidin in both synaptosomes and in PDSs. In the absence of the toxin, each purified protein sedimented at the lowest sucrose concentration suggesting that either each sedimented as a low molecular weight to that neither entered the gradient. In toto, these studies established not only that GAPDH and actin were components of each neuronal structure but also demonstrated a selective GAPDH localization as part of a GAPDH–F-actin protein complex.

Given the localization of GAPDH in neuronal structures, its function was then examined (Wu et al., 1997). For that purpose, its activity, along with a series of glycolytic enzymes was determined in the PSD. Those studies indicated that, lactate dehydrogenase (LDH), along with the previously reported GAPDH and PGK, exhibited considerable activity. As these three enzymes are required for ATP synthesis (LDH converting the NADH formed into NAD for subsequent reactions), the porcine PSD preparation was incubated with ADP, glyceraldehyde-3-phosphate, NAD, and $^{32}$P. Quantitation of ATP indicated that it was synthesized endogenously with an average yield of 361 fmol ATP formed/mg PSD protein. Similar results were observed with rat cortex PSD indicating the generality of this finding. As will be described in detail in Chapter 8, nitric oxide binding to GAPDH is intrinsic to its moonlighting functions. In this study, it was determined not only that PSD contained an endogenous nitric oxide synthase but also that its production of nitric oxide (NO) inhibited GAPDH activity. As such, these studies demonstrate the glycolytic function of neuronal and begin to indicate the role of nitric oxide in the regulation of GAPDH function.

It may be considered counterintuitive, considering the role of mitochondria in ATP production, that synaptic structures contain their own "regional" mechanisms for glycolytic ATP production. However, as will be also considered in Section V, "the unique role of sperm-specific GAPDH," there is an intriguing rationale for the presence of this energy producing mechanism, which is directly related to cell/tissue function. In particular, in this instance, a recent investigation suggested that the glycolytic production of ATP in synaptic vesicles may be directly related to the function of glutamate as a neurotransmitter (Ikemoto et al., 2003).

In that investigation, synaptic vesicles were prepared from bovine cerebellum. Preliminary studies confirmed the presence of both GAPDH and 3-phosphoglycerate kinase (3-PGK) as defined by immunoblot analysis. Coimmunoprecipitation/immunoblot studies identified their physical association using antibodies to both in the former and in the latter. Intriguingly, a subtraction/readdition protocol indicated that binding of GAPDH to the synaptic vesicle was a prerequisite for 3-PGK binding, i.e., binding of the latter to salt washed synaptic vesicles required the addition of GAPDH. In contrast, 3-PGK addition was not required for GAPDH binding.

The functional consequences of "regional" glycolytic ATP production were examined by quantitating the accumulation of the neurotransmitter glutamate into synaptic vesicles. The experimental paradigm was to incubate synaptic vesicles with [$^3$H]glutamate in the presence or absence of the reaction components required for ATP production (GAP, NAD, ADP, and $P_i$). Uptake of [$^3$H] glutamate was quantitated by a filter-binding assay.

These studies demonstrated that the accumulation of [$^3$H]glutamate into synaptic vesicles was dependent on the presence of the GAPDH/3-PGK substrates. Omission of one of the components abolished [$^3$H]glutamate uptake. Kinetic analysis indicated that a concentration of 12 μM ATP was sufficient for endogenous [$^3$H]glutamate uptake. This was an order of magnitude lower that the apparent $K_m$ for exogenously supplied ATP (0.8–1.2 mM).

Accordingly, it was suggested that this vesicle-bound glycolytic complex was more effective in stimulating glutamate uptake than exogenously supplied ATP. Inhibitor studies were undertaken to confirm that supposition. In those studies, exposure to iodoacetate, a specific GAPDH inhibitor, reduced glutamate uptake proportionally in a dose–response manner. In contrast, it did not affect glutamate uptake when exogenous ATP was supplied.

The physiological significance of these findings was examined by determining the release of glutamate from synaptic vesicles subsequent to depolarization. Pretreatment with iodoacetate decreased the amount of glutamate released. In contrast, treatment with iodoacetate subsequent to glutamate accumulation into synaptic vesicles did not significantly affect its release by exocytosis. From these studies, it was concluded that GAPDH fulfilled a required role in glutamate-mediated neuronal firing.

## 2. THE ROLE OF MOONLIGHTING GAPDH AS A PROTEIN PHOSPHORYLASE IN NEURONAL STRUCTURE AND FUNCTION

As described in previous chapters, initial studies on GAPDH provided the first indications that GAPDH may not be simply a glycolytic housekeeping protein. In that regard, such an investigation reported that rabbit skeletal muscle GAPDH was capable not only of autophosphorylation but also exhibited an intrinsic phosphorylase activity (Kawamoto and Caswell, 1986). In this study, GAPDH autophosphorylation was dependent on MgATP with rapid saturation at c.1 mM MgATP. Phosphorylated GAPDH exhibited a pronounced stability and instability at acidic versus neutral and basic pH, respectively. Substrate analysis demonstrated that the extent of phosphorylation was reduced by >90%, >40%, and >68% by NADH, NAD+, and glyceraldehyde-3-phosphate, respectively. Incubation of the phosphorylated protein with each of the metabolites resulted in its dephosphorylation. The latter could be observed after a 30 s incubation. The physiological significance of this finding was suggested by the demonstration that purified rabbit muscle GAPDH contained phosphotyrosine (Sergienko et al., 1992).

A second intriguing finding was the observation that phosphorylated GAPDH was capable of transferring its phosphate group to other proteins. The experimental paradigm was to use [$^{32}$P]ATP as the phosphate donor in the GAPDH reaction and terminal cisternae/triad proteins as the acceptor. Two phosphorylated proteins of $M_r$ = 72 and 80 kDa were detected. Reactions that included unlabeled ATP in the GAPDH phosphorylation reaction followed by the addition of [32]ATP in the transferase reaction resulted in a decrease in the amount of radiolabel in the acceptor proteins. This indicated that the phosphate group was derived from GAPDH and that a two-step reaction was responsible for the phosphorylation of the acceptor proteins, i.e., GAPDH autophosphorylation followed by the transfer of the phosphate group to the acceptor protein.

This two-step phosphorylation reaction is similar to that observed for GAPDH as a transnitrosylase (Chapter 8; Kornberg et al., 2010). Subsequent studies verified this new neuronal GAPDH activity (Wu et al., 1997) demonstrated that GAPDH could phosphorylate the hepatitis B virus core protein (Duclos-Vallée et al., 1998) and functioned as a phosphorylase in mammalian cells (Engel et al., 1996).

## 2.1 GAPDH Phosphorylation of GABA$_A$R: Regulation of Receptor-Mediated Neuronal Cell Transmission

GABA ($\gamma$-aminobutyric acid) is the principle inhibitory brain neurotransmitter. The GABA receptor is a complex structure containing five subunits some of which are phosphorylated. As is the case for many GAPDH studies, an investigation was begun to define both the GABA$_A$R protein phosphorylase and the mechanisms through which it modifies GABA$_A$R (Laschet et al., 2004).

In those investigations, isolation of the GABA$_A$R indicated the presence of a 38 kDa protein, which was constantly observed copurifying with the receptor subunit. Incubation of the purified receptor subunit with 1 mM Mg$^{++}$ (physiological concentration) and with [$^{32}$P]ATP resulted in the phosphorylation of the GABA$_A$ 51 kDa $\alpha$1 subunit. At 10 $\mu$M Mg$^{++}$, phosphorylation of the 38 kDa protein was observed while little phosphorylation of GABA$_A$R was detected. In contrast, at 100–1000 $\mu$M Mg$^{++}$, significant GABA$_A$R and little 38 kDa protein phosphorylation was observed. It was concluded not only that the 38 kDa protein was an endogenous protein kinase responsible for GABA$_A$R phosphorylation but also that a two-step mechanism involved in this posttranslational modification.

Sequence analysis of the p38 protein identified it as GAPDH. To verify its identity, it was recognized by an anti-GAPDH monoclonal antibody (which did not recognize any other protein in the purified protein preparation). Coimmunoprecipitation of phosphorylated GABA$_A$R was observed using the anti-GAPDH monoclonal antibody. Further, colocalization analysis using double fluorescence immunocytochemistry demonstrated the association of GABA$_A$R and GAPDH at the plasma membrane in rat hippocampal neuronal cells. The localization of the GABA$_A$R phosphorylation site was examined using recombinant GABA$_A$R constructs. Those studies indicated that GAPDH phosphorylated $\alpha$1thr$^{337}$ and $\alpha$1ser$^{416}$.

Intriguingly, the origin of the ATP used in the phosphorylation reaction was examined using the purified receptor preparation to which was added glyceraldehyde-3-phosphate, NAD$^+$, ADP, Mg$^{++}$, and $^{32}$P, i.e., the requirements for GAPDH glycolytic activity. Under those conditions, phosphorylation of GABA$_A$R was observed. Further, modulation of the reaction components (deletion, increase, or decrease) affected phosphorylation. Thus, it was concluded that neuronal glycolysis was required for GABA$_A$R phosphorylation.

The physiological significance of GAPDH-mediated $GABA_AR$ phosphorylation was examined using whole-cell recordings of GABA-elicited currents in dissociated neurons. In this experimental paradigm, using 7 mM MgATP, it was noted that the response of $GABA_AR$ to increasing applications of GABA diminished (termed rundown). Addition of GAPDH to the reaction reduced the level of $GABA_AR$ rundown by 34%. Addition of GAPDH plus glyceraldehyde-3-phosphate reduced rundown by 79%. Inactivation of GAPDH by iodoacetamide abolished this preventive effect. In toto, these studies demonstrate that GAPDH is responsible for $GABA_AR$ phosphorylation, the mechanism involves two steps, i.e., GAPDH autophosphorylation followed by its transfer of the phosphate group to the $GABA_AR$ $\alpha 1$ subunit, and that GAPDH-mediated $GABA_AR$ phosphorylation affects neuronal signal transduction.

## 2.2 GAPDH Deficiency of $GABA_AR$ Phosphorylation in the Etiology of Epilepsy

A common theme of investigations examining the functional diversity of GAPDH involves the analysis of moonlighting GAPDH activity in relation to human pathology. Again, regrettably, investigation of GAPDH structure and function in neuronal disease is no exception (Chapter 12; Moonlighting GAPDH and Age-Related Neurodegenerative Disease: Diversity of Protein Interactions and Complexity of Function).

That being said, GAPDH phosphorylation of $GABA_AR$ may be involved in the pathology of epilepsy as well (Laschet et al., 2007). In this investigation, the degree of $GABA_AR$ $\alpha 1$-subunit phosphorylation was examined in cortical tissue from epileptic patients as compared to nonepileptic controls. The experimental paradigm was to incubate isolated patient membrane fractions with [$^{32}$P]ATP and 1 mM $MgCl_2$ in 50 mM Hepes-Tris buffer (pH 7.3). Following the isolation of the protein pellet, the sample was analyzed by autoradiography after SDS-PAGE.

Although both the patient and the control samples demonstrated endogenous phosphorylation of the 51 kDa $\alpha 1$ subunit of $GABA_AR$, there was a noticeable reduction in the extent of modification in the patient samples. Statistical analysis demonstrated a $P = .0002$ (patients, n = 23, controls, n = 5). As a control, the binding of an agonist, in this case, [$^3$H] flunitrazepam, was determined by photoaffinity labeling. No statistical difference was observed between the patient and the control samples. This finding indicated that there was no reduction in the levels of the subunit in the patient population. Further addition of the GAPDH reaction mixture, [$^{32}$P] phosphate, ADP, $Mg^{++}$, $NAD^+$, and glyceraldehyde-3-phosphate increased $\alpha 1$ phosphorylation. This result indicated that there was no deficiency in the ability of patient samples to produce ATP.

Physiological differences in the patient versus control samples were examined using the experimental paradigm described above, i.e., rundown of $GABA_A$ currents. Comparison of each group indicated that the rundown in the patient population was increased and that this difference was statistically significant ($P<.001$). Intriguingly, the addition of glyceraldehyde-3-phosphate protected the patient population neurons from diminution in $GABA_A$ currents. This effect could be abolished by the addition of the specific GAPDH inhibitor iodoacetamide. Accordingly, it was postulated that these findings suggest that there may be a relationship between the marked difference in neuronal function in epileptic patients as compared to controls and their reduction of GADPH-mediated phosphorylation of $GABA_A R$.

## 2.3 GAPDH Phosphorylation of L1: Regulation of Neuronal Cell Growth

Another role for GAPDH as a phosphorylase is indicated by a recent study defining it as a cell adhesion factor L1 binding partner and determining the physiological significance of L1–GAPDH protein binding (Makhina et al., 2009). The experimental paradigm was to use coimmunoprecipitation from synaptosomal membrane fractions using an L1 antibody as a means to detect such binding partners. Immunoblot analysis detected a 37 kDa binding protein that was identified as GAPDH by mass spectroscopy. For confirmation, chemical cross-linking studies were performed using recombinant L1 (extracellular domain). Those investigations identified GAPDH as an L1 binding partner as well. Reversing the paradigm in ELISA demonstrated that soluble L1 (extracellular domain) recognized bound GAPDH.

The L1 GAPDH binding site was defined by ELISA using recombinant L1 fragments derived from its extracellular domain. Intriguingly, the L1 domain containing its entire immunoglobulin-like (Ig-like) domain (I–VI) was required for binding as fragments containing segments of that domain failed to bind to GAPDH. Similarly, the L1 fragment containing its fibronectin-homologous domains 1–5 and 4–5 bound GAPDH. This binding in vitro was corroborated by confocal microscopy studies demonstrating the colocalization of GAPDH and L1 on the neuronal cell surface in vivo.

Sequence analysis of L1 indicated the presence of the $\alpha 1$ subunit $GABA_A R$ GAPDH binding motif within its Ig-like domain. Accordingly, the Ig-like domain (I–VI) was incubated with rabbit muscle GAPDH and [$^{32}$P]ATP followed by autoradiography after SDS-PAGE. A marked phosphorylation of the Ig-like domain (I–VI) was observed. No phosphorylation was detected in the absence of GAPDH. Curiously, these studies revealed a complex phosphorylation pattern. The extracellular domains of two proteins, L1 and neural cell adhesion molecule were capable of autophosphorylation yet, in the presence of GAPDH, neither were phosphorylated nor GAPDH was autophosphorylated. Further the fibronectin-homologous domain 1–5 was autophosphorylated in the presence or absence of GAPDH. Thus, in contrast to the findings with $GABA_A R$, it would appear that GAPDH phosphorylation of adhesion factor L1 is part of a complex regulatory mechanism.

The physiological relevance of L1 phosphorylation by GAPDH was then examined. The experimental paradigm was to quantitate neurite outgrowth of cerebellar neurons on poly-L-lysine coated plates. Pretreatment of the poly-L-lysine plates with GAPDH and ATP enhanced L1-dependent neurite outgrowth while addition of a protein kinase inhibitor blocked that effect. Similarly, treatment with alkaline phosphatase after GAPDH exposure in the presence of ATP reduced neurite outgrowth. Anti-GAPDH antibodies reduced L1-induced neurite outgrowth but only in the presence of ATP. It was concluded that GAPDH phosphorylated L1 and that posttranslational modification increased L1-mediated neurite outgrowth.

## 3. MOONLIGHTING GAPDH AND THE REGULATION OF LIGAND-GATED ION CHANNEL SIGNALING

Receptor-mediated cell signaling is an a priori requirement for function in most, if not all, cell organs and tissue. As such, neurotransmission is no exception. As indicated in Table 7.2, moonlighting GAPDH activity may fulfill an important role in receptor-mediated neurotransmission and thus in neuronal function.

### 3.1 ATP-Mediated Release of GAPDH From Microglial Cells

The studies described above, indicating the role of extracellular GAPDH on neurite outgrowth (Makhina et al., 2009), raise a fundamental question, i.e., what is the origin of the extracellular GAPDH? One possibility is an interesting report describing the ability of extracellular ATP to release GAPDH from microglial cells (Takenouchi et al., 2015). The latter are located in the brain and appear to provide a protective immune function.

The experimental paradigm used in that investigation was to examine GAPDH localization as a function of extracellular ATP concentration in lipopolysaccharide treated microglial cells in culture. Using that protocol, extracellular GAPDH was detected in the culture media as defined by both immunoblot analysis and by ELISA. This effect was ATP concentration dependent. Kinetic analysis indicated a rapid effect of ATP with demonstrable extracellular GAPDH observed 20 min after ATP addition. Enzymatic assay indicated that the released GAPDH was catalytically active. As the P2X7 receptor may be involved in this process (Dubyak, 2012), the effect of its antagonist, $CuCl_2$, on ATP-mediated GAPDH release was determined. The study indicated that $CuCl_2$ addition in the presence of 2 mM ATP inhibited GAPDH extracellular release.

Morphological analysis indicated a potential mechanism through which extracellular GAPDH may be released from microglial cells. In particular, the latter are known to form both microvesicle-like structures as well as exosomes each of which may be produced as a consequence of ATP exposure. Immunostaining with an anti-GAPDH antibody indicated the presence of GAPDH in the microvesicle-like structures, while immunoblot analysis

**TABLE 7.2 Role of GAPDH in Ligand-Gated Ion Channel Cell Signaling[a]**

| Ion | Receptor | Agonist | GAPDH Function/Property | References |
|---|---|---|---|---|
| Cl[-] | GABA$_A$ | GABA (γ-aminobutyric acid) | Generation of ATP; phosphorylation of α1 subunit of GABA$_A$ | Laschet et al. (2004) |
| K[+] | K$_{ATP}$ | ATP | GAPDH identified as binding to the Kir6.2 pore channel subunit | Jovanović et al. (2005)[b] and Dhar-Chowdhury et al. (2005)[b] |
| K[+] | P2X7 (microglial cells) | ATP | Release of extracellular GAPDH | Takenouchi et al. (2015) |
| Na[+] | SCN1A, SCN3A | N.C. | GAPDH binds to 3'-UTR of both SCN1A and SCN3A mRNA differentially affecting mRNA stability; GAPDH$^{thrP}$ may be active binding GAPDH species | Lin et al. (2017) |

N.C., Not cited
[a]Chronological order.
[b]Studies performed in cardiac cells.

demonstrated its presence in exosomes. Of note, microvesicle-like structures may contain F-actin, which had previously been demonstrated to bind to GAPDH in synaptic structures (Rogalski-Wilk and Cohen, 1997).

As indicated in Table 7.2, the P2X7 receptor is a $K^+$ ion channel. Addition of high $K^+$ concentrations to the microglial cell cultures diminished ATP-mediated extracellular GAPDH release. Further analysis indicated that a $K^+/H^+$ ionophore, nigericin, could trigger the release of GAPDH in the absence of ATP. This effect was abolished by the presence of high extracellular $K^+$ concentrations. In toto, as these studies demonstrate that ATP may stimulate the release of GAPDH from microglial cells, it is tempting to speculate that this phenomenon produces the extracellular GAPDH, which binds to extracellular L1 stimulating neurite outgrowth (Makhina et al., 2009).

## 3.2 Moonlighting GAPDH and the Posttranscriptional Control of Sodium Channel Protein Expression

The study of receptor-mediated cell signaling is focused usually on the analysis of agonist and antagonist effects and the mechanisms through which they affect receptor action. That being said, the control of receptor gene expression may represent also an important theme in the analysis of signal transduction.

As said before, a general theme throughout this work is that the study of human pathologies has resulted in the elucidation of many GAPDH moonlighting activities. In this instance, clinical genetic analyses indicated the dysregulation of two $Na^+$ ion–gated channel genes, *SCN1A* and *SCN3A*, in the epilepsy. In particular, *SCN1A* exhibits a loss of function and *SCN3A* appears to be upregulated.

In a recent investigation, the posttranscriptional regulation of each sodium channel gene was examined (Lin et al., 2017). The experimental paradigm was to use 3′-UTR constructs from each gene to define protein-binding partners and then to determine whether perturbations in those protein–nucleic interactions may underlie specific alterations in *SCN1A* and *SCN3A* gene expression.

Three synthetic oligoribonucleotides probes were used in RNA–EMSA to identify those binding partners in HEK-293 cells. Each oligonucleotide contained a conserved region (termed CS1, CS2, and CS3) located in the 3′-UTR mRNA domains of both genes. These studies revealed that a shift band was observed solely using the CS1 oligonucleotide probe. Binding was reduced in a dose-dependent manner using an unlabeled probe, indicating the specificity of the RNA–protein interaction. In a pull down assay, a 36 kDa protein was identified. Incubation with an anti-GAPDH antibody reduced the amount of protein bound. The use of purified GAPDH in the RNA–EMSA protocol confirmed the identity of the protein as GAPDH.

Recombinant gene experiments using determination of luciferase activity as the experimental paradigm was used to examine the effect of GAPDH–CS1 binding on mRNA stability. Luciferase activity in CS-1 deletion constructs of the *SCN1A* gene was higher than that observed in the wild type control. In contrast, no difference in luciferase activity was observed in the CS1-deleted construct of the *SCN3A* gene. In GAPDH knock down studies, increased reporter activity was observed in cells containing the *SCN1A* 3'-UTR as compared to the wild type control. In contrast, GAPDH knock down decreased luciferase activity in cells containing the *SCN3A* 3'-UTR. The effect of GAPDH on the stability of each mRNA was examined by treating transfected cells with Actinomycin D then determining residual luciferase reporter activity. Those studies demonstrated that cells transfected with the *SCN1A* 3'-UTR exhibited lower luciferase activity than control cells. In contrast, cells transfected with the *SCN3A* 3'-UTR displayed higher luciferase levels than the control cells. From these cumulative studies, it was concluded that binding of GAPDH to the *SCN1A* CS-1 3'-UTR decreased the stability of the *SCN1A* mRNA, while its binding to the *SCN3A* CS-1 3'-UTR increased its mRNA stability. In toto, these studies suggested a selective regulation of brain sodium channel genes by GAPDH.

The physiological significance of this differential regulation was indicated by studies in normal, seizure and seizure mice treated with a ketogenic diet (seizure induced by injection of kainic acid; KD used to treat childhood epilepsy). In the mouse hippocampus, initial studies demonstrated that *SCN1A* and *SCN3A* mRNA levels were reduced in seizure mice by 30% in the former and increased by 80% in the latter. KD treatment restored normal mRNA levels for both genes.

Quantitation of GAPDH levels by immunoblot analysis in the mouse hippocampus indicated that there was a 2.5-fold increase in the seizure animals, which was reduced to normal levels by KD treatment. Analysis of GAPDH phosphorylation (thr$^{244}$) demonstrated that, in the seizure mice, GAPDH$^{thrP}$ increased 3.5-fold and was also reduced to normal by KD treatment. RNA–EMSA indicated increased binding of GAPDH purified from the seizure mouse, while normal binding was observed in GAPDH purified from the KD-treated animals. Treatment of GAPDH$^{seizure}$ with alkaline phosphatase reduced binding. In toto, these studies suggest that not only does GAPDH regulate differentially two brain sodium channel genes but also that its phosphorylation is required for its effect on mRNA stability.

## 3.3 Role of Moonlighting GAPDH in K$_{ATP}$ Channel Structure and Function

As with most receptors, the K$_{ATP}$ channel is a multiprotein complex that functions in a variety of tissues including the pancreas, brain, and heart. Recent evidence suggested that moonlighting GAPDH associates with that receptor complex and may help to regulate its function (Jovanović et al., 2005; Dhar-Chowdhury et al., 2005).

Coimmunoprecipitation was performed using a membrane fraction with an antibody directed against the SUR2A regulatory subunit of the K$_{ATP}$ channel

(Jovanović et al., 2005). The analysis detected a p42 protein that was identified as GAPDH by MALDI–TOF analysis. As a control, the coimmunoprecipitation protocol was repeated using an anti-Kir6.2 antibody. The latter is considered as a pore-forming $K_{ATP}$ subunit. It also detected GAPDH as a binding protein. Lastly, the reverse study using an anti-GAPDH antibody coimmunoprecipitated both SUR2A and Kir6.2.

To define the specificity of GAPDH binding, recombinant gene constructs were prepared, which contained either the SUR2A or the Kir6.2 $K_{ATP}$ channel subunit. Each was transfected into A549 cells, which does not contain either subunit. Coimmunoprecipitation studies demonstrated that an anti-GADPH antibody precipitated the Kir6.2 protein but not the SUR2A subunit. Reversing the coimmunoprecipitation protocol demonstrated that the anti-Kir6.2 antibody but not the anti-SUR2A antibody precipitated GAPDH. Further analysis demonstrated that GAPDH bound to the C-terminal portion of the Kir6.2 protein.

In a concurrent study, two-hybrid screening using the Kir6.2 cytosolic carboxyl terminus as bait was utilized to probe a rat cardiac cDNA library (Dhar-Chowdhury et al., 2005). Two potential interacting proteins were detected. Sequence analysis identified them as GAPDH and triosephosphate isomerase (TPI). For confirmation, coimmunoprecipitation assays were performed in COS7L cells transfected with full-length Kir6.2-HA (hemagglutinin) using an anti-GAPDH antibody. Immunoblot analysis demonstrated that the Kir6.2-HA protein bound to GAPDH. Reversing the antibody sequence using a polyclonal anti-HA antibody demonstrated the immunoprecipitation of GAPDH. Similarly, using a TPI-GFP fusion protein and anti-GFP antibodies, coimmunoprecipitation was observed for Kir6.2-HA as well. Although it was not detected in the initial two-hybrid screen, further coimmunoprecipitation studies indicated that pyruvate kinase (PK) also bound to Kir6.2. The physical association of Kir6.2 with GAPDH and PK was examined in vivo by colocalization studies using dual labeled immunocytochemical protocols. Those investigations demonstrated that each glycolytic protein exhibited a strikingly similar subcellular localization as was detected for Kir6.2. In toto, these two independent reports identify GAPDH, as well as other glycolytic proteins, as constituents of the $K_{ATP}$ channel. In accord with the synaptic studies described in this chapter and those described in Section V for sperm GADPH, these findings suggest that the compartmentalization of glycolytic pathways in specific organs and tissues may be a general phenomenon in higher organisms.

## 4. SUMMARY

The investigations described in this chapter highlight three activities of moonlighting GAPDH in neuronal structure and function. These include its singular synaptic distribution enabling regional ATP formation, the importance of its phosphorylase activity in cell signaling as well as its significance with respect to ligand-gated ion channel signaling. That being said, of particular importance may be its role in the regulation of two opposite neuronal signals, i.e., glutamate as the major excitatory neurotransmitter and GABA as the major inhibitory neurotransmitter. Each depends on a specific moonlighting GAPDH activity. Further, the variety of mechanisms through which GAPDH exhibits its neuronal functional diversity (protein–protein interactions, posttranscriptional regulation of gene expression, posttranslational protein modification) illustrates its characteristic diversity, which is one of the major themes of the present chapter and Chapters 1–6.

## REFERENCES

Dhar-Chowdhury, P., Harrell, M., Han, S., et al., 2005. The glycolytic enzymes, glyceraldehyde-3-phosphate dehydrogenase, triose phosphate isomerase, and pyruvate kinase are components of the $K_{ATP}$ channel macromolecular complex and regulate its function. J. Biol. Chem. 280, 38464–38470.

Dubyak, G., 2012. P2X7 receptor regulation of non-classical secretion from immune effector cells. Cell. Microbiol. 14, 1697–1706.

Duclos-Vallée, J.-C., Capel, F., Mabin, H., Petit, M.-A., 1998. Phosphorylation of the hepatitis B virus core protein by glyceraldehyde-3-phosphate dehydrogenase protein kinase activity. J. Gen. Virol. 79, 1665–1670.

Engel, M., Seifert, M., Theisinger, B., Seyfert, U., Welter, C., 1996. Glyceraldehyde-3-phosphate dehydrogenase and Nm23-H1/nucleoside diphosphate kinase A: two old enzymes combine for the novel Nm23 protein phosphotransferase function. J. Biol. Chem. 273, 20058–20065.

Ikemoto, A., Bole, D., Ueda, T., 2003. Glycolysis and glutamate accumulation into synaptic vesicles: role of glyceraldehyde-3-phosphate dehydrogenase and 3-phosphoglycerate kinase. J. Biol. Chem. 276, 5929–5940.

Jovanović, S., Du, Q., Crawford, R., et al., 2005. Glyceraldehyde-3-phosphate dehydrogenase serves as an accessory protein of the cardiac sarcolemmal $K_{ATP}$ channel. EMBO Rep. 6, 848–852.

Kawamoto, R., Caswell, A., 1986. Autophosphorylation of glyceraldehyde-3-phosphate dehydrogenase and phosphorylation of protein from skeletal muscle microsomes. Biochemistry 25, 656–661.

Kornberg, M., Sen, N., Hara, M., et al., 2010. GAPDH mediates nitrosylation of nuclear proteins. Nat. Cell Biol. 12, 1094–1102.

Laschet, J., Kurcewicz, I., Minier, F., et al., 2007. Dysfunction of $GABA_A$ receptor glycolysis-dependent modulation in human partial epilepsy. Proc. Natl. Acad. Sci. U.S.A. 104, 3472–3477.

Laschet, J., Minier, F., Kurcewicz, I., et al., 2004. Glyceraldehyde-3-phosphate dehydrogenase is a $GABA_A$ receptor kinase linking glycolysis to neuronal inhibition. J. Neurosci. 24, 7614–7622.

Leung, T., Hall, C., Monfries, C., Lim, L., 1987. Trifluoperazine activates and releases latent ATP-generating enzymes associated with the synaptic plasma membrane. J. Neurochem. 49, 232–238.

Lin, G.-W., Lu, P., Zeng, T., et al., 2017. GAPDH-mediated posttranscriptional regulations of sodium channel *Scn1a* and *Scn3a* genes under seizure and ketogenic diet conditions. Neuropharmacology 113, 480–489.

Makhina, T., Loers, G., Schulze, C., et al., 2009. Extracellular GAPDH binds to L1 and enhances neurite outgrowth. Mol. Cell. Neurosci. 41, 206–218.

Rogalski-Wilk, A., Cohen, R., 1997. Glyceraldehyde-3-phosphate dehydrogenase activity and F-actin associations in synaptosomes and postsynaptic densities of porcine cerebral cortex. Cell. Mol. Neurobiol. 17, 51–70.

Schläfer, M., Wolknandt, W., Zimmermann, H., 1994. Putative synaptic vesicle nucleoside transporter identified as glyceraldehyde-3-phosphate dehydrogenase. J. Neurochem. 63, 1924–1931.

Sergienko, E., Kharitonenkov, A., Bulargina, T., Muronetz, V., Nagradova, N., 1992. D-glyceraldehyde-3-phosphate dehydrogenase purified from rabbit muscle contains phosphotyrosine. FEBS Lett. 304, 21–23.

Stadler, H., Fenwick, E., 1983. Cholinergic synaptic vesicles from *Torpedo marmorata* contain an atractyloside-binding protein related to the mitochondrial ADP/ATP carrier. Eur. J. Biochem. 136, 377–382.

Takenouchi, T., Tsukimoto, M., Iwamaru, Y., et al., 2015. Extracellular ATP induces unconventional release of glyceraldehyde-3-phosphate dehydrogenase from microglial cells. Immunol. Lett. 167, 116–124.

Wu, K., Aoki, C., Elste, A., Rogalski-Wilk, Siekevitz, P., 1997. The synthesis of ATP of glycolytic enzymes in the postsynaptic density and the effect of endogenously generated nitric oxide. Proc. Natl. Acad. Sci. U.S.A. 94, 13273–13278.

## FURTHER READING

Huo, J., Zhu, X.-L., Ma, R., Dong, H.-L., Bin-Xiao, B., 2016. GAPDH/Siah1 cascade is involved in traumatic spinal cord injury and could be attenuated by sivelestat sodium. Neuorscience 330, 171–180.

Kumar, S., Kim, Y., 2016. Glyceraldehyde-3-phosphate dehydrogenase is a mediator of hemocyte-spreading behavior and molecular target of immunosuppressive factor CrV1. Dev. Comp. Immunol. 54, 97–108.

Lubec, G., Labudova, O., Cairns, N., Fountoulakis, M., 1999. Increased glyceraldehyde-3-phosphate dehydrogenase levels in the brain of patients with Down's syndrome. Neurosci. Lett. 260, 141–145.

Rudkouskaya, A., Sim, V., Shah, A., et al., 2010. Long-lasting inhibition of presynaptic metabolism and neurotransmitter release by protein S-nitrosylation. Free Radic. Biol. Med. 49, 757–769.

Yamaji, R., Chatani, E., Harada, N., et al., 2005. Glyceraldehyde-3-phosphate dehydrogenase in the extracellular space inhibits cell spreading. Biochim. Biophys. Acta 1726, 261–271.

Zhai, D., Lee, F., D'Souza, C., et al., 2015. Blocking GluR2-GAPDH ameliorates experimental autoimmune encephalomyelitis. Ann. Clin. Transl. Neurol. 2, 388–400.

# Physiological Stress and GAPDH Functional Diversity

# Chapter 8

# The Significance of Nitric Oxide–Modified GAPDH: Regulation of Apoptosis, Cell Signaling, and Heme Metabolism

*A Star is Born*

Moss Hart

In spite of many exemplary performances over a number of years, many actors and actresses seek that one role which marks them as a star. One need only think of John Wayne in "Stagecoach," Marilyn Monroe in "Some Like It Hot," Humphrey Bogart in "Casablanca," and Marlin Brando in "On the Waterfront" as actors and actresses with considerable pedigrees prior to those films yet who, by virtue of those performances, were catapulted to that elite level reserved for very few thespians.[1]

Similarly, early studies on the functional diversity of GAPDH had been reported for a number of years establishing it as a moonlighting protein (rev. in Sirover, 1996). Yet, in spite of those diverse reports, it remained a protein of interest perhaps noted as an "intellectual curiosity." Using the analogy described above, it could be termed a "character actor," i.e., an individual who appeared regularly on stage and screen but never obtained top billing.

Accordingly, in this chapter, we shall consider those studies, which marked GAPDH as a major protein of significant interest, fulfilling a critical role in normal cell function. These studies relate to the nitric oxide (NO)–mediated formation of SNO-GAPDH (GAPDH[cys149-NO])[2]; its role as an initiator of apoptosis through its protein interaction with Siah1, subsequent nuclear proteasome degradation, changes in nuclear protein structure and function, downstream gene regulation, the transnitrosylase activity of SNO-GAPDH, and finally, the intriguing role of SNO-GAPDH in heme regulation. Conversely, its binding to

---

1. For those who are unfamiliar with these actors, actresses, and movies of yesteryear, a Google search is recommended.
2. Although GAPDH is a highly conserved protein, there is some variance in the location of its active site cysteine (149, 150, or 152).

Glyceraldehyde-3-Phosphate Dehydrogenase (GAPDH). http://dx.doi.org/10.1016/B978-0-12-809852-3.00008-X
**131**

GOSPEL may prevent apoptosis. In addition, unmodified GAPDH may function in autophagy, a mechanism to ensure cell survival. In toto, these investigations defined GAPDH as an elite protein worthy of its "top billing" in the pantheon of important cell molecules.[3]

## 1. GAPDH REGULATION IN APOPTOSIS

As indicated in Table 8.1, a series of studies established that GAPDH expression was increased during programmed cell death. For the most part, these studies were performed using neuronal cell primary cultures in which apoptosis was induced merely by time in cell culture. The latter was termed days in vitro (DIV). In particular, following isolation, neuronal cells exhibited increased cell death approximately 16–17 DIV (Sunaga et al., 1995). That this cell death was due to apoptosis was indicated by electron microscopic analysis and by determination of DNA fragmentation (Ishitani et al., 1996a). Further analysis revealed that treatment with cytosine arabinoside also induced apoptosis in these neuronal cells (Ishitani and Chuang, 1996; Chen et al., 1999).

Accordingly, this system was particularly useful as an experimental paradigm in that it permitted a detailed temporal analysis of those events, which were a part of the reproducible apoptotic program initiated by these neuronal cells in culture. For example, it was noted that inhibition of transcription and translation by actinomycin D and cycloheximide, respectively, reduced the rate of neuronal cell apoptosis (Ishitani et al., 1996a,b, 1997; Ishitani and Chuang, 1996).

As with many GAPDH studies, it was noted initially that neuronal cell apoptosis was associated with the appearance of a 38-kDa protein, which comigrated with commercially available GAPDH (Sunaga et al., 1995). Curiously, each protein was immunoreactive with a monoclonal antibody prepared an Alzheimer's disease plaque yet was not recognized by antibodies prepared against the β-amyloid protein (Sunaga et al., 1995).

That 38-kDa protein was identified as GAPDH by protein sequence analysis (Ishitani et al., 1996a). As with apoptosis, suppression of GAPDH expression was effected by treatment with either actinomycin D or cycloheximide (Ishitani et al., 1996a,b, 1997; Ishitani and Chuang, 1996). Temporal analysis demonstrated increased expression of GAPDH mRNA at 11 DIV, which was maximal at 15 DIV, while that of the GAPDH protein reached its highest levels at 17–19 DIV (Ishitani et al., 1996a). Similarly, both GAPDH mRNA and protein were enhanced in a temporal manner following exposure to cytosine arabinoside (Ishitani and Chuang, 1996).

---

3. This is not intended to diminish the significance of the host of GAPDH studies presented in this book. Each is important in its own right. Rather, it is the personal opinion of the author that these investigations, relating GAPDH and its interaction with nitric oxide to apoptosis, were the first to alter permanently the perception of GAPDH by the scientific community. The latter has been cemented by the diverse, detailed analyses described in this text.

**TABLE 8.1 GAPDH Expression in Apoptosis**[a]

| System | Apoptotic Stimulus | GAPDH Expression | "Unique Finding" | References |
|---|---|---|---|---|
| Cerebellar granule cells | Days in culture | Increased protein levels | AD plaque monoclonal antibody recognition | Sunaga et al. (1995) |
| Cerebellar granule cells | Days in culture | Increased protein levels | Prior increase in GAPDH mRNA expression | Ishitani et al. (1996a) |
| Cerebellar granule cells | Cytosine arabinoside | Increased mRNA and protein levels | Antisense GAPDH prevents apoptosis | Ishitani and Chuang (1996) |
| Cerebellar cortical cells | Days in culture | Increased mRNA and protein levels | Antisense GAPDH prevents apoptosis | Ishitani et al. (1996b) |
| Neuroblastoma/glioma hybrid cells | Lipopolysaccharide | Inhibition of activity | Nitric oxide GAPDH modification | Nomura et al. (1996) |
| Cerebellar granule cells | Low potassium | Increased mRNA and protein levels | Antisense GAPDH prevents apoptosis | Ishitani et al. (1997) |
| Neuroblastoma/glioma hybrid cells | Koningic acid | Inhibition of activity | Comparable effect of NO donors | Nakazawa et al. (1997) |
| Cerebellar granule cells, PC12 cells | Cytosine arabinoside | Increased mRNA and protein levels | p53 gene dependent | Chen et al. (1999) |

AD, Alzheimer's disease; NO, nitric oxide.
[a]In chronological order.

As with all moonlighting protein studies, it was necessary to establish the physiological significance of the respective protein's new function. In this instance, antisense studies were utilized to consider the effect of GAPDH diminution on the rate of programmed cell death (Ishitani and Chuang, 1996; Ishitani et al., 1996b, 1997). These studies demonstrated that irrespective of the mechanism through which apoptosis was initiated (DIV or cytosine arabinoside), addition of antisense GAPDH prevented the induction of programmed cell death.

The generality of this new GAPDH function was examined using other experimental paradigms. In neuroblastoma/glioma hybrid cells, apoptosis was induced by either liposaccharide or koningic acid (Nomura et al., 1996; Nakazawa et al., 1997, respectively). In each case, changes in cell morphology and the induction of DNA fragmentation were used to validate the initiation of programmed cell death. However, in contrast to the findings described above, each of these studies reported an inhibition of GAPDH catalytic activity. In the Nomura et al.'s (1996) study, it was noted that liposaccharide is a potent inducer of NO synthase that stimulated the covalent attachment of ($^{32}$P) NAD$^+$ to GAPDH. In the Nakazawa et al.'s (1997) report, it was noted not only that koningic acid is a specific inhibitor of GAPDH, binding to its active site cysteine, but also that a similar effect was observed using exogenous NO donors.

Apoptosis represents a complex program of gene expression. For that reason, the cerebellar granule cells model was used to examine the interrelationship between cytosine arabinoside–induced apoptosis, GAPDH expression, and that of apoptotic-related genes (Chen et al., 1999). In those studies, increased levels of GAPDH, p53, and Bax mRNA were observed. It was noted that GAPDH expression persisted for the longest interval (up to 24 h), while p53 mRNA levels were increased for the shortest time period (1–4 h after drug exposure). Bax mRNA was in the middle with increases observed between 2 and 12 h.

Antisense studies demonstrated that both p53 and GAPDH constructs protected the cerebellar granule cells from drug-induced programmed cell death. Intriguingly, in antisense p53-treated cells, GAPDH mRNA and protein levels were diminished. This suggested that GAPDH may be a downstream target of p53. To test this possibility, cerebellar granule cells derived from p53$^{-/-}$ transgenic mice were treated with cytosine arabinoside to induce apoptosis. There was a fourfold increase in cell survival in those cells as compared to that of cells from the p53$^{+/+}$ control. Further, GAPDH expression was only slightly increased in the p53$^{-/-}$ cells. Accordingly, these findings suggest that p53 is an upstream regulator of GAPDH during apoptosis in cerebellar granule cells.

## 2. NUCLEAR TRANSLOCATION OF GAPDH IN APOPTOSIS

In toto, the studies described above and presented in Table 8.1 establish increased GAPDH expression as an a priori requirement for apoptosis. That being said, they do not demonstrate that a moonlighting GAPDH activity is involved. In particular, as neuronal cells depend on glycolysis for ATP production, it would be reasonable to suggest that the increase in GAPDH is simply due to an elevated requirement for energy generation during programmed cell death.That being said, as indicated in Table 8.2, in a series of elegant reports, it was established that the nuclear translocation of GAPDH was also an a priori requirement for the initiation of apoptosis. In an initial study (Sawa et al., 1997), subcellular fractionation demonstrated a time-dependent increase in GAPDH in the low-speed particulate fraction as a function of apoptotic cell death. This was noteworthy for five reasons: First, these studies were performed in a variety of cell types including S49, HEK 293 and PC12 cells, primary thymocytes, and cortical neurons. As such, this study demonstrated convincingly the generality of GAPDH expression during apoptosis. Second, the appearance of GAPDH in the low particulate fraction was time-dependent coordinate with a decrease in cell viability. Third, immunocytochemical analysis demonstrated the presence of nuclear GAPDH as a function of apoptotic stimuli. In contrast, in untreated cells, GAPDH displayed a cytosolic localization. Fourth, exposure of cells to antisense GAPDH not only prevented apoptosis but also resulted in a purely cytosolic GAPDH distribution. Fifth, analysis of nuclear GAPDH revealed that it was tightly associated with DNA, i.e., treatment with DNase I or washing with high salt (3 M or 5 M NaCl) failed to disrupt the GAPDH–DNA linkage.

In another study (Saunders et al., 1997), differential sucrose density centrifugation analysis (1,000, 20,000, and $200,000 \times g$) was used to isolate nuclear, mitochondria, microsomal, and cytoplasmic fractions in cerebellar granule cells exposed to cytosine arabinoside in culture. Pellets from each centrifugation were designated as $P_1$, $P_2$, and $P_{200}$, respectively. Immunoblot analysis demonstrated that drug treatment increases GAPDH levels by 2.9- and 4.5-fold in the $P_1$ and the $P_{20}$ fractions, respectively. Introduction of antisense GAPDH prior to drug treatment eliminated the increases in immunoreactive GAPDH in each fraction. Intriguingly, GAPDH-specific activity decreased in the $P_1$ and the $P_{20}$ fractions despite the increase in immunoreactive GAPDH protein.

In a further analysis, an antibody was prepared against the overexpressed protein (Ishitani et al., 1998). Control studies demonstrated that it reacted with commercially obtained chicken muscle GAPDH. Immunological analysis of subcellular protein fractions in cytosine arabinoside–treated CGC cells revealed that a majority of the overexpressed GAPDH protein was located in the nuclear $(P_1)$ fraction. Intriguingly, there appeared to a small but significant amount of immunoreactive protein in the mitochondria. Electron microscopic analysis confirmed the nuclear localization. Treatment with either antisense GAPDH or cycloheximide blocked both the increase in immunoreactive GAPDH and its nuclear localization. In a subsequent study, the nuclear localization of GAPDH isoforms was examined (Saunders et al., 1999). The presence of isoforms was well established in early GAPDH studies (rev. in Sirover, 1999). Again, the cytosine arabinoside treatment of CGC cells was used as the experimental paradigm. Significant findings in this study included the time-dependent increase

**TABLE 8.2** Nuclear Localization of Moonlighting GAPDH in Apoptosis[a]

| System | Apoptotic Stimulus | Nuclear Translocation | "Unique Finding" | References |
|---|---|---|---|---|
| PC12, HEK 293,[b] S49,[b] primary thymocytes,[b] cerebral cortical cells | Normal growth factor, serum withdrawal, Dex[b] stimulation, days in culture, respectively | Cytosol-to-nuclear translocation following apoptotic stimuli | GAPDH tightly bound to DNA; antisense GAPDH blocks apoptosis and nuclear translocation | Sawa et al. (1997) |
| Cerebellar granule cells | Cytosine arabinoside | Increased immunoreactive GAPDH | Reduced levels of GAPDH nuclear catalysis | Saunders et al. (1997) |
| Cerebellar neurons | Cytosine arabinoside, low $K^+$ in serum-free medium | Major immunoreactive protein localization | Minor mitochondrial localization | Ishitani et al. (1998) |
| Cerebellar granule cells | Cytosine arabinoside | Time-dependent localization | Coordinate nuclear decrease in GAPDH and UDG activities[c] | Saunders et al. (1999) |
| COS-1, HEK 293, PC 12 cells transfected with rat brain GAPDH fusion construct | Serum withdrawal | Cytoplasmic-to-nuclear translocation of GAPDH–GFP fusion protein | Time dependent; nuclear translocation observed after $H_2O_2$ exposure | Shashidharan et al. (1999) |
| Rat ventral prostate | Androgen withdrawal | Time-dependent localization | Nuclear GAPDH localization following androgen replacement | Epner et al. (1999) |
| COS-7 cells | GAPDH transfection | Immunolocalization of transfected GAPDH | Apoptosis induction independent of GAPDH source | Tajima et al. (1999) |
| Neuroblastoma, fibroblasts | Staurosporine, MG132,[d] $H_2O_2$, FeCN | Nuclear translocation except in Bcl2 cells | GAPDH transfection eliminates Bcl2 inhibition | Dastoor and Dreyer (2001) |

[a]Chronological order.
[b]Dexamethasone.
[c]Uracil DNA glycosylase.
[d]Proteasome inhibitor.

in the nuclear GAPDH protein (804%) of the control at 16 h after drug exposure; the coordinate, time-dependent decrease of both nuclear GAPDH glycolytic activity and an identical time-dependent reduction in nuclear uracil DNA glycosylase activity (Fig. 3, p. 928); and the selective increase of three nuclear GAPDH basic isoforms. As expected, treatment with antisense GAPDH eliminated the increase in GAPDH basic proteins. In toto, this study suggests not only a selectivity in the form of nuclear GAPDH expressed during apoptosis but also an inverse relationship between the nuclear expression of the GAPDH protein and those enzyme activities attributed to it.

Concurrent with the studies described above, the role of GAPDH during apoptosis in prostate cells was examined as a function of androgen withdrawal and replacement (Epner et al., 1999). In these studies, as defined by immunohistochemical analysis, nuclear GAPDH was observed 5–7 days after androgen withdrawal. It was stated that, at this time, approximately 20% of the ventral prostate cells were undergoing apoptosis per day. Of particular significance, it was noted that, in the prostate, androgen withdrawal induces apoptosis only in the ventral lobe with minimal programmed cell death occurring in the dorsal or lateral lobes. Unfortunately, it does not appear that a similar control analysis was performed in the latter prostate regions.

The dynamics of apoptotic-induced GAPDH nuclear location was examined using a transfected GAPDH–GFP fusion protein in COS-1, HEK 293, and PC12 cells (Shashidharan et al., 1999). The former was transiently transfected, while the latter two were stable transfectants. In PC12 cells following serum withdrawal, histochemical analysis demonstrated that the GAPDH–GFP construct was detected in the nucleus as compared to its cytoplasmic localization in the PC12 control. Parallel histochemical changes were observed in the endogenous GAPDH protein. Similarly, exposure to stably transfected HEK 293 or transiently transfected cells to $H_2O_2$ resulted in a time-dependent nuclear relocation of the GAPDH–GFP fusion protein. In accord with previous findings (Sawa et al., 1997), the GAPDH–GFP protein was associated with nuclear DNA.

A further study revealed an intriguing finding (Tajima et al., 1999). In this investigation, a GAPDH clone was prepared from mRNA isolated from age-induced CGC cells (14–15 DIV). The resultant clone was then transfected into COS-7 cells. Surprisingly, in the absence of any apoptotic inducer, not only was the transfected protein localized in the nucleus but also there was a demonstrative increase in apoptotic cell death. The latter was confirmed by DNA fragmentation and morphological analyses. As a control, transfection of a cloned GAPDH gene from nonapoptotic CGCs yielded the same findings. Accordingly, it was postulated that GAPDH may represent a "killer protein" in programmed cell death.

As noted, apoptosis is not only a complex program of gene expression but is also tightly regulated by effectors and inhibitors. One such inhibitor is Bcl2. To examine that interrelationship, Dastoor and Dreyer (2001) examined apoptotic-related GAPDH nuclear translocation as a function of Bcl2 expression. The experimental paradigm was a neuroblastoma cell line and two R6 fibroblast cell lines, an R6 control and one overexpressing Bcl2 (R6-Bcl2) as well as trans-fection analysis using a GAPDH–GFP construct. Apoptosis was induced by exposure to either staurosporine or MG132. Oxidative stress was quantitated after exposure to either $H_2O_2$ or FeCN. GAPDH expression was monitored using immunocytochemical localization of endogenous GAPDH or that of the GAPDH–GFP construct.

The critical question posed in this study was the effect of Bcl2 expression on apoptosis generally and on GAPDH specifically. Examination of endoge-nous GAPDH in neuroblastoma cells or in the control R6 fibroblast cell line as a function of staurosporine- or MG132-induced apoptosis demonstrated the expected increase in programmed cell death, increased GAPDH expression, and its nuclear translocation. In contrast, as expected following drug exposure, apoptosis was decreased in the R6-Bcl2 cell line. Similarly, the nuclear trans-location of GAPDH was reduced significantly. Intriguingly, GAPDH appeared to be localized in the perinuclear region as if Bcl2 prevented it from accessing a "door," which enabled it to enter the cell nucleus.

In contrast, a different result was observed when cells were transfected with the GAPDH–GFP construct. In R6-Bcl2-transfected cells following drug expo-sure, not only was a significant level of apoptosis observed but also there was considerable GAPDH–GFP nuclear translocation. This finding suggests that Bcl2 inhibition of apoptosis can be overcome by overexpression of GAPDH.

## 3. ROLE OF NITRIC OXIDE IN GAPDH-MEDIATED APOPTOSIS

As discussed above, significant studies demonstrated not only that GAPDH was involved in apoptosis but also that its nuclear translocation was an a pri-ori requirement for programmed cell death. This presented a paradox in that, although sequence analysis demonstrated the presence of a nuclear export signal (Brown et al., 2004), it also indicated the absence of a nuclear local-ization domain. Thus, there did not appear to be a mechanism through which GAPDH, by itself as a tetramer, could redistribute from the cytoplasm to the nucleus. As evidence was presented that GAPDH could be dissociated into a dimer or a monomer (Park et al., 2009), it was possible that a sequential change in its oligomeric structure could mediate its apoptotic intracellular movement. However, it was reported that such a change in GAPDH oligomeric structure resulted in a reduction of apoptosis (Carlile et al., 2000). Accordingly, a differ-ent mechanism would presumably be required for apoptotic-related GAPDH nuclear translocation.

### 3.1 Nitric Oxide–Mediated GAPDH Nuclear Translocation

The salient study that provided the answer to this apparent dilemma not only clarified the mechanism for GAPDH nuclear translocation during apoptosis, but also demonstrated again the utility of GAPDH as the quintessential moonlighting protein (Hara et al., 2005; comment by Benhar and Stamler, 2005). Significant findings reported in this study are the cytosolic binding of GAPDH to Siah1; the nuclear translocation of this protein–protein complex; the role of NO modification of GAPDH as a facilitator of complex formation and intracellular redistribution; the catalysis of nuclear protein degradation by Siah1; and the functional specificity of nuclear GAPDH–Siah1 nuclear translocation, i.e., its restriction to the initiation of programmed cell death (rev. in Hara and Snyder, 2006).

Siah1 (seven in absentia homolog 1) is an E3 ubiquitin ligase that catalyzes protein degradation and is involved in apoptosis (Hu et al., 1997; Li et al., 1997; Tang et al., 1997). Its binding to GAPDH was identified by yeast two-hybrid analysis. The latter used both N-terminal and C-terminal GAPDH constructs. Siah1 was the only protein identified by the C-terminal GAPDH domain. GAPDH–Siah1 binding was confirmed by coimmunoprecipitation in transfected HEK 293 cells. The Siah1–GAPDH binding site was identified as a 19-amino acid sequence in the C-terminal region (220–238). Mutational analysis identified GAPDH$^{lys225}$ as required for binding.

Apart from its apoptotic function, sequence analysis demonstrated that Siah1 contained a nuclear localization signal, which could provide the requisite vehicle for GAPDH nuclear transport. This was confirmed by transfection analysis in which cells were transfected with GAPDH by itself or in combination with Siah1. The former was characterized by a predominantly cytosolic localization with weak nuclear staining, while the latter was characterized by a GAPDH nuclear localization in approximately 70%–80% of the cells. Of significance, cotransfection of a Siah1 construct lacking its nuclear localization signal (Siah1$^{\Delta NLS}$) reduced markedly GAPDH nuclear transport.

The role of NO in GAPDH–Siah1 nuclear transport was demonstrated by the treatment of HEK 293 cells with the NO donors S-nitroso-glutathione (GSNO) or with sodium nitroprusside (SNP). Each significantly increased GAPDH nuclear translocation as well as its coimmunoprecipitation. The role of GAPDH S-nitrosylation was confirmed through its inhibition with dithiothreitol (DTT) or by its reversal by ascorbate. Antibody analysis indicated the formation of a sulfonated derivative at GAPDH$^{cys150}$ following staurosporine treatment of HEK 293 cells. Although redundant, it is again important to emphasize that the physiological significance of any putative moonlighting GAPDH function needs to be established. In this instance, a well-documented NO experimental paradigm was used, i.e., exposure to macrophages (RAW264.7 cells) to lipopolysaccharide (LPS), a component of endotoxin. This results in NO generation as a consequence of inducible nitric oxide synthetase (iNOS) activation (this will be further discussed in Chapter 13).

Using this protocol, at 24 h after LPS exposure, there was an increase not only in the formation of S-nitrosylated GAPDH but also in the binding of GAPDH to Siah1. Previous studies demonstrated significant apoptotic cell death at that interval. The initiation of apoptosis is eliminated by the addition of an iNOS inhibitor, which also prevents GAPDH nuclear translocation. Further, depletion of GAPDH by siRNA prevents apoptosis subsequent to LPS exposure. In addition, the use of macrophages isolated from wild-type (wt) and iNOS knockout mice demonstrated GAPDH S-nitrosylation in the former but not in the latter. Further, the apoptotic specificity of SNO-GAPDH–Siah1 nuclear translocation was examined using the GAPDH[lys225] mutant, which can be S-nitrosylated but does not bind to Siah1. Further, GAPDH is not S-nitrosylated during S phase and is not located in the nuclear matrix. Accordingly, a different mechanism is required for the S phase localization of GAPDH during cell proliferation (see Chapter 7).

Quantitative analysis indicated, at 24h, that 4.3% of total cellular GAPDH is translocated to the nucleus. Simultaneous determination of total GAPDH glycolytic activity revealed no diminution in that activity until 72h after LPS treatment. The significant effect of that small amount of SNO-GAPDH formed and transported is in accord with the hypothesis that a "pool" of GAPDH exists in vivo from which a small portion may be recruited for each of its moonlighting functions (Sirover, 2014).

The physiological significance of SNO-GAPDH–Siah1 nuclear translocation is exemplified further by determination of SNO-GAPDH mediated increases in Siah1 stability. Pulse-chase experiments indicated that, in the absence of complex formation, Siah1 has a $T_{1/2}$ of approximately 5 min with no detectable ($^{35}$S) radiolabel at 40 min. In contrast, significant ($^{35}$S) Siah1 is detected over the 40 min kinetic analysis. This enhanced stability is reflected in its enhanced cytotoxicity in vivo mediated by its degradation of nuclear proteins, which is illustrated in Fig. 8.1.

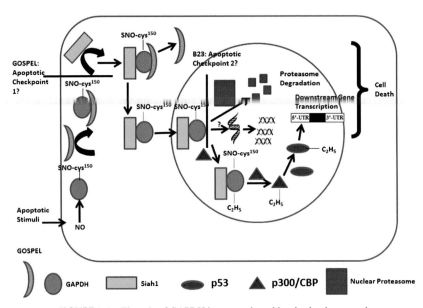

**FIGURE 8.1** The role of GAPDH in apoptosis and its checkpoint control.

## 3.2 SNO-GAPDH Gene Regulation During Apoptosis

The studies described above provided the requisite information to define not only the mechanism of GAPDH nuclear translocation during apoptosis but also the requirement of SNO-GAPDH for one of the basic apoptotic pathways, i.e., nuclear proteasome degradation mediated by Siah1. Subsequent studies demonstrated the basic requirement of SNO-GAPDH for the program of gene expression observed during apoptosis (Sen et al., 2008).

These investigations revealed a number of significant findings. In their temporal order, these included the acetylation of SNO-GAPDH by p300/CBEB binding protein (CBP); autoacetylation of p300/CBP; downstream regulation of p53; p53-directed expression of several apoptosis regulated genes, p53-upregulated modulator of apoptosis (PUMA), Bax, and p21; and lastly, the requirement of each acetylation modification for the initiation of programmed cell death.

LPS-activated and interferon-$\gamma$ (IFN$\gamma$)-activated macrophages were used as the experimental paradigm to identify GAPDH acetylation by p300/CBP. The latter was determined by immunoprecipitation with an antiacetyl lysine antibody. As defined by iNOS inhibition, GAPDH acetylation was dependent on its initial SNO modification. Depletion of p300 and CBP by RNA interference resulted in the absence of GAPDH acetylation. Mutational analysis demonstrated that the alteration of lys$^{160}$ (GAPDH$^{lysK160R}$) eliminated GAPDH acetylation identifying that amino acid residue as the site of modification. The interaction of SNO-GAPDH with p300/CBP was demonstrated both in vitro and in vivo by affinity chromatography, coimmunoprecipitation, and by colocalization in macrophage nuclei. Each was prevented by the GAPDH$^{lysK160R}$ mutant. Accordingly, these studies established that apoptosis required a double GAPDH posttranslational modification. Further, they indicated the utility of the GAPDH$^{lysK160R}$ mutant as a means to dissect this aspect of apoptotic GAPDH function.

That being said, a subsequent study examined the subcellular localization and functional consequences of a GAPDH construct containing a nuclear localization signal (NLS-GAPDH). In activated macrophages, that construct bound to p300/CBP either in iNOS deleted cells or following mutation of the NLS-GAPDH active site cysteine (NLS-GAPDH$^{C150S}$). Based on these findings, it was postulated that NO may not be involved in GAPDH–p300/CBP binding. Instead, it may function as a "transport mechanism" to facilitate GAPDH nuclear transport. As such, it would be the catalyst for all that followed.

Subsequent autoacetylation of p300/CBP in macrophage-activated cells was demonstrated by immunoprecipitation with a p300 antibody followed by immunoblot analysis using an acetylation-specific p300/CBP antibody. Macrophage iNOS inhibition by the addition of an iNOS inhibitor eliminated autoacetylation as did the use of macrophages from iNOS deficient mice. Similarly, knockdown of GAPDH abolished autoacetylation as did transfection of the GAPDH$^{lysK160R}$ mutant into macrophages prior to the addition of LPS/IFN$\gamma$. Thus, these studies defined autoacetylation of p300/CBP as the next step in this NO-initiated apoptotic cell pathway.

Downstream regulation was analyzed first by determining p53 acetylation as a function of NO exposure. In these studies, GAPDH-transfected HEK 293 cells were treated with the NO donor GSNO. Subsequently, following p53 acetylation was determined in cell lysates by immunoprecipitation with an antiacetyl lysine antibody followed by immunoblot analysis using an anti-p53 antibody. The observed p53 acetylation could be blocked by transfection of the GAPDH$^{lysK160R}$ mutant. Further, this posttranslational modification was prevented by the use of an iNOS inhibitor or by the addition of GAPDH RNAi.

The downstream effect of p53 acetylation was determined by quantitation of PUMA, Bax, and p21 gene expression. Each is involved in apoptosis and each is regulated by p53. PUMA protein and mRNA levels were increased in activated macrophages. Those inductions of PUMA gene expression were blocked by iNOS inhibition. RNA levels of Bax and p21 were increased in GSNO-treated cells. With respect to PUMA, chromatin immunoprecipitation (ChIP) analysis demonstrated the physical association of SNO-GAPDH, p300, and p53 at the PUMA promoter subsequent to GSNO exposure. The formation of this protein–DNA complex was blocked by GAPDH RNAi and by transfection of the GAPDH$^{lysK160R}$ mutant. As it did not appear that Siah1 was present in the protein–DNA complex, these findings suggest that there may be a dissociation of the GAPDH and Siah1 complex as a function of SNO-GAPDH gene regulation. It may be reasonable to speculate that separation may occur as a result of GAPDH acetylation, which presumably alters GAPDH tertiary structure.

As with all such moonlighting GAPDH studies, it was necessary to establish the physiological significance of these findings. That was demonstrated using both activated macrophages and neuroblastoma SH-SY5Y cells. GAPDH transfection increased apoptosis that was blocked by the GAPDH$^{lysK160R}$ mutant; the former increased PUMA levels but the latter did not; p53 knockdown in the former or the latter decreased apoptosis. Accordingly, as illustrated in Fig. 8.1, these findings suggest a sequential regulatory pathway, which begins with NO modification of GAPDH and ends with apoptosis.

## 4. SNO-GAPDH AS A TRANSNITROSYLASE

Regulation of cell function by NO presents an intriguing paradox. Its rapid diffusion rate makes it an effective regulatory molecule, yet its short half-life limits its effectiveness. Accordingly, mechanisms might exist to resolve this paradox, thereby enhancing its signaling potential. Recent evidence suggests that cells utilize transnitrosylation of NO from protein to protein as a means to maximize its regulatory role (rev. in Nakamura and Lipton, 2013). As indicated below, SNO-GAPDH can function as such a "carrier," transferring its NO group to target proteins.

The initial report of SNO-GAPDH as a transnitrosylase focused on its ability to nitrosylate nuclear proteins (Kornberg et al., 2010; commented on by Stamler and Hess, 2010). In this study, HEK 293 cells exposed to GSNO were used as the experimental paradigm. Coimmunoprecipitation demonstrated the physical association of SNO-GAPDH with SIRT1. In addition, transnitrosylation was demonstrated in an in vitro assay in which purified SNO-GAPDH was incubated with SIRT1. Further, the GAPDH active site cysteine was required for this reaction as indicated by the lack of SIRT1 nitrosylation using a GAPDH$^{C150S}$ mutant. Kinetic analysis indicated that SNO-GAPDH was more effective in nitrosylating SIRT1 than the NO donor, GSNO. Further mutational studies indicated the requirement for GAPDH$^{thr152}$. RNA interference depleting GAPDH in HEK 293-nNOS cells demonstrated its necessity for SIRT1 nitrosylation. Prevention of SNO-GAPDH nuclear localization through the use of a Siah1$^{\Delta NLS}$ mutant eliminated SIRT1 nitrosylation yet had no effect on the nitrosylation of β-tubulin that was used as a control.

The physiological significance of SNO-GAPDH–mediated transnitrosylation was demonstrated by enzymatic analysis and by downstream gene regulation. For the former, SNO-GAPDH–inhibited SIRT1 enzyme activity in an in vitro histone deacetylation assay. For the latter, SIRT1-mediated deacetylation of PGC1α was dependent on GAPDH expression as defined by its depletion by GAPDH RNAi. Further analysis demonstrated that SNO-GAPDH can transfer its NO moiety to HDAC2 and DNA–PK. As with SIRT1, their nitrosylation was eliminated by both GAPDH RNAi and by the Siah1$^{\Delta NLS}$ mutant. In toto, these cumulative studies not only identify GAPDH as a transnitrosylase but also demonstrate its role in the regulation of three important cell proteins.

The second report describing the transnitrosylase activity of SNO-GAPDH provided new information on its role in the regulation of basic cell metabolic pathways (Rodriguez-Ortigosa et al., 2014). In particular, this investigation involved not only the identification of the mechanisms through which the liver regulates bile salt synthesis but also the characterization of its dysregulation in human clinical disorders. As such, it provides a further example of the breadth of moonlighting GAPDH activity.

A classical characteristic of most metabolic pathways is their regulation through modulation of the first committed enzyme in that pathway, i.e., their inhibition by the end product of the pathway. In this instance, bile salts affect cholesterol 7α-hydroxylase whose gene is termed *CYP7A1*. Repression of the latter involves histone deacetylation in the *CYP7a1* promoter region by HDAC 1/2. As recent evidence interrelated liver bile salts and nitric oxide synthesis (Rodriguez-Ortigosa et al., 2010), it seemed reasonable to determine whether or not that interrelationship involved SNO-GAPDH.

The experimental paradigm involved the use of cholate-rich diets with or without injection of an NOS inhibitor. The use of the diet by itself results in repression of bile salt synthesis including the repression of *CYP7A1* expression. Initial studies without the inhibitor demonstrated induction of hepatic iNOS and endothelial NOS mRNAs. Addition of the inhibitor altered the regulatory mechanisms, which regulate bile salt pool size as defined by increases in serum and intrahepatic bile salt concentrations. Quantitation of *CYP7A1* expression revealed a diminution in its down regulation in NOS inhibitor–treated animals.

Subsequently, the nuclear localization of GAPDH was examined and identified in primary rat hepatocytes incubated with cholic acid. This change in subcellular distribution was not only time and dose dependent but also was diminished in cultures exposed to the NOS inhibitor. GAPDH RNAi eliminated cholic acid–induced diminution in *CYP7A1* expression. Deprenyl, which inhibits GAPDH nuclear transport, and the use of a Siah1$^{\Delta NLS}$ mutant blocked cholic acid–induced repression of *CYP7A1* expression as well. Finally, the production of SNO-GAPDH was determined. It was identified in the nuclei of primary hepatocytes exposed to cholic acid but not in hepatocytes exposed to both cholic acid and the NOS inhibitor.

Quantitation of SNO-GAPDH, SNO-HDAC-2, and SNO-SIRT1 indicated an association between the former and both of the latter proteins. Inhibition of GAPDH nuclear translocation by deprenyl or by transfection of Siah1$^{\Delta NLS}$ diminished the formation of all three nitrosylated proteins. Finally, SNO-GAPDH and SNO-HDAC2 formation were determined in cholestatic livers from patients as well as normal patients. Each was detected in the former but not in the latter. Similarly, in an animal model (bile duct ligation), each was also present. In toto, these findings demonstrate a new SNO-GAPDH transnitrosylase function, i.e., the metabolic control of bile salt synthesis through the nitrosylation of two proteins, HDAC2 and SIRT1, which are involved in controlling bile salt homeostasis.

The third study examined the role of SNO-GAPDH as a mitochondrial transnitrosylase (Kohr et al., 2014). The experimental paradigm utilized was an ischemic preconditioning protocol involving repeat intervals of ischemia/reperfusion (GAPDH and ischemia are discussed in Chapter 10). Using that protocol, it was observed that mitochondrial GAPDH levels were increased fourfold. Subsequently, the uptake of both GAPDH and SNO-GAPDH into isolated mitochondria was determined in vitro. Each appeared to be imported into the mitochondria.

The critical experiment was to determine not only whether perfusion with the NO donor, GSNO, resulted in GAPDH binding to mitochondrial proteins but also whether that binding resulted in their S-nitrosylation. As reported, in the absence of GSNO, GAPDH bound to tubulin and actin in the cytosol with little binding to mitochondrial proteins. The cytosolic binding to tubulin was one of the early examples of GAPDH protein interactions (rev. in Sirover, 1999). In contrast, following GSNO perfusion, GAPDH bound to mitochondrial aconitate hydratase, OPA1, and the mitochondrial phosphate carrier. However, in isolated heart mitochondria, incubation with GSNO resulted in a large number of S-nitrosylated proteins, while incubation with SNO-GAPDH transnitrosylation was observable, but significantly reduced as compared to the GSNO sample.

Subsequently, an S-nitrosylation resin–assisted capture (SNO-RAC) protocol was used to determine protein S-nitrosylation (Kohr et al., 2011) as a function of GAPDH overexpression or its reduction by GAPDH RNAi in HepG2 cells. The latter expresses both nNOS and eNOS. Using this protocol, a number of S-nitrosylated proteins were determined both at baseline, while others were detected as a result of GAPDH overexpression, most notably acetyl-CoA acetyltransferase, long-chain fatty acid CoA ligase as well as a 2.5-fold increase in the formation of SNO-Hsp60. Reduction of GAPDH resulted in a marked reduction in the number of mitochondrial SNO proteins that were identified. Similarly, the expression of a GAPDH mutant (GAPDH$^{C150S}$) resulted in the reduction of mitochondrial S-nitrosylated proteins. Accordingly, these correlations suggest that GAPDH functions as a mitochondrial transnitrosylase.

## 5. SNO-GAPDH AS A CHECKPOINT FOR APOPTOSIS

Although the studies described above identified SNO-GAPDH as a transnitrosylase, they appear to present a logical inconsistency with respect to the role of SNO-GAPDH in vivo. In other words, if SNO-GAPDH is involved in both programmed cell death and in transnitrosylation, once it is formed, how does the cell determine which pathway it will follow? This suggests the possibility that some control mechanism is necessary to "decide" whether SNO-GAPDH will function in apoptosis or will exhibit its transnitrosylase activity. As discussed in this section, recent studies suggest the possibility that cells might employ a two-step control mechanism to regulate SNO-GAPDH function.

## 5.1 Cytosolic GOSPEL as an Apoptotic Checkpoint

As with Siah1, a yeast two-hybrid analysis using the GAPDH N-terminal domain identified a 52-kDa protein as a GAPDH binding partner (Sen et al., 2009; commented on by Nakamura and Lipton, 2009). The protein was not recognized using the GAPDH C-terminal domain as a probe. It was expressed in a number of tissues with similar mRNA and protein levels. Subcellular localization indicated it was localized in the cytoplasm. Mapping studies using deletion constructs coupled to immunoprecipitation analyses demonstrated that their binding to GAPDH required their amino acids from positions 160–200 while GAPDH amino acids 80–120 were also necessary.

A series of studies were performed examining its role in NO-mediated GAPDH functions. These investigations demonstrated that S-nitrosylation of the protein was required for GAPDH binding; mutation of its cys$^{47}$ (C47S) abolished its NO modification; it competed successfully with Siah1 for binding to GAPDH as defined by inhibition analysis (IC$_{50}$ = 10 nM) as compared to an IC$_{50}$ = 100 nM for Siah1; its depletion increased GAPDH–Siah1 binding in neuronal cell cultures; it prevented nuclear translocation of GAPDH as defined by its overexpression in cultured cells; and it prevented NMDA-induced neuronal toxicity in vivo.

Considering these properties, the protein was named GOSPEL (GAPDH's competitor of Siah protein enhances life). It was proposed that a competition may exist between GOSPEL and Siah1 for GAPDH binding to provide a control mechanism to regulate both moonlighting GAPDH functions and the cellular response to oxidative or other stress. Further, it was noted that the pronounced differences in GOSPEL and Siah1 affinity for GAPDH coupled with their differences in GAPDH binding sites (N-terminal vs. C-terminal, respectively) might infer the presence of an intermediate tertiary protein complex, i.e., GOSPEL–GAPDH–Siah1 (Sirover, 2011). As illustrated in Fig. 8.1, formation of the complex would result in the dissociation of GOSPEL followed by nuclear translocation of the GAPDH–Siah1 complex. Considering their relative affinities, it was suggested that this displacement reaction would require significant levels of GAPDH S-nitrosylation, "tipping the balance" from the GOSPEL–GAPDH to GAPDH–Siah1. By definition, as illustrated in Fig. 8.1, this could provide a cytosolic checkpoint for determining not only SNO-GAPDH function but also the "decision" to initiate apoptosis.

## 5.2 Nuclear B23/Nucleophosmin as an Apoptotic Checkpoint

As with the yeast two-hybrid protocol, proteomic analysis, especially coimmunoprecipitation studies, has proved of great utility in the analysis of protein: protein interactions. In such a study, the use of a GST-B23/nucleophosmin (B23) antibody identified its binding partners as GAPDH and as Siah1 (Lee et al., 2012). B23 is also a multifunctional protein involved in cell proliferation and in cell survival exhibiting antiapoptotic activity (Ahn et al., 2005). Domain analysis indicated that Siah1 is bound with B23 amino acid residues 117–186, while GAPDH bound to residues 4–186. Analysis in vivo demonstrated that, in the absence of oxidative stress, there was a colocalization of B23 with Siah1 in both the cytoplasm and the nucleus. There did not appear to be a similar colocalization with GAPDH. However, under nitrosative stress, their nuclear colocalization was observed, consistent with that change in GAPDH intracellular distribution.

Subsequent analysis revealed that, under those conditions, the formation of SNO-B23 was observed. Intriguingly, that modification was accompanied by a coordinate decrease in the amount of SNO-GAPDH. Knock down of GAPDH followed by GSNO exposure did not result in the formation of SNO-B23. Mutational studies indicated that $B23^{cys275}$ as its nitrosylation site. The use of the mutant $GAPDH^{C150S}$ blocked SNO-B23 formation. As this suggested a GAPDH-mediated transnitrosylation reaction, the in vitro modification of B23 by SNO-GAPDH was determined and confirmed.

Studies were then performed to examine the physiological significance of B23/GAPDH/Siah1 interactions. Four such studies were performed either in neuronal cells in culture or in vivo. First, B23 disrupted GAPDH–Siah1 binding, i.e., overexpression of B23 coordinate with GSNO exposure resulted in enhanced formation of an SNO-B23–Siah1 complex, which was not observed using a mutant B23$^{C275S}$ mutant. Said overexpression diminished GAPDH–Siah1 binding, while depletion of B23 (knockdown) increased GAPDH–Siah1 interaction. Second, B23 protected neuronal cells in culture from NMDA-induced cytotoxicity as defined by increases in cell viability subsequent to NMDA exposure. In those studies, B23 wt or the B23$^{C275S}$ mutant were introduced into the neuronal cells by adenovirus infection. Introduction of an shB23-expressing adenovirus increased NMDA-induced cell toxicity. Third, the in vivo neuroprotective effects of B23 were examined by administration of a series of adenovirus constructs into mouse cerebral cortex. Infection with B23 wt–containing virus decreased lesion size by 82% and decreased lesion volume 10-fold. Mice injected with shB23 exhibited an increase in lesion size presumably due to depletion of the endogenous protein. Fourth, B23 appeared to inhibit Siah1 E3 ligase activity. In particular, overexpression of B23 wt inhibited the degradation of nuclear receptor corepressor (NCoR1), which is a Siah1 E3 ligase target. Further, depletion of endogenous B23 or the overexpression of the B23$^{C275S}$ mutant increased Siah1 self-ubiquitination, which is known to stabilize Siah1. In toto, as illustrated in Fig. 8.1, these findings suggest that nuclear B23 may function as a nuclear apoptotic checkpoint permitting the cell to make a final "go or no go" decision with respect to the initiation of apoptosis. Cumulatively, the experiments in this section describing SNO-GAPDH–GOSPEL or SNO-GAPDH–B23 binding demonstrate the complexity of moonlighting GAPDH activity not only in relation to its interaction with NO but also as it pertains to apoptosis.

## 6. SNO-GAPDH AND THE REGULATION OF HEME METABOLISM

As indicated above, moonlighting GAPDH as SNO-GAPDH may play a significant role in bile salt regulation through classical feedback inhibition mechanisms. Another basic mechanism of enzyme regulation is the modulation of catalysis through the addition or removal of protein cofactors that are essential for activity, i.e., the addition of the cofactor to the apoenzyme may be a rate-limiting step in catalysis.

In that regard, heme is a required component of many proteins. As such, its metabolism, transport, and subcellular localization are a priori requirements not only for its function but also for cell viability. For that reason, a novel experimental paradigm was developed through which the mechanisms of heme control into a specific apoprotein could be determined (Waheed et al., 2010). Through this procedure, it was demonstrated that NO could block heme addition to a variety of heme proteins. These included eNOS, iNOS, two cytochrome p450 enzymes as well as catalase, and hemoglobin. The specificity of this effect was established by determining that NO did not affect ATP levels, guanylate cyclase, gene induction, or protein expression. The significance of these negative findings is that they highlighted the uniqueness of the NO mediated effect on protein structure and function.

Through those studies, it was hypothesized that a common cellular constituent may modulate the effect of NO on heme insertion to these diverse proteins. Accordingly, using a macrophage-like cell line (RAW 264.7), hemin-agarose affinity chromatography coupled with 2D gel chromatography and mass spectroscopy was used in an attempt to determine not only did such a common element exist but also to identify and to characterize that cellular molecule (Chakravarti et al., 2010).

In a familiar scenario, one of the proteins bound was identified as GAPDH. However, unfamiliarly, GAPDH binding to hemin agarose was diminished when the cells were treated first with an NO donor, Noc18. Intriguingly, DTT could restore GAPDH–heme binding, thereby suggesting that a reversible cysteine modification was responsible for the observed GAPDH reduction in binding. The latter was confirmed as SNO-GAPDH. Accordingly, an active site cysteine mutation was constructed, in this case GADPH$^{C152S}$. As expected, the latter was not nitrosylated in NO-generating cells and did not display any catalytic activity yet bound heme normally in cells or in vitro. Coimmunoprecipitation analysis demonstrated that GAPDH and iNOS bound in cells and that blocking NO synthesis increased that binding. In vitro studies indicated a strong physical association ($K_d = 0.6\,nM$); that it was eliminated by GAPDH S-nitrosylation and that eNOS and nNOS also bound to GAPDH. Thus, this was the first suggestion that GAPDH–heme binding can be regulated by NO.

Using the experimental apoenzyme protocol described above, it was noted that GSNO-inhibited heme insertion into iNOS in a dose-dependent manner. Depletion of GAPDH (si-GAPDH) resulted in a virtually complete reduction in heme incorporation into iNOS. Overexpression of GAPDH increased heme–iNOS binding while a similar experiment using the GAPDH$^{C152S}$ mutant resulted in the heme–iNOS insertion becoming resistant to NO exposure. Finally, analysis of endogenous NO formed by iNOS was examined to ascertain whether there was a correlation between the formation of SNO-GAPDH and diminution of heme insertion into iNOS. The latter diminished over time, which could be reversed by addition of an iNOS inhibitor. Of note, overexpression of either wt GAPDH or the GAPDH$^{C125S}$ mutant partially or completely protected transfected cells against the effect of endogenous NO.

Analyses of intermediary metabolic pathways indicate usually their cyclic nature. As such, as the studies above demonstrated the role of SNO-GAPDH formation in the inhibition of heme insertion into iNOS thereby diminishing its activity, it would seem reasonable to suggest that a mechanism may exist to remove the SNO moiety from GAPDH to facilitate its incorporation of heme into iNOS, thereby recovering its activity.

To analyze the latter, the Noc18/macrophage cell model was used (Chakravarti and Stueher, 2012). The recovery of iNOS activity was measured following Noc18 removal in cell cultures, which contained L-NAME to inhibit

NO production from iNOS. These studies revealed that, subsequent to Noc18 removal, there was a burst of heme insertion into apo-iNOS with most of the recovery detected within the first 30 min, after Noc18 removal. This recovery of function was coordinate with the rapid denitrosation of GAPDH (80% in 20 min).

The rapid degradation of SNO-GAPDH suggests the possibility that it is a substrate for a cellular denitrosylase. For that reason, thioredoxin-1 (Trx1) expression was varied to determine if it was responsible for the removal of the SNO group from SNO-GAPDH. These studies demonstrated that its knockdown by its siRNA diminished cellular ability to recover heme insertion into iNOS following Noc18 removal. In contrast, diminished levels of SNO-GAPDH subsequent to Noc18 exposure was observed in cells overexpressing Trx1 expression. Accordingly, it is reasonable to suggest that these cumulative findings describe a novel GAPDH metabolic cycle involving its S-nitrosylation, which inhibits heme insertion into iNOS, while its denitrosylation permits the recovery of iNOS activity due to GAPDH-mediated heme insertion. As such, as with bile salts, SNO-GAPDH functions as a feedback inhibitor to modulate iNOS function. Further, these studies are a further demonstration of the multifaceted utility of GAPDH as a moonlighting protein. Lastly, the reader is referred to an elegant analysis detailing the stoichiometry and parameters of GAPDH–heme binding (Hannibal et al., 2012).

## 7. MOONLIGHTING GAPDH AND APOPTOSIS: AN ALTERNATIVE MECHANISM

There is an old maxim that states that, "Life is never simple, then it gets complicated." So far, the role of GAPDH in apoptosis was relatively simple even though the pathway it initiated was quite detailed and complex. GAPDH was S-nitrosylated at its active site cysteine and everything else followed in a straightforward manner.

Recently, an interesting study suggests a more complicated but intriguing role for GAPDH in the initiation of programmed cell death (Tristan et al., 2015). In this investigation the role of apoptosis signal-regulating kinase (ASK1) in mediation of apoptotic GAPDH function was probed. Coimmunoprecipitation analysis using mouse brain extracts demonstrated the formation of a ternary protein complex composed of ASK1–Siah1–GAPDH, which was enhanced by exposure to $H_2O_2$. Subsequently, complex formation was probed using HEK 293 cells in culture. Those studies indicated that ASK1 phosphorylation was correlated with complex formation.

To probe the mechanisms of complex formation, recombinant proteins were utilized in in vitro binding assays. These studies demonstrated that ASK1 formed a complex with Siah1 but not with GAPDH. However, once the ASK1–Siah1 complex was formed, GAPDH enhanced their binding suggesting sequential Siah1 and GAPDH binding to form the ternary complex. Further studies in vitro and in vivo demonstrated that ASK1 phosphorylated Siah1 at several threonine residues, which were required for GAPDH binding. It was noted that the magnitude of the ASK1 increases in Siah1$^P$–GAPDH binding was greater than that observed by sodium nitroprusside, an NO donor. Mutation of the Siah1 residues to alanine that eliminated Siah1 phosphorylation reduced GAPDH binding.

The physiological significance of the ASK1–Siah1$^{Thr-P}$–GAPDH was then examined. Nuclear translocation of GAPDH was increased by wt ASK1 but was not observed using a "kinase dead" ASK1 mutant (KD ASK1) or the Siah1$^{thr\rightarrow ala}$ mutant. Use of the mutants also revealed a diminution in GAPDH–p300 binding, in the acetylation of p300 as well as in the reduction of *PUMA* expression. In toto, several conclusions may be drawn from these studies: ASK1 may activate the GAPDH–Siah1 apoptotic signaling cascade; such activation requires the formation of a ternary ASK1–Siah1$^P$–GAPDH complex; and, lastly, that ASK1 activation of the GAPDH–Siah1 apoptotic signaling cascade occurs in the absence of NO, i.e., it does not require the formation of SNO-GAPDH. In consideration of the latter conclusion, it was hypothesized that there may be a number of stressors, which may activate the GAPDH–Siah1 signaling cascade to initiate programmed cell death. This is a novel finding and has the potential to expand the utility of moonlighting GAPDH as an apoptotic protein of significance.

## 8. GAPDH AND AUTOPHAGY: THE DUALITY OF MOONLIGHTING GAPDH FUNCTION

It has been said of politicians that they can be on both sides of an issue in the same sentence. Similarly, at times, GAPDH can exhibit properties, which appear to be directly in opposition to each other. In that regard, the apoptotic role of GAPDH is cell destructive in nature (Fig. 8.1). In contrast, recent evidence suggests that GAPDH may also play a significant role in cell survival through its role in autophagy.

The primary study that suggests the latter indicated both a cytoplasmic and a nuclear role for GAPDH (Colell et al., 2007; commented on by Rathmell and Kornbluth, 2007; Song and Finkel, 2007). This investigation used an intriguing experimental paradigm. A retroviral cDNA library was introduced into Jurkat cells, which were then exposed to lethal concentrations of staurosporine, a known inducer of apoptosis. Insightfully, the cells were cultured in media, which contained a pancaspase inhibitor. Cells that survived were "recloned" by inducing apoptosis in media, which did or did not contain the inhibitor. The retroviral inserts in the former were then analyzed. Through this procedure, GAPDH was identified as a protein whose selective expression was observed in the selected cells.

Subsequently, the role of GAPDH in cell survival was examined in a variety of cell culture models. Expression of GAPDH did not prevent apoptosis but did prevent caspase-independent cell death (CICD). Subcellular studies indicated a protective role for GAPDH through its ability to enable cells to recover from mitochondrial outer-membrane permeabilization (MOMP). Of note, again indicating a duality of moonlighting GAPDH function, this appears to be in direct contrast to the apoptotic role of GAPDH in promoting proapoptotic membrane permeabilization (Tarze et al., 2007).

The mechanisms that may underlie this new moonlighting GAPDH activity were then examined (Colell et al., 2007). Overexpression of GAPDH resulted in elevated ATP levels in cells cultured in high glucose (10 mM) but not in low glucose (0.1 mM). Mutational analysis yielded two intriguing mutants, GAPDHΔ2 that elevated ATP levels and GAPDH glycolytic activity and GAPDH$^{149S176F}$. The latter failed to increase ATP levels but displayed a nuclear localization.

Surprisingly, transfection of each mutant individually failed to rescue HeLa cells from CICD. However, cell survival was observed following their cotransduction, indicating that both the cytoplasmic and nuclear functions of GAPDH were required. Examination of the latter indicated that nuclear GAPDH was required for the transcription of the autophagic gene Atg12. More recent analysis examining autophagy subsequent to glucose starvation indicates a role for AMPK-dependent GAPDH phosphorylation resulting in Sirt1 activation (Chang et al., 2015), in the autophagic effect of mitochondrial uncoupling proteins (Dando et al., 2013) and of rotenone-induced mitochondrial injury (Liang et al., 2015). In toto, these cumulative studies indicate a basic yet complex role for GAPDH in autophagy.

## 9. SUMMARY

Paradoxically, this chapter describes the function of moonlighting GAPDH in a series of complex, multistep cell pathways, all of which are based on its single posttranslational modification by NO. In particular, the formation of SNO-GAPDH initiates a pleiotropic apoptotic cascade, regulates the structure and function of important heme proteins, maintains the homeostasis of the bile salt metabolic pathway, and provides a mechanism to nitrosylate acceptor cell proteins. As such, these studies highlight not only the diversity of GAPDH activity but also its physiological significance. That being said, the mechanism through which the cell modulates SNO-GAPDH function remains unknown, i.e., once it is formed, which activity does it exhibit? In addition, these studies highlight the importance of protein–protein interactions in moonlighting GAPDH activity, most notably the role of tertiary protein complexes, i.e., GOSPEL–GAPDH–Siah1; B23–GAPDH–Siah1; and ASK1–Siah1–GAPDH. Each plays a major role in apoptosis. A new study suggests that GAPDH, in the absence of its NO modification, may also be of significance with respect to the initiation of programmed cell death. Lastly, GAPDH may function in autophagy both in the mitochondria and in the nucleus. The latter emphasizes not only the role of subcellular localization in GAPDH function but also the duality of its activity, in this case being involved not only in cell destruction but also in its survival.

# REFERENCES

Ahn, J., Liu, X., Cheng, J., et al., 2005. Nucleophosmin/B23, a nuclear PI(3,4,5)P$_3$ receptor, mediates the anti-apoptotic effects of NGF inhibiting CAD. Mol. Cell 18, 435–445.

Benhar, M., Stamler, J., 2005. A central role for S-nitrosylation in apoptosis. Nat. Cell Biol. 7, 645–646.

Brown, V., Krynetski, E., Krynetskaia, N., et al., 2004. A novel CRM1-mediated nuclear export signal governs nuclear accumulation of glyceraldehyde-3-phosphate dehydrogenase following genotoxic stress. J. Biol. Chem. 279, 5984–5992.

Carlile, G., Chalmers-Redman, M., Tatton, N., et al., 2000. Reduced apoptosis after nerve growth factor and serum withdrawal: conversion of tetrameric glyceraldehyde-3-phosphate dehydrogenase to a dimer. Mol. Pharmacol. 57, 2–12.

Chakravarti, R., Aulak, K., Fox, P., Stueher, D., 2010. GAPDH regulates cellular heme insertion into inducible nitric oxide synthetase. Proc. Natl. Acad. Sci. U.S.A. 104, 18004–18009.

Chakravarti, R., Stueher, D., 2012. Thioredoxin-1 regulates cellular heme insertion by controlling S-nitrosation of glyceraldehyde-3-phosphate dehydrogenase. J. Biol. Chem. 287, 16179–16186.

Chang, C., Su, H., Zhang, D., et al., 2015. AMPK-dependent phosphorylation of GAPDH triggers Sirt1 activation and is necessary for autophagy upon glucose starvation. Mol. Cell 60, 930–940.

Chen, R.-W., Saunders, P., Wei, H., et al., 1999. Involvement of glyceraldehyde-3-phosphate dehydrogenase (GAPDH) and p53 in neuronal apoptosis: evidence that GAPDH is upregulated by p53. J. Neurosci. 19, 9654–9662.

Colell, A., Ricci, J.-E., Tait, S., et al., 2007. GAPDH and autophagy preserve survival after cytochrome c release in the absence of caspase activation. Cell 129, 983–997.

Dando, I., Fiorini, C., Pozza, E., et al., 2013. UCP2 inhibition triggers ROS-dependent nuclear translocation of GAPDH and autophagic cell death in pancreatic adenocarcinoma cells. Biochim. Biophys. Acta 1833, 672–679.

Dastoor, Z., Dreyer, J.-L., 2001. Potential role of nuclear translocation of glyceraldehyde-3-phosphate dehydrogenase in apoptosis and oxidative stress. J. Cell Sci. 114, 1643–1653.

Epner, D., Sawa, A., Isaacs, J., 1999. Glyceraldehyde-3-phosphate dehydrogenase expression during apoptosis and proliferation of rat ventral prostate. Biol. Reprod. 61, 687–691.

Hannibal, L., Collins, D., Brassard, J., et al., 2012. Heme binding properties of glyceraldehyde-3-phosphate dehydrogenase. Biochemistry 51, 8514–8529.

Hara, M., Agrawal, N., Kim, S., et al., 2005. S-nitrosylated GAPDH initiates apoptotic cell death by nuclear translocation following Siah1 binding. Nat. Cell Biol. 7, 665–674.

Hara, M., Snyder, S., 2006. Nitric oxide-GAPDH-Siah: a novel cell death cascade. Cell. Mol. Neurobiol. 26, 527–538.

Hu, G., Chung, Y.-L., Glover, T., Valentine, V., Look, A., Fearon, E., 1997. Characterization of human homologs of the *Drosophila* seven in absentia (sina) gene. Genomics 46, 103–111.

Ishitani, R., Chuang, D.-M., 1996. Glyceraldehyde-3-phosphate dehydrogenase antisense oligodeoxynucleotides protect against cytosine arabinonucleotide-induced apoptosis in cultured cerebellar neurons. Proc. Natl. Acad. Sci. U.S.A. 93, 9937–9941.

Ishitani, R., Sunaga, K., Hirano, A., et al., 1996a. Evidence that glyceraldehyde-3-phosphate dehydrogenase is involved in age-induced apoptosis in mature cerebellar neurons in culture. J. Neurochem. 66, 928–935.

Ishitani, R., Kimura, M., Sunaga, K., et al., 1996b. An antisense oligodeoxynucleotide to glyceraldehyde-3-phosphate dehydrogenase blocks age-induced apoptosis of mature cerebrocortical neurons in culture. J. Pharmacol. Exp. Ther. 278, 447–454.

Ishitani, R., Sunaga, K., Tanaka, M., Aishita, H., Chuang, D.-M., 1997. Overexpression of glyceraldehyde-3-phosphate dehydrogenase is involved in low K+-induced apoptosis but not necrosis of cultured cerebellar granule cells. Mol. Pharmacol. 51, 542–550.

Ishitani, R., Tanaka, M., Sunaga, K., Katsube, N., Chuang, D.-M., 1998. Nuclear localization of overexpressed glyceraldehyde-3-phosphate dehydrogenase in cultured cerebellar neurons undergoing apoptosis. Mol. Pharmacol. 53, 701–707.

Kohr, M., Sun, J., Aponte, A., et al., 2011. Simultaneous measurement of protein oxidation and S-nitrosylation during preconditioning and ischemia/reperfusion injury with resin-assisted capture. Circ. Res. 108, 418–426.

Kohr, M., Murphy, E., Steenbergen, C., 2014. Glyceraldehyde-3-phosphate dehydrogenase act as a mitochondrial trans-S-nitrosylase in the heart. PLoS One 9, e111448.

Kornberg, M., Sen, N., Hara, M., et al., 2010. GAPDH mediates nitrosylation of nuclear proteins. Nat. Cell Biol. 12, 1094–1102.

Lee, S., Kim, C., Leem, K.-H., Ahn, J.-Y., 2012. S-nitrosylation of B23/nucleophosmin by GAPDH protects cells from the Siah1-GAPDH death cascade. J. Cell Biol. 199, 65–76.

Li, S., Li, Y., Carthew, R., Lai, Z.-C., 1997. Photoreceptor cell differentiation requires regulated proteolysis of the transcriptional repressor Tramtrack. Cell 90, 469–478.

Liang, S., Figtree, G., Aiqun, M., Ping, Z., 2015. GAPDH-knockdown reduce rotenone-induced H9C2 cells death via autophagy and anti-oxidative stress pathway. Toxicol. Lett. 234, 162–171.

Nakazawa, M., Uehara, T., Nomura, Y., 1997. Koningic acid (a potent glyceraldehyde-3-phosphate dehydrogenase inhibitor)-induced fragmentation and condensation of DNA in NG-108-15 cells. J. Neurochem. 68, 2493–2499.

Nakamura, T., Lipton, S., 2009. According to GOSPEL: filling in the GAP(DH) of NO-mediated neurotoxicity. Neuron 63, 3–6.

Nakamura, T., Lipton, S., 2013. Emerging role of protein-protein transnitrosylation in cell signaling pathways. Antioxid. Redox Signal. 18, 239–249.

Nomura, Y., Uehara, T., Nakazawa, M., 1996. Neuronal apoptosis by glial NO: involvement of the inhibition of glyceraldehyde-3-phosphate dehydrogenase. Hum. Cell 9, 203–214.

Park, J., Han, D., Kim, K., Kang, Y., Kim, Y., 2009. O-GlcNAcylation disrupts glyceraldehyde-3-phosphate dehydrogenase homo-tetramer formation and mediates its nuclear translocation. Biochim. Biophys. Acta 1794, 254–262.

Rathmell, J., Kornbluth, S., 2007. Filling a GAP(DH) in caspase-independent cell death. Cell 129, 861–863.

Rodriguez-Ortigosa, C., Banales, J., Olivas, I., et al., 2010. Biliary secretion of S-nitrosoglutathione is involved in the hypercholeresis induced by ursodeoxycholic acid in the normal rat. Hepatology 52, 667–677.

Rodriguez-Ortigosa, C., Celay, J., Olivas, I., et al., 2014. A GAPDH-mediated trans-nitrosylation pathway is required for feedback inhibition of bile salt synthesis in rat liver. Gastroenterology 147, 1084–1093.

Saunders, P., Chalecka-Franaszek, E., Chuang, D.-M., 1997. Subcellular distribution of glyceraldehyde-3-phosphate dehydrogenase in cerebellar granule cells undergoing cytosine arabinoside-induced apoptosis. J. Neurochem. 69, 1820–1828.

Saunders, P., Chen, R.-W., Chuang, D.-M., 1999. Nuclear translocation of glyceraldehyde-3-phosphate dehydrogenase isoforms during neuronal apoptosis. J. Neurochem. 72, 925–932.

Sawa, A., Khan, A., Hester, L., Snyder, S., 1997. Glyceraldehyde-3-phosphate dehydrogenase: nuclear translocation participates in neuronal and nonneuronal cell death. Proc. Natl. Acad. Sci. U.S.A. 94, 11669–11674.

Sen, N., Hara, M., Kornberg, M., et al., 2008. Nitric oxide-induced nuclear GAPDH activates p300/ CBP and mediates apoptosis. Nat. Cell Biol. 10, 866–873.

Sen, N., Hara, M., Ahmad, A., et al., 2009. GOSPEL: a neuroprotective protein that binds to GAPDH upon S-nitrosylation. Neuron 63, 81–91.

Shashidharan, P., Chalmers-Redman, R., Carlile, G., et al., 1999. Nuclear translocation of GAPDH-GFP fusion protein during apoptosis. NeuroReport 10, 1149–1153.

Sirover, M., 1996. Emerging new functions of the glycolytic protein, glyceraldehyde-3-phosphate dehydrogenase, in mammalian cells. Life. Sci. 58, 2271–2277.

Sirover, M., 1999. New insights into an old protein: the functional diversity of mammalian glyceraldehyde-3-phosphate dehydrogenase. Biochim. Biophys. Acta 1432, 159–184.

Sirover, M., 2011. On the functional diversity of glyceraldehyde-3-phosphate dehydrogenase: biochemical mechanisms and regulatory control. Biochim. Biophys. Acta 1810, 741–751.

Sirover, M., 2014. Structural analysis of glyceraldehyde-3-phosphate dehydrogenase functional diversity. Int. J. Biochem. Cell Biol. 57, 20–26.

Song, S., Finkel, T., 2007. GAPDH and the search for alternative energy. Nat. Cell Biol. 9, 869–870.

Stamler, J., Hess, D., 2010. Nascent nitrosylases. Nat. Cell Biol. 12, 1024–1026.

Sunaga, K., Takahashi, H., Chuang, D., Ishitani, R., 1995. Glyceraldehyde-3-phosphate dehydrogenase is over-expressed during apoptotic death of neuronal cultures and is recognized by a monoclonal antibody against amyloid plaques from Alzheimer's brain. Neurosci. Lett. 200, 133–136.

Tajima, H., Tsuchiya, K., Yamada, M., et al., 1999. Over-expression of GAPDH induces apoptosis in COS-7 transfected with cloned GAPDH cDNAs. NeuroReport 10, 2029–2033.

Tang, A., Neufeld, T., Kwan, E., Rubin, G., 1997. PHYL Acts to down-regulate TTK88, a transcriptional repressor of neuronal cell fates, by a SINA-dependent mechanism. Cell 90, 459–467.

Tarze, A., Deniaud, A., Le Bras, M., et al., 2007. GAPDH, a novel regulator of the pro-apoptotic mitochondrial membrane permeabilization. Oncogene 26, 2606–2620.

Tristan, C., Ramos, A., Shahani, N., et al., 2015. Role of apoptosis signal-regulating kinase 1 (ASK1) as an activator of the GAPDH-Siah1 stress-signaling cascade. J. Biol. Chem. 290, 56–64.

Waheed, S., Ghosh, A., Chakravarti, R., et al., 2010. Nitric oxide blocks cellular heme insertion into a broad range of heme proteins. Free Radic. Biol. Med. 48, 1548–1558.

## FURTHER READING

He, C., Klionsky, D., 2009. Regulation and signaling pathways of autophagy. Annu. Rev. Genet. 43, 67–93.

Katsube, N., Sunaga, K., Aishita, H., Chuang, D.-M., Ishitani, R., 1999. ONO-1603, a potential antidementia drug, delays age-induced apoptosis and suppresses overexpression of glyceraldehyde-3-phosphate dehydrogenase in cultured central nervous system neurons. J. Pharmacol. Exp. Ther. 288, 6–13.

Chapter 9

# GAPDH and Hypoxia: Mechanisms of Cell Survival During Oxygen Deprivation

*Necessity is the Mother of Invention*

Plato

Hypoxia, a.k.a. oxygen deficiency, presents an existential threat to cells and organisms. The program of gene regulation developed in response results in the increased expression of what have been termed hypoxia associated proteins (HAPs), including GAPDH (Zimmerman et al., 1991). Although conventional wisdom considered increased GAPDH expression as involved predominantly in anaerobic energy generation, the elucidation of its functional diversity suggests that one or more of those activities may also be required as hypoxia survival mechanisms.

Accordingly, the goal of this chapter is to consider the mechanisms through which cells increase GAPDH as part of their hypoxic response. In particular, this includes increases in GAPDH mRNA and protein, subcellular localization of hypoxia-induced GAPDH, the cell specificity of hypoxia-related GAPDH expression, and, perhaps most notably, the complex transcriptional mechanisms through which cells modulate GAPDH expression, including both the role of specific transcription factors and that of GAPDH promoter sequences in GAPDH regulation during hypoxia. With respect to the latter, recent evidence suggests that specific cell types utilize distinct transcriptional mechanisms to increase GAPDH function during hypoxia.

## 1. UPREGULATION OF GAPDH EXPRESSION DURING HYPOXIA

As indicated in Table 9.1, initial studies demonstrated GAPDH induction as a characteristic of endothelial cell response to hypoxia (Graven et al., 1994, 1998). Increased GAPDH expression comprised both transcriptional and translational increases in GAPDH mRNA and protein, respectively. Other studies extended this observation both to human (Zhong and Simons, 1999) and to brain capillary endothelial cells (Yamaji et al., 2003). Other cell types exhibiting

Glyceraldehyde-3-Phosphate Dehydrogenase (GAPDH). http://dx.doi.org/10.1016/B978-0-12-809852-3.00009-1

**TABLE 9.1** Upregulation of GAPDH in Hypoxia

| Cell Type | Species | Method of GAPDH Quantitation | "Unique" Finding | References |
|---|---|---|---|---|
| Endothelial cells | Bovine | mRNA, nuclear runoff, protein, subcellular fractionation | Increase in both cytoplasmic and nuclear GAPDH | Graven et al. (1994) |
| Endothelial cells | Bovine | mRNA/protein | Nitric oxide generation blocks hypoxic GAPDH up regulation | Graven et al. (1998) |
| Endothelial cells | Human | mRNA | Misuse of GAPDH as an internal standard | Zhang and Simons (1999) |
| Alveolar epithelial cells | Rat | RNase protection, GAPDH protein synthesis | Increase in GAPDH enzymatic activity | Escoubet et al. (1999) |
| Prostate cancer cells | Human | mRNA | Misuse of GAPDH as an internal standard | Zhong and Simons (1999) |
| Hepatoma | Human | mRNA | HIF-2$\alpha$ regulation of hypoxic GAPDH | Graven et al. (2003) |
| Brain capillary endothelial cells | Mouse | mRNA/protein | Ca$^{++}$ control of GAPDH hypoxic expression | Yamaji et al. (2003) |
| Melanoma melanocytes | Human | mRNA/protein | Downregulation of GAPDH in normal melanocytes | Silverthorne and Alani (2010) |
| Breast cancer cells | Human | mRNA/protein | Sp1 control of GAPDH regulation | Higashimura et al. (2011) |

hypoxia-induced GAPDH expression included rat alveolar epithelial cells (Escoubet et al., 1999), human prostate cancer (Zhong and Simmons, 1999), hepatoma (Graven et al., 2003), melanoma (Silverthorne and Alani, 2010), and breast cancer cells (Higashimura et al., 2011). The relationship between hypoxia-related changes in GAPDH expression and tumorigenesis is discussed in Chapter 16. Although not included in Table 9.1, hypoxia-induced increases in GAPDH mRNA and protein were detected in IMR-90 human embryonic lung fibroblasts and in bovine pulmonary artery smooth muscle cells (Graven et al., 1993, 1999).

However, these hypoxia-dependent responses were lesser in magnitude and may reflect the increased sensitivity of those cells to hypoxia. Endothelial cells survive at least 96h in 0% $O_2$. In contrast, IMR-90 and smooth muscle cells survive between 24 and 48h in 0% $O_2$, while mouse renal tubular epithelial cells survive for only 8h in 0% $O_2$ (Graven and Farber, 1998).

A number of cell types do not express GAPDH during hypoxia. These include human glioblastoma cells (Said et al., 2007) as well as mouse and human hepatoma, rat lung adenocarcinoma, and human colon cancer cells (Said et al., 2009). The observation that each of these tumor cell lines does not induce GAPDH during hypoxia suggests that this regulatory control mechanism is not characteristic of cancer cells (the role of hypoxia in tumorigenesis is considered in Chapter 16). Further, during hypoxia, GAPDH expression is decreased in normal melanocytes (Silverthorne and Alani, 2010) and in skeletal muscle cells (Webster, 1987). In toto, these cumulative studies emphasize the cell selectivity of hypoxia-induced GAPDH expression, indicating its designated role in cell survival (Graven et al., 1993).

This data set also reveals several other intriguing aspects of GAPDH regulation during hypoxia (termed "unique findings" in Table 9.1). In particular, they shed light not only on the basic mechanisms of GAPDH expression but also on its physiological significance in response to hypoxia. This is indicated by subcellular changes in GAPDH localization, the effect of nitric oxide on its regulation, the role of $Ca^{++}$ in hypoxic GAPDH expression, and the complex regulatory mechanisms through which hypoxia-induced increases in GAPDH gene transcription are effected.

## 2. SUBCELLULAR LOCALIZATION OF HYPOXIA-INDUCED GAPDH

As previously discussed, GAPDH displays a characteristic subcellular localization, which may depend on the particular function exhibited (Tristan et al., 2011; Sirover, 2012). Further, it was postulated that mechanisms exist through which its intracellular localization may be altered in response to its in vivo functional requirements. Accordingly, it may be of interest that not only is the GAPDH protein increased in endothelial cytoplasm, membrane, and nuclear fractions but also GAPDH glycolytic activity is differentially increased in each fraction, i.e., nuclear GAPDH enzyme activity appears to be increased more than that observed in either the cytoplasmic or membrane fractions (Graven et al., 1994). That being said, the observed increases, although statistically significant, do not appear to be proportional to the magnitude with which GAPDH mRNA or protein is increased during hypoxia. Similarly, in mouse capillary epithelial cells, hypoxia-related increases in the GAPDH protein were observed in the cytoplasmic, particulate, and nuclear fractions (Yamaji et al., 2003). Accordingly, these cumulative findings suggest that the observed increases in hypoxia-induced GAPDH increase may not be due to solely its role in anaerobic energy generation.

## 3. ROLE OF NITRIC OXIDE IN HYPOXIC REGULATION OF GAPDH EXPRESSION

These studies also suggest that nitric oxide (NO) affects hypoxic GAPDH upregulation (Graven et al., 1998). In particular, NO-blocked GAPDH gene transcription and translation as evidenced by analysis of Northern blot and SDS-PAGE of $^{35}$S methionine–labeled proteins, respectively. It was suggested that this may be mediated by a guanylate cyclase mechanism. That being said, the NO-generating agent used, sodium nitroprusside (SNP), is known to modify the GAPDH protein at its active site cysteine. As discussed in Chapter 8, NO modification at that GAPDH residue may be the critical initiating event in the pleiotropic apoptotic cascade in which GAPDH plays a major part. Accordingly, it would be of interest to determine the extent of SNO-GAPDH formed during hypoxia as well as that formed when hypoxic cells are exposed to SNP.

In contrast, addition of $N^6$-nitro-L-arginine (L-NNA), an inhibitor of nitric oxide synthetase, did not affect GAPDH induction during hypoxia (Graven et al., 1998), suggesting that this effect is independent of the production of endogenous NO presumably formed by cNOS as no evidence has been presented for any induction of iNOS in endothelial cells during hypoxia. Historically, NO was first identified as "endothelial cell relaxing factor" (Furchgott, 1999). Thus, the observed effect of SNP-produced NO might mimic that observed in vivo by exogenous NO, which acts within endothelial cells in vivo.

Other studies suggest another mechanism through which NO be involved in GAPDH function during hypoxia. In particular, recent investigations indicate that GAPDH may be involved in the regulation of endothelian-1 (ET-1) levels in endothelial cells (Rodriguez-Pascual et al., 2008). As discussed in Chapter 2, these studies indicate that GAPDH binds to ET-1 mRNA, which results both in a decrease in its stability and subsequent reduction in ET-1 protein levels. As ET-1 functions as a vasoconstrictor, its reduction may result in vasodilation in vivo. Those studies show also that NO modification of GAPDH prevents its binding to ET-1 mRNA, which would result in the persistence of ET-1 protein facilitating vasoconstriction. Accordingly, it may be of interest to determine the potential interrelationship between this new GAPDH function, hypoxia, and the role of NO in endothelial cells.

## 4. ROLE OF Ca$^{++}$ IN HYPOXIA GAPDH EXPRESSION

Ca$^{++}$ is a well-established regulator of intra- and intercellular function. Recent evidence suggests that its regulatory activity extends to the control of GAPDH expression during hypoxia (Yamaji et al., 2003). Their studies in mouse brain capillary epithelial cells used inhibitor analysis not only to establish the effect of Ca$^{++}$ on hypoxic GAPDH regulation but also to suggest upstream elements through which Ca$^{++}$ mediates GAPDH expression during hypoxia.

A number of inhibitors were used, which included an endogenous Ca$^{++}$ chelator, a nonselective cation channel blocker and an inhibitor of Na$^{+}$/Ca$^{++}$ exchangers. Each inhibited hypoxic induction of GAPDH expression in a dose-dependent manner. In contrast, the lack of effect of nifedipine and verapamil suggested that L-type Ca$^{++}$ channels are not required for the action of Ca$^{++}$ on hypoxic GAPDH regulation. Further studies analyzed the effect of KN-93 and curcumin in these inhibition studies. Each reduced hypoxic GAPDH expression in a dose-dependent manner.

It was noted that KN-93, a Ca$^{++}$/calmodulin-dependent kinase inhibitor may participate in the activation of c-Fos during hypoxia. Curcumin is a known inhibitor of c-Jun activity. Previous studies indicated that c-Fos and c-Jun can form a protein–protein complex, which may bind to AP-1 elements (Angel and Karin, 1991). Accordingly, the action of each inhibitor could be due to the prevention of protein complex formation precluding its binding to AP-1 sequences. Intriguingly, analysis of GAPDH gene structure indicated the presence of three such sequences within the GAPDH 5′-flanking sequence. Regrettably, studies have yet to be performed to define the significance of those elements in relation to hypoxic GAPDH regulation.

## 5. MISUSE OF GAPDH AS AN INTERNAL STANDARD

The role of GAPDH as a control for transcriptional or translational studies will be considered in detail in Section 7. Historically, as a "housekeeping" protein of little interest, it was used extensively in such studies based on the presumption that it was invariant under a wide variety of physiological conditions. Therefore, unbeknownst to the investigators, variation of GAPDH expression was an unintentional consequence of their studies complicating the interpretation of their findings.

Analysis of gene expression during hypoxia was no exception. Investigation of GAPDH mRNA as well as that of β-actin mRNA and cyclophilin mRNA revealed that each was increased during hypoxia (Zhong and Simons, 1999). In contrast, they found that 28S RNA expression was invariant indicating its utility as an internal standard. A strength of this study was the use of four different cell lines with varied biological characteristics. Of particular interest in this study was the observation that was the danger in the use of GAPDH as an internal standard under conditions in which upregulation of hypoxia-inducible factor-1α (HIF-1α) was observed (see below).

## 6. TRANSCRIPTIONAL REGULATION OF GAPDH IN HYPOXIA

It is trite to state, but nevertheless true, that gene and/or protein expression under physiological stress (or other stimuli) does not occur in a vacuum. Instead, it is usually part of a rigorously controlled cell signaling pathway. In this manner, cells may exhibit the phenotypic expression of their genetic information.

So it is with GAPDH during hypoxia. Structural analysis of the GAPDH promoter revealed six hypoxia-inducible factor-1α (HIF-1α) binding sites which were termed hypoxia response elements (HREs). HIF-1α is a transcription factor, which is present in vivo as HIF-1, a heterodimer in which it is bound to HIF-1β previously identified as the aryl hydrocarbon nuclear translocator (Semenza, 1999). The presumption is that the latter contains the nuclear localization signal, which transports HIF-1α into the nucleus.

Initial studies used GAPDH promoter constructs linked to a chloramphenicol acetyltransferase (CAT) reporter gene as well as electrophoretic mobility shift assays (EMSAs) to localize critical sequences within the GAPDH promoter region (Graven et al., 1998). Cells were transiently transfected with gene constructs, which were then exposed to 0% $O_2$. Based on CAT activity, it was possible to exclude specific sequences within the GAPDH promoter ultimately localizing the HIF-1 binding site to a 19 nucleotide sequence (−130 to −112) within the GAPDH promoter region. However, those studies indicated that a 5′-flanking sequence was also required for binding. Transfection of the construct into either endothelial smooth muscle cells or into IMR-90 fibroblasts demonstrated CAT activity. However, the magnitude of that response was considerably lower than that observed for the endothelial cells. This may correlate with the lower level of GAPDH induction detected in those cells during hypoxia.

EMSA revealed binding of a cellular protein in transfected endothelial cells. However, that protein was not identified as HIF-1α. Nevertheless, in comparison with HIF-1 sequences in other hypoxia-associated proteins, it was designated as an HIF-1 binding site. Mutational analysis revealed that altering either the 19 nucleotide sequences or the 5′-flanking region reduced CAT activity, thereby demonstrating that each was required for HIF-1 binding.

Subsequent studies examined the structural dynamics of GAPDH HREs in prostate cancer cells (Lu et al., 2002). In these investigations, three different prostate cancer cell lines of varying differentiation characteristics were used. The expression model utilized was transfection analysis of luciferase reporter gene constructs. As with the previous study, GAPDH promoter sequences of varying length and composition were used.

As expected, hypoxia-induced expression was readily observed using a construct containing a 1112 GAPDH promoter sequence. However, varying degrees of induction were detected, which appeared to be dependent on the degree of differentiation, i.e., the more highly differentiated cell line exhibited the greatest induction, while the lesser differentiated cell lines exhibited reduced levels of luciferase reporter activity.

Surprisingly, through the use of shorter GAPDH promoter constructs, an HRE binding site was identified, which was clearly distinct from the previously located HIF-1 site (−125 to −121). In particular, this novel site consisted not only of two HIF-1 consensus sequences (−217 to −203) but also which were inverted and separated by five nucleotides. Constructs containing this sequence by itself increased expression of the luciferase reporter gene during hypoxia. Mutation of this novel HRE site eliminated reporter gene expression during hypoxia.

Although HIF-1α is a major hypoxia-related transcription factor (as HIF-1), recent evidence suggests a role for HIF-2α as well (Ema et al., 1997; Flamme et al., 1997). The latter forms a heterodimeric (HIF-2) in which it is bound to Ah receptor nuclear translocator as well. For that reason, studies were initiated to examine the relative roles of HIF-1α and HIF-2α in hypoxic GAPDH regulation (Graven et al., 2003). Further, the relative contribution of each was determined in two different cell strains, human lung microvascular endothelial cells, and human hepatoma (Hep3B) cells.

Initial studies demonstrated the identical time course for hypoxic GAPDH mRNA in both cell strains. Both started to increase at 4–8 h, were maximal 24 h, and declined significantly at 48 h. However, the extent of induction was much reduced in the hepatoma cells. As before, GAPDH promoter constructs linked to the CAT reporter gene were used in transient transfection assays to demonstrate that the identical HRE used in endothelial cells was also used in the hepatoma cells. Again, mutation of that sequence eliminated CAT activity during hypoxia.

The question then arises as to the relative contribution of HIF-1α and HIF-2α to the observed hypoxic induction of GAPDH. To answer this question, in each cell type, immunoblot analysis was performed quantitating their expression as a function of the hypoxia time course. These studies yielded a surprising result. In the hepatoma cells, both HIF-1α and HIF-2α expression were increased to the same degree over the time course. Each started to increase at 2–4 h, were maximally increased at 18 h, and then declined precipitously at 24 h. It should be noted that the observed increase in GAPDH mRNA peaked at 24 h, at which time HIF-1α and HIF-2α expression was minimal. Accordingly, this suggests that the information transmitted by each protein had been received and acted upon.

In contrast, different results were observed in endothelial cells. Although HIF-1α expression was identical to that observed in the hepatoma cells, the levels of the HIF-2α protein continued to increase progressively to the 24 h interval. At that time, endothelial cell HIF-1α expression had diminished to basal levels. These results demonstrate that either the half-life of the endothelial cell HIF-2α is greater than that observed for the endothelial cell HIF-1α protein or that transcription of the former continues when the transcription of the latter has ceased in epithelial cells. Similarly, in hepatoma cells, either the half-life of both proteins is similar or the temporal sequence of their transcription is identical. In toto, these findings suggest that the increased response of endothelial cells, i.e., induction of GAPDH, may be due to the increased stability of HIF-2α during hypoxia.

Lastly, the hypoxic regulation of GAPDH was compared in breast cancer versus prostate cancer cells (Higashimura et al., 2011). In the former, GAPDH mRNA and protein were increased. Time course analysis revealed a twofold increase in GAPDH mRNA at 12 h with a sixfold increase at 24 h. At the latter interval, a twofold increase in the GAPDH protein was observed.

Transcriptional activation was studied using GAPDH constructs coupled to a luciferase reporter model in transient transfection studies. Equivalent results were observed in both breast and prostate cancer cells using the complete promoter sequence (−1091 to +25). However, transfection of a −1091 to −231 construct resulted in a 50% reduction in breast cancer GAPDH mRNA induction with no decrease observed in the prostate cancer cell lines. As this sequence contains three HREs, mutation studies were used to identify the critical breast cancer HRE. Those investigations identified HRE1 as necessary in breast cancer cells. In contrast, only HRE6 was functional in prostate cancer cells.

To determine the relative roles of HIF-1α and HIF-2α, in hypoxic GAPDH regulation, siRNA studies were performed in breast cancer cells. Knockdown of the former but not the latter resulted in the reduction of hypoxic GAPDH gene expression. Further, analysis of the HRE1 sequence revealed the presence of a potential Sp family binding site (5′-GGGCGC-3′) at positions -983 to -978. Mutation of that sequence reduced luciferase reporter gene activity. Consistent with this finding, not only is basal Sp1 expression higher in breast cancer versus prostate cancer cells but also hypoxia increases its expression. Finally, Sp1 knockdown reduced HRE1 in breast cancer hypoxia GAPDH regulation as well as its control by both Sp1 and HIF-1α.

## 7. GENETIC REGULATION OF HYPOXIC GAPDH EXPRESSION

As indicated in the last section, not only are distinct HRE sequences involved in the regulation of hypoxia-induced GAPDH expression but also other diverse upstream elements function in the control of GAPDH activity during oxygen deprivation. In particular, it is reasonable to suggest that these are significantly more complicated than the "rather simple" transcription of the GAPDH gene and its translation into protein.

Accordingly, as indicated in Fig. 9.1, it may be possible to devise a current model detailing those genetic factors and the GAPDH sequences with which they interact. These include not only the role of HIF-1α and HIF-2α but also

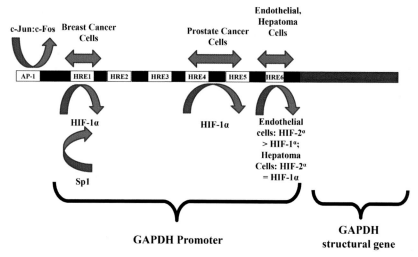

**FIGURE 9.1** Genetic regulation of GAPDH activity during hypoxia.

the roles of Sp1, c-Jun, and c-Fos. It is also may be of interest that different cell types utilize distinct HREs to regulate hypoxia-related changes in GAPDH expression. Although the reason for this is unknown at the present time, the separate mechanisms through which three different cancers (breast, prostate, and hepatoma) control GAPDH expression could be potentially related to cancer development. Similarly, the potential roles of c-Jun and c-Fos in GAPDH gene expression as each relates to tumorigenesis, and the ability of cancer cells to survive in a hypoxic environment may be of interest.

Finally, the model describes an intriguing interrelationship between the extent of GAPDH induction during hypoxia and the relative expressions of HIF-1$\alpha$ and HIF-2$\alpha$ within the cell. As illustrated, the relative increases in the extent of GAPDH expression detected in endothelial cells and in hepatoma cells, respectively, may depend on differences in the half-lives, rates of synthesis, or rates of degradation of each factor, which influence their binding to HRE6 (Graven et al., 2003).

## 8. SUMMARY

It is hoped that, from this discussion, the reader will gain insight into the complexity of GAPDH function in hypoxia. In particular, it is not a simple enhancement of a glycolytic gene to ensure adequate energy generation. Instead, not only is it a highly regulated signaling pathway but also the subcellular increases in GAPDH may indicate a role for its moonlighting activities during hypoxia. The physiological relevance of GAPDH function in hypoxia may be reflected in its cell specificity, i.e., this induction is detected only in specific cell types. Regrettably, the rationale for that specificity appears to be unknown as are GAPDH moonlighting activities, which may function in hypoxia. Each may be a fertile area for future research.

# REFERENCES

Angel, P., Karin, M., 1991. The role of Jun, Fos and the AP-1 complex in cell-proliferation and transformation. Biochim. Biophys. Acta 1072, 129–157.

Ema, M., Taya, S., Yokotani, N., et al., 1997. A novel bHLH-PAS factor with close sequence similarity to hypoxia-inducible factor 1a regulates the *VEGF* expression and is potentially involved in lung and vascular development. Proc. Natl. Acad. Sci. U.S.A. 94, 4273–4278.

Escoubet, B., Planès, C., Clerici, C., 1999. Hypoxia increases transcription in rat alveolar epithelial cells. Biochem. Biophys. Res. Commun. 266, 156–161.

Flamme, I., Fröhlich, T., von Reutern, M., et al., 1997. HRF, a putative basic helix-loop-helix-PAS-domain transcription factor is closely related to hypoxia-inducible factor-1a and developmentally expressed in blood vessels. Mech. Dev. 63, 51–60.

Furchgott, R.F., 1999. Endothelium-derived relaxing factor: discovery, early studies, and identification as nitric oxide. Biosci. Rep. 19, 235–251.

Graven, K.K., Bellur, D., Klahn, B.D., Lowrey, S.L., Amerger, E., 2003. HIF-2α regulates glyceraldehyde-3-phosphate dehydrogenase expression in endothelial cells. Biochim. Biophys. Acta 1626, 10–18.

Graven, K.K., Farber, H.W., 1998. Endothelial cell hypoxic stress proteins. J. Lab. Clin. Med. 132, 456–463.

Graven, K.K., McDonald, R.J., Harrison, H.W., 1998. Hypoxic regulation of endothelial glyceraldehyde-3-phosphate dehydrogenase. Am. J. Physiol. 274, C347–C355.

Graven, K.K., Troxler, R.F., Kornfeld, H., Panchenkov, M.V., Farber, H.W., 1994. Regulation of endothelial glyceraldehyde-3-phosphate dehydrogenase expression by hypoxia. J. Biol. Chem. 269, 24446–24453.

Graven, K.K., Yu, Q., Pan, D., Roncarati, J.S., Farber, H.W., 1999. Identification of an oxygen responsive enhancer element in the glyceraldehyde-3-phosphate dehydrogenase gene. Biochim. Biophys. Acta 1447, 208–218.

Graven, K.K., Zimmerman, L.H., Dickson, E.W., Weinhouse, G.L., Farber, H.W., 1993. Endothelial cell hypoxia associated proteins are cell and stress specific. J. Cell. Physiol. 157, 544–554.

Higashimura, Y., Nakajima, Y., Yamaji, R., et al., 2011. Up-regulation of expression by HIF-1 activity depending on Sp1 in hypoxic breast cancer cells. Arch. Biochem. Biophys. 509, 1–8.

Lu, S., Gu, X., Hoestje, S., Epner, D.E., 2002. Identification of an additional hypoxia responsive element in the glyceraldehyde-3-phosphate dehydrogenase gene promoter. Biochim. Biophys. Acta 1574, 152–156.

Rodriguez-Pascual, F., Redondo-Horcajo, M., Magan-Marchal, N., Lagares, D., Martinez-Ruiz, A., et al., 2008. Glyceraldehyde-3-phosphate dehydrogenase regulates endothelin-1 expression by a novel, redox-sensitive mechanism involving mRNA stability. Mol. Cell. Biol. 28, 7139–7155.

Said, H.M., Hagemann, C., Stojic, J., et al., 2007. GAPDH is not regulated in human glioblastoma under hypoxic conditions. BMC Mol. Biol. 8, 55.

Said, H.M., Polat, B., Hagemann, C., et al., 2009. Absence of GAPDH regulation in tumor-cells of different origin under hypoxic conditions in-vitro. BMC Res. Notes 2, 8.

Semenza, G., 1999. Regulation of mammalian $O_2$ homeostasis by hypoxia-inducible factor 1. Annu. Rev. Cell. Biol. Dev. Biol. 15, 551–578.

Silverthorne, C.F., Alani, R.M., 2010. Glyceraldehyde-3-phosphate dehydrogenase expression is altered by hypoxia in melanoma cells and primary melanocytes. Melanoma Res. 20, 61–65.

Sirover, M., 2012. Subcellular dynamics of multifunctional protein regulation: mechanisms of GAPDH intracellular translocation. J. Cell. Biochem. 113, 2193–2200.

Tristan, C., Shahani, N., Sedlak, T., Sawa, A., 2011. The diverse functions of GAPDH: views from different subcellular compartments. Cell. Signal. 23, 317–323.

Webster, K.A., 1987. Regulation of glycolytic enzyme RNA transcriptional rates by oxygen availability in skeletal muscle cells. Mol. Cell. Biochem. 77, 19–28.

Yamaji, R., Fujita, K., Takahashi, S., et al., 2003. Hypoxia up-regulates glyceraldehyde-3-phosphate dehydrogenase in mouse brain capillary endothelial cells: involvement of $Na^+/Ca^{++}$ exchanger. Biochim. Biophys. Acta 1593, 269–276.

Zhong, H., Simons, J.W., 1999. Direct comparison of GAPDH, $\beta$-actin, cyclophilin and 28S rRNA as internal standards for quantifying RNA levels under hypoxia. Biochem. Biophys. Res. Commun. 259, 523–526.

Zimmerman, L.H., Levine, R.A., Farber, H.W., 1991. Hypoxia induces a specific set of stress proteins in cultured endothelial cells. J. Clin. Invest. 87, 908–914.

# Chapter 10

# Moonlighting GAPDH and Ischemia: Cellular and Molecular Effects of Oxygen Deprivation and Reperfusion

*Actions have consequences*

An old proverb

Ischemia can be considered as a clinical presentation of hypoxia, i.e., oxygen deprivation to organs, tissues, and cells. As such, in the form of stroke, myocardial damage, or kidney failure, it represents an existential threat to life itself. That being said, considerable evidence suggests that the response to this life-threatening challenge evokes a quite different cellular response than that discussed in the previous chapter. In particular, this may be reflected in ischemic specific changes in GAPDH expression and function.

Accordingly, in this chapter, we will discuss those ischemic specific changes in two parts. First, we shall consider the role of GAPDH in the physiological response of the organism to an ischemic insult including those events that occur during reperfusion. Primarily, this is focused on neuronal, cardiovascular, and renal perturbations. However, an intriguing case report on retinal ischemia as well as age-related ischemic responses in skeletal muscle will also be considered. Second, molecular mechanisms through which moonlighting GAPDH performs an essential function fundamental to ischemia-related cell activity will be considered. This includes the interrelationship between GAPDH and glutamate receptor–mediated cell signaling, a novel ischemic-dependent interrelationship between GAPDH and poly (ADP) ribose polymerase-1 (PARP-1) as well as the interrelationship between nitric oxide (NO)-induced GAPDH posttranslational modification and ischemia. In toto, these reports will bear further testament to the physiologically significant role of a protein, which was once considered as a classical housekeeping protein of little interest.

## 1. SIGNIFICANCE OF GLYCOLYSIS AND GAPDH IN ISCHEMIA AND REPERFUSION

Early studies indicated the importance of both glycolysis and GAPDH in an organism's response to ischemia. With respect to the former, those

Glyceraldehyde-3-Phosphate Dehydrogenase (GAPDH). http://dx.doi.org/10.1016/B978-0-12-809852-3.00010-8

investigations focused primarily on quantitation of glucose utilization as well as the determination of glycolytic intermediates, ATP, and lactate concentrations as a function of ischemia (Rovetto et al., 1975; Neely and Grotyohann, 1984; Morgan et al., 1984; Mallet et al., 1990; Owen et al., 1990; Jeremy et al., 1993). As such, these studies were among the first to indicate the role of GAPDH as a rate-limiting step in ischemia-related changes in glycolysis. In particular, that early analysis of glycolytic intermediates noted that GAPDH limits glycolysis in ischemia (Rovetto et al., 1975). Further, these early studies demonstrated that iodoacetate, a specific inhibitor of GAPDH, diminished significantly the myocardial functional recovery (Jeremy et al., 1993). Investigation on the role of oxidative stress during ischemia indicated the preferential S-glutathiolation of GAPDH at its active site cysteine (Eaton et al., 2002). The latter modification inhibits GAPDH activity (Mohr et al., 1991).

Subsequently, detailed analysis revealed significant ischemic-related changes in the regulation of GAPDH transcription and translation. Northern blot analysis demonstrated that GAPDH mRNA increased using a variety of now classical ischemic or ischemic/reperfusion protocols (Feldhaus and Liedtke, 1998). Six different protocols were used, which may be considered as either "severe" or "mild." Intriguingly, GAPDH mRNA was increased only under the "severe" conditions. In a subsequent study, ischemia-related increases in GAPDH mRNA were compared to that observed for pyruvate kinase and for pyruvate dehydrogenase (Liedtke and Lynch, 1999). Similar increases in each mRNA were observed, consistent with the ischemia-related requirement for glycolysis. Kinetic analysis indicated that increases in GAPDH mRNA could be detected 40 min after ischemic treatment as well as for 4 days subsequent to partial coronary stenosis. Analysis of GAPDH enzymatic activity revealed that GAPDH catalysis was reduced 3 h after ischemia onset but returned to normal at 24 h. A similar pattern of $V_{max}$ reduction followed by recovery was also observed (Knight et al., 1996).

In toto, these studies are in accord with the classical view of GAPDH as a mere glycolytic protein of seemingly little interest. That being said, there was an initial hint that ischemia-related GAPDH expression may be indicative of the expression of its moonlighting activities as well. Immunoreactive analysis of GAPDH subcellular localization indicated that ischemia induced a defined nuclear GAPDH translocation (Tanaka et al., 2002). It was reported that GAPDH nuclear translocation was observed soon after reperfusion (3 h) and persisted up to 48 h. In a more recent study, using transgenic mice that overexpressed GAPDH (Tg GAPDH), it was reported that retinal injury subsequent to ischemia was reduced in Tg GAPDH animals as compared with the wt control (Cai et al., 2015). This included diminution of vascular degeneration; reduction of apoptosis and necrosis; reduced expression of DNA damage-related genes; and, lastly, elimination of ischemia-related decreases in ATP levels. As described below, detailed analyses revealed the scope of GAPDH moonlighting activities during ischemia/reperfusion.

## 2. GAPDH AND AMPAR EXCITOTOXICITY IN ISCHEMIA

α-amino-3-hydroxy-5-methylisoxazole-4-propionic acid subtype glutamate receptor (AMPAR) is a transmembrane ligand-gated ion channel, implicated deeply in neuronal ischemia. In particular, its activation by glutamate is an a priori requirement not only for the complex cellular response to neuronal ischemia but also for the toxicity associated with such events.

Recent investigations indicate an intriguing role for GAPDH in this neuronal ischemic response. Initially, as with many such moonlighting protein studies, GAPDH was identified as an AMPAR-binding protein by an exploratory affinity protocol using AMPAR recombinant constructs (GST–GluR1$_{NT}$ and GST–GluR2$_{NT}$) and solubilized rat hippocampal tissue (Wang et al., 2011). GAPDH was identified by SDS-PAGE of the precipitated rat hippocampal proteins followed by mass spectroscopy. It was noted that GAPDH was bound to GST–GluR2$_{NT}$ but not to the GST–GluR1$_{NT}$ construct. The GAPDH binding site on GluR2 was identified as the Y142-K172N-terminus region. As GluR2 is an extracellular component of the AMPAR transmembrane ion channel, further cellular studies revealed not only that GAPDH was bound to GluR2 on the cell surface but also that glutamate activation increased the formation of this protein–protein complex. As noted, GAPDH is not only an internal cellular protein but also is extracellular in nature where it exhibits multiple functions (Chapters 3 and 13).

In any study of moonlighting proteins such as GAPDH, it is necessary to determine the physiological significance of the putative new function (rev. in Sirover, 1999, 2011, 2014). In this instance, a GluR2 inhibitory peptide (GluR2$_{NT1-3-2}$) was used to assess the consequences of preventing GAPDH–GluR2 binding. These studies demonstrated that preincubation of cells in culture with the peptide not only prevented formation of the GAPDH–GluR2 complex but also inhibited AMPAR-mediated excitotoxicity. Finally, subcellular localization studies suggested that the GADPH–GluR2 complex was internalized following its formation subsequent to AMPAR activation. In those studies, the latter was inferred by the decrease in both GAPDH and GluR2 plasma membrane concentration. The use of a dynamin mutant suggested that this internalization occurred through a dynamin-mediated process.

As most of the studies described above were performed in vitro or in cultured cells, further analysis was conducted in vivo (Zhai et al., 2013). In these investigations, transient global cerebral ischemia (10 min occlusion of the carotid arteries) resulted in the formation of the GAPDH–GluR2 complex as defined by coimmunoprecipitation of GluR2 and GAPDH using a GluR2 antibody. This suggested that this protein–protein interaction was an early event following ischemic insult. Notably, preincubation with the inhibitory peptide reduced complex formation by approximately threefold.

Further studies using a 90-min transient ischemic model demonstrated that the inhibitory peptide prevented neuronal loss whether it was injected 1 h before or 2 h after ischemia induction. Subcellular localization analysis suggested that, in vivo, the peptide prevented the internalization of the GAPDH/GluR2 complex. As defined by analysis of plasma membrane and intracellular content, there appeared to be no decrease in GAPDH and GluR2 concentration in the former following inhibitor treatment. In contrast, without pretreatment, a statistical decrease in each was observed. GAPDH shRNA in a lentiviral vector was used to define the role of GAPDH in neuronal ischemia in vivo. These studies conclusively demonstrated that diminution of GAPDH prevented neuronal toxicity following ischemic insult. In particular, there was a lack of noticeable cerebral infarct area and volume in GADPH shRNA–treated animals as compared to the control.

Lastly, intracellular analysis defined the molecular mechanisms, which may underlie this new moonlighting GAPDH function (Zhai et al., 2014). Glutamate stimulation of HEK-293 cells expressing GluR1/GluR2 as well as rat hippocampal and cortical neurons in culture resulted in GAPDH nuclear localization. Consistent with previous investigations (Chapter 8), Siah1 appeared to mediate GAPDH nuclear translocation as defined by siRNA Siah1 studies. Further, as previous studies indicated a role for p53 function in GAPDH-mediated apoptosis (Chapter 8), coimmunoprecipitation studies were performed using a p53 antibody. Those studies demonstrated the formation of a GAPDH–p53 complex. The use of constructs containing GAPDH fragments localized the p53 binding site to $I_{221}$–$E_{250}$. Intriguingly, a GST–GAPDH antibody but not a GST–GluR2$_{NT}$ coprecipitated the GAPDH–p53 complex. This suggests that, once in the nucleus, GAPDH dissociates from GluR2.

The physiological significance of the GAPDH– p53 protein complex in ischemia was examined in cells in culture and in vivo. Preincubation of the HEK-293 cells expressing GluR1/GluR2 with a p53 inhibitor diminished glutamate-induced cell death. Further, coimmunoprecipitation studies in vivo demonstrated not only the ischemia-dependent formation of the GAPDH–p53 protein but also its diminution in ischemic animals preincubated with a GAPDH inhibitor peptide. The latter inhibited AMPAR receptor-mediated cell death as measured by increased neuronal survival and by a smaller infarct volume.

In summary, as illustrated in Fig. 10.1, these studies describe the binding of extracellular GAPDH to AMPAR through its GluR2 component, its endocytic intracellular transfer, the cytoplasmic transport of the GAPDH–GluR2 complex; its nuclear translocation, the formation of a nuclear GAPDH–p53 protein complex; the downstream transcriptional activation of p53 regulated genes followed by neuronal toxicity.

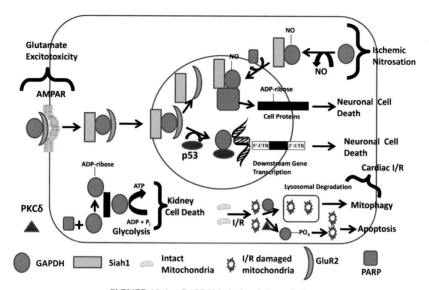

**FIGURE 10.1**    GAPDH in ischemia/reperfusion.

## 3. GAPDH AND NITRIC OXIDE-MEDIATED ISCHEMIC CHANGES IN CELL STRUCTURE AND FUNCTION

As indicated, detailed analysis revealed an intimate interrelationship between NO function and moonlighting GAPDH. This includes NO-mediated apoptosis (Chapter 8), its role in infection and immunity (Chapter 13), and in the regulation of mRNA function (Chapter 2). In virtually all such instances, there was a coordinate relationship between stimulation of NO and the resultant effect on GAPDH expression, structure, and function. That being said, initial ischemic studies interrelating NO and GAPDH revealed a surprising twist, i.e., an inverse relationship between the stimulation of NO production and GAPDH expression (Galea et al., 1998).

In these studies, neuronal ischemia was induced by occlusion of the middle cerebral artery (MCAO). The temporal sequence of inducible NO synthase (iNOS) mRNA and protein were examined postocclusion in the ischemic core. The latter displayed a not unusual regulatory pattern, i.e., iNOS was undetectable in intact tissue, was initially observed 8 h after MCAO, peaked between 14 and 24 h then decreased to near control levels 48–72 h after MCAO. Quantitation of immunoreactive iNOS protein revealed parallel increases in immunoreactive protein. The latter implied there was no translational delay in iNOS protein synthesis. From these studies it was concluded that neuronal ischemia elicited an inflammatory response, which presumably facilitated cell necrosis.

Simultaneous quantitation of GAPDH mRNA demonstrated a significant decrease of GAPDH mRNA levels 14 h after MCAO. In particular, the most notable reduction (approximately fivefold) was in the core area itself, whereas a fourfold decrease in GAPDH mRNA expression was noted in the penumbra of the ischemic brain. No diminution of expression was observed in the contralateral area. This may be of special significance in that ischemia affected vesicular iNOS and NO, whereas the effect on GAPDH was neuronal in nature.

The rationale for this inverse relationship may relate not only to the ischemic-dependent posttranslational modification of GAPDH but also to the consequences of those modifications. Using a transient four-vessel occlusion (4-VO) model or the transient focal MCAO procedure followed by reperfusion, the formation and subcellular localization of S-nitrosylated GAPDH (SNO–GAPDH) were examined (Li et al., 2012). In the 4-VO model, SNO–GAPDH was elevated 6 h after reperfusion, which, by inhibitor analysis, appeared to be NMDA receptor dependent. Subcellular localization analysis indicated that, at the 6 h interval, SNO–GAPDH was localized in the nucleus. Coimmunoprecipitation studies demonstrated this was mediated by Siah1. Inhibitor studies using nNOS inhibitor 7-nitroindazole indicated the latter's neuroprotective effect reducing cerebral ischemia thereby, demonstrating that endogenous NO was responsible for the formation of SNO–GAPDH. A similar protective effect was observed in the MCAO model using the monoamine oxidase-B inhibitor deprenyl hydrochloride.

Other studies, although not specifically neuronal in nature, also indicated the significance of ischemia-related GAPDH posttranslational modification. However, in these instances, they documented the formation and consequences of nitrotyrosine GAPDH formation. The first involved a case study of a woman blinded as a consequence of ophthalmic artery ischemia (Kawaji et al., 2011). Using surgically obtained tissue, immunohistochemical studies were performed using an antinitrotyrosine or anti-GAPDH antibody. These investigations demonstrated GAPDH nuclear localization in retinal pigment epithelial (RPE) cells and in choroid endothelial cells in the ischemic area. This subcellular distribution was not observed in RPE cells and choroid not exhibiting visible ischemia. Correspondingly, a similar immunological profile was observed using the antinitrotyrosine antibody, i.e., significant immunoreactivity in locales exhibiting apoptotic changes with little nitrotyrosine immunoreactivity in unaffected regions.

To define the interrelationship between these in vivo immunological findings, human RPE cells in culture were exposed to oxidative stress (200 μM $H_2O_2$ for 24 h) followed by immunocytochemical analysis. These studies revealed the identical findings, i.e., in $H_2O_2$-treated cells, nuclear GAPDH was detected as was positive nuclear localization of nitrotyrosine. It would now be of interest not only to determine whether the nuclear localization of nitrotyrosine represented its modification of nuclear GAPDH, but also whether the modification occurred in the cytoplasm facilitating GAPDH nuclear translocation.

In the second investigation, the effect of ischemia followed by reperfusion (I/R) on GAPDH nitration was examined in skeletal muscle as a function of age (Bailey et al., 2011). In this instance, tourniquet-induced ischemia was utilized for a duration of 2 h. GAPDH mRNA expression was quantitated using PCR while GAPDH protein levels were determined by immunoblot analysis in young (6–7 months) and old (24–27 months) mice.

Analysis of mRNA levels demonstrated a significant decrease in GAPDH mRNA. However, there did not appear any significant difference as a function of age. In contrast, significant age-related differences were observed in immunoreactive GAPDH protein levels. In young mice, there was an initial decrease in GAPDH protein (days 1 and 2 I/R) followed by an increase at days 5 and 7 following I/R. In old animals, the initial decrease in GAPDH protein was observed, which did not increase at days 5 and 7 after I/R.

Immunoblot analysis using an antinitrotyrosine antibody demonstrated significant GAPDH modification in young mice at 3 days, which subsequently declined. In contrast, nitrotyrosine modification of old mice continued to increase with maximal levels observed at 5–7 days. In toto, these two studies indicate that, in addition to ischemic-induced S-nitrosylation of GAPDH, as well its S-thiolation and hyperoxidation (Eaton et al., 2002; Hwang et al., 2007, respectively), there is also a considerable nitrotyrosine modification as well. The functional significance of this is unknown. In addition, the age effect reported in skeletal muscle may have clinical significance given the age-related incidence of stroke in the general population.

## 4. GAPDH AND POLY (ADP) RIBOSE POLYMERASE IN ISCHEMIA

PARP is a nuclear protein involved intimately in the cellular response to DNA damage. Its mechanism of action involves its ADP-ribosylation of numerous proteins through which it modulates cell viability as well as apoptosis. Of note, its enzymatic function requires $NAD^+$ as a cofactor, which not only can have a drastic effect on many enzymes that also require $NAD^+$ as a cofactor but also on cellular ATP concentration. The latter is due to the stoichiometric requirement of 4 ATP consumed per molecule of $NAD^+$ formed.

Recent genetic evidence indicates a significant role for PARP in ischemia (Eliasson et al., 1997; Endres et al., 1997). In these studies, neuronal ischemia was induced in $PARP^{+/+}$, $PARP^{+/-}$, and $PARP^{-/-}$ mice as well as in $PARP^{+/+}$ mice treated with 3-aminobenzamide, a specific PARP inhibitor. Subsequently, following reperfusion, physiological parameters were utilized to determine whether ablation of PARP or its enzymatic inhibition could affect the animal's ischemic response.

In both studies, dramatic effects were observed. In particular, in both reports, $PARP^{-/-}$ mice exhibited a pronounced resistance to the harmful effects of neuronal ischemia. This included a significant decrease in cell death as measured in cerebral cortical cells in culture, in vivo diminution of infarct volume and area as well as the in vivo absence of ADP-ribose polymer formation as compared to the PARP$^{+/+}$ animals. Further, pretreatment of PARP$^{+/+}$ with 3-aminobenzamide reduced the infarct volume and area. Quantitation of NAD$^+$ levels were significantly higher in ischemic PARP$^{-/-}$ animals than their PARP$^{+/+}$ counterparts in accord with the demonstrative deficient PARP activity. In toto, these findings indicate a significant role for PARP in the pathology of neuronal ischemia commensurate with its established role in nuclear DNA damage and repair.

That being said, more recent investigations suggest an ischemic-dependent role for PARP in glycolysis generally and its interaction with GAPDH specifically. Ischemic renal injury (IRI) in the PARP$^{+/+}$ and PARP$^{-/-}$ model was used to determine the interrelationship between PARP-1, glycolysis, and GAPDH (Devalaraja-Narashimha and Padanilam, 2009). Ischemia was induced by bilateral renal pedicle clamping for 37 min followed by their removal to allow reperfusion. In addition, some studies were performed using a porcine-derived proximal tubular cell line.

Following IRI, GAPDH activity was determined in PARP$^{+/+}$ and PARP$^{-/-}$ renal outer medullary proteins. GAPDH activity was severely diminished in PARP$^{+/+}$ kidney when examined 3 or 6 h after IRI. In contrast, although inhibition of activity was observed in PARP$^{-/-}$ kidney, that diminution of activity was significantly reduced as compared with that observed in ischemic PARP$^{+/+}$ animals. As defined by immunoprecipitation with an anti-PAR antibody followed by immunoblot analysis, GAPDH was so modified in PARP$^{+/+}$ cells as a function of IRI. An in vitro assay was used to demonstrate that the latter was responsible for the observed reduction in GAPDH activity. The functional significance of PARP modification of GAPDH was demonstrated by measuring lactate and ATP levels in PARP$^{+/+}$ and PARP$^{-/-}$ animals. In both cases lactate and ATP levels were significantly higher in the PARP$^{-/-}$ IRI–treated kidney as compared with that of the PARP$^{+/+}$ IRI–treated kidney. Accordingly, as illustrated in Fig. 10.1, PARP performs an additional function contributing to ischemic toxicity, i.e., inhibition of anaerobic energy production through its modification of GAPDH.

Further analysis revealed an intriguing aspect of the ischemic interrelationship between GAPDH and PARP-1 (Nakajima et al., 2015). These studies were initiated to identify proteins which bound to GAPDH as a function of exposure to the NO generator NOC18. The model system was human neuronal SH–SY5Y cells expressing a myc-tagged GAPDH. The latter permitted coimmunoprecipitation analysis using an antimyc antibody. Through that analysis, exposure to the NO generator resulted in the coimmunoprecipitation of a 116-kDa protein subsequently identified as PARP-1 by mass spectroscopy.

Intriguingly, this association was observed only in SH–SY5Y nuclei indicating that NO exposure resulted in the formation of SNO–GAPDH, which was bound to Siah1, thereby inducing its nuclear translocation. This was confirmed by mutational analysis in which GAPDH$^{C152A}$ eliminated GAPDH–PARP-1, whereas no effect was detected using a GAPDH$^{K162R}$ mutant. The former alters the active site GAPDH cysteine, which is required for its S-nitrosylation, whereas the latter is required for its binding to p300/CBP. Similarly, mutagenesis protocols were used to map the relative binding sites on each protein. These studies demonstrated that the 20 N terminus of GAPDH bound to the C-terminal region of PARP-1. Further, use of GAPDH$^{G10A}$ or GAPDH$^{G12A}$ abolished GAPDH–PARP-1 binding, indicating that both glycine residues are required for this protein–protein interaction.

As these studies were performed in cell culture, their relationship to neuronal ischemia was determined in vivo. The model system used was ischemia induced in rat striatum by the MCAO protocol followed by reperfusion for 1 h. Nitrosative stress and oxidative stress were measured by the determination of nitrotyrosine and protein carbonyl formation, respectively. Relevant findings of this study were first, that ischemia-related nuclear GAPDH could be detected, and, second, that its binding to PARP-1 increased approximately sixfold. Of note, it appeared that PARP-1 activity was increased as a function of GAPDH binding.

To determine the latter, a novel strategy was employed. Rat GAPDH (rGAPDH) expression was reduced using rGAPDH siRNA and replaced by a coexpressing human GAPDH (hGAPDH). The hGAPDH protein bound PARP-1 while a mutant, hGAPDH$^{G10A}$, did not. In addition, the former stimulated PARP-1 activity while the latter did not. Quantitation of physiological parameters (23 h after reperfusion) indicated significant increases in infarct volume in rats expressing hGAPDH but an approximate threefold reduction of infarct volumes in rats expressing hGAPDH$^{G10A}$. In toto, these studies indicate not only that the formation of a nuclear GAPDH–PARP-1 nuclear protein complex in vivo is required for neuronal ischemic toxicity but also that the formation of this complex results in the activation of PARP-1 (Fig. 10.1). As noted in the study, these findings suggest that PARP-1 stimulation may require both DNA strand breakage and its binding to nuclear GAPDH. If this assertion is correct, this report identifies a new function for moonlighting GAPDH.

## 5. GAPDH AND THE REGULATION OF MITOPHAGY DURING ISCHEMIA/REPERFUSION

As previously described (Chapter 8), evidence suggests that moonlighting GAPDH may be intimately involved in autophagic cell survival (Colell et al., 2007). Studies suggest an additional role for GAPDH in basic mechanisms through which cardiac cells enhance their survival during reperfusion following an ischemic insult (Yogalingam et al., 2013). In particular, they indicated that GAPDH facilitated mitophagy, the lysosomal mediated elimination of damaged mitochondria to ensure cell survival. Further, these studies suggest that the regulation of this new GAPDH function may be a focal point at which cardiac cell survival is ensured or cardiac cell death is initiated.

As with many moonlighting protein studies, serendipity played a major role in the elucidation of this new GAPDH function. In ischemia/reperfusion, protein kinase Cδ (PKCδ) translocates from the cytoplasm to the mitochondria where it is implicated in cell death based on its deleterious effects, including reduced ATP synthesis and an increased mitochondrial-induced apoptosis (Churchill et al., 2005; Churchill and Szweda, 2005). In a further analysis of PKCδ prevention of mitochondrial mitophagy during cardiac ischemia/reperfusion, it was noticed that levels of the loading control, GAPDH, increased in the mitochondrial fraction (Yogalingam et al., 2013). This prompted an analysis not only of GAPDH function but also its interaction with PKCδ during ischemia/reperfusion.

These studies included a determination not only that GAPDH associates with mitochondria during ischemia/reperfusion facilitating mitophagy but also that its enzymatic activity was unnecessary for that association. Further, subcellular immunocytochemical analysis indicated that catalytically inactive GAPDH (termed iGAPDH) facilitated the formation of lysosomal-like structures (LL), which was considered as suggestive of mitochondrial uptake.

Subsequently, it was determined that mitochondria-associated cardiac GAPDH was phosphorylated both in vivo and in cardiac cells in culture. Accordingly, as measured in in vitro assays, both endogenous rat heart GAPDH and purified recombinant GAPDH were phosphorylated by PKCδ. Phosphopeptide mapping indicated modification at both Ser-241 and Thr-246. However, the majority (70%) of phosphorylation occurred at Thr-246. Of note, introduction of a GAPDH$^{T246A}$ mutant prevented PKCδ phosphorylation. Analysis of the wt versus the mutant-transfected cells demonstrated the presence of cytosolic cytochrome c following ischemia/reperfusion but not in cells expressing the T246A GAPDH mutant. Further, using transmission electron microscopy, it was observed using the iGAPDH$^{T246A}$ mutant that mitochondria were engulfed by the LL structures but not in the wt T246 iGAPDH$^P$–transfected cells.

In toto, it was concluded that, following ischemia/reperfusion, GAPDH facilitates mitophagy by lysosomal uptake of damaged mitochondria. PKCδ-mediated GAPDH phosphorylation abolishes this new GAPDH function, resulting in damaged mitochondrial release of cytochrome c inducing mitochondrial-mediated apoptosis. As such, as illustrated in Fig. 10.1, it was hypothesized that the formation of GAPDH$^{T246P}$ is a regulatory control point for determining the balance between cardiac cell survival or death, i.e., at a certain level of GAPDH$^{T246P}$ formation, that balance shifts toward apoptosis.

## 6. SUMMARY

As with other GAPDH moonlighting activities, it is hoped that, from this discussion, the reader will gain insight into the complexity of GAPDH function in ischemia. In particular, changes in its expression or structure are not, as the early studies indicated, related solely to the role of glycolysis in ischemia/reperfusion. Instead, as indicated in Fig. 10.1, these studies present significant evidence demonstrating an a priori requirement for GAPDH in a number of complex intracellular pathways, including ischemia-related receptor-mediated cell signaling/p53 response to ischemic stress, NO-related changes in PARP-1 structure and function, and the role of the GAPDH and its interactions with PARP-1 as a focal point for mitophagic control. In addition, these investigations provide a means to delineate GAPDH-specific differences between an organism's response to ischemia and cellular responses to hypoxia.

## REFERENCES

Bailey, C., Hammers, D., DeFord, J., et al., 2011. Ischemia-reperfusion enhances GAPDH nitration in aging skeletal muscle. Aging 3, 1003–1017.

Cai, R., Xue, W., Liu, S., et al., 2015. Overexpression of glyceraldehyde-3-phosphate dehydrogenase prevents neurovascular degeneration after retinal injury. FASEB J. 29, 2749–2758.

Churchill, E., Murriel, C., Chen, C.-H., Mochly-Rosen, D., Szweda, L., 2005. Reperfusion-induced translocation of δPKC to cardiac mitochondria prevents pyruvate dehydrogenase reactivation. Circ. Res. 97, 78–85.

Churchill, E., Szweda, L., 2005. Translocation of δPKC to mitochondria during cardiac reperfusion enhances superoxide anion production and induces loss in mitochondrial function. Arch. Biochem. Biophys. 439, 194–199.

Colell, A., Ricci, J.-E., Tait, S., et al., 2007. GAPDH and autophagy preserve survival after cytochrome c release in the absence of caspase activation. Cell 129, 983–997.

Devalaraja-Narashimha, K., Padanilam, B., 2009. PARP-1 inhibits glycolysis in ischemic kidneys. J. Am. Soc. Nephrol. 20, 95–103.

Eaton, P., Wright, N., Hearse, D., Shattock, M., 2002. Glyceraldehyde-3-phosphate dehydrogenase oxidation during cardiac ischemia and reperfusion. J. Mol. Cell. Cardiol. 34, 1549–1560.

Eliasson, M., Sampei, K., Mandir, A., et al., 1997. Poly (ADP-ribose) polymerase gene disruption renders mice resistant to cerebral ischemia. Nat. Med. 3, 1089–1095.

Endres, M., Wang, Z.-Q., Namura, S., Waeber, C., Moskowitz, M., 1997. Ischemic brain injury is mediated by the activation of poly (ADP-ribose) polymerase. J. Cereb. Blood Flow Metab. 17, 1143–1151.

Feldhaus, L., Liedtke, A., 1998. mRNA expression of glycolytic enzymes and glucose transporter proteins in ischemic myocardium with and without reperfusion. J. Mol. Cell. Cardiol. 30, 2475–2485.

Galea, E., Golanov, E., Feinstein, D., et al., 1998. Cerebellar stimulation reduces inducible nitric oxide synthase expression and protects brain from ischemia. Am. J. Physiol. 274, H2035–H2045 (Heart Circ. Physiol. 43).

Hwang, I., Yoo, K.-Y., Kim, D., et al., 2007. Hyperoxidized peroxiredoxins and glyceraldehyde-3-phosphate dehydrogenase immunoreactivity and protein levels are changed in the gerbil hippocampal CA1 region after transient forebrain ischemia. Neurochem. Res. 32, 1530–1538.

Jeremy, R., Ambrosio, G., Pike, M., Jacobus, W., Becker, L., 1993. The functional recovery of post-ischemic myocardium requires glycolysis during early reperfusion. J. Mol. Cell. Cardiol. 25, 261–276.

Kawaji, T., Elner, V., Yang, D.-L., Clark, A., Petty, H., 2011. Ischemia-induced nitrotyrosine formation and nuclear translocation of glyceraldehyde-3-phosphate dehydrogenase in human retinal pigment epithelium in vivo. Redox. Rep. 16, 24–26.

Knight, R., Kofoed, K., Schelbert, H., Buxton, D., 1996. Inhibition of glyceraldehyde-3-phosphate dehydrogenase in post-ischaemic myocardium. Cardiovasc. Res. 32, 1016–1023.

Li, C., Feng, J.-J., Wu, Y.-P., Zhang, G.-Y., 2012. Cerebral ischemia-reperfusion induces GAPDH S-nitrosylation and nuclear translocation. Biochem. (Mosc.) 77, 671–678.

Liedtke, A., Lynch, M., 1999. Alteration of gene expression for glycolytic enzymes in aerobic and ischemic myocardium. Am. J. Physiol. 277, H1435–H1440 (Heart. Circ. Physiol. 46).

Mallet, R., Hartman, D., Bünger, R., 1990. Glucose requirement for postischemic recovery of perfused working heart. Eur. J. Biochem. 188, 481–493.

Mohr, S., Hallak, H., de Boitte, A., Lapetina, E., Brüne, 1991. Nitric oxide-induced S-glutathionylation and inactivation of glyceraldehyde-3-phosphate dehydrogenase. J. Biol. Chem. 274, 9427–9430.

Morgan, H., Neely, J., Kira, Y., 1984. Factors determining the utilization of glucose in rat hearts. Basic Res. Cardiol. 79, 292–299.

Nakajima, H., Kubo, T., Ihara, H., et al., 2015. Nuclear-translocated glyceraldehyde-3-phosphate dehydrogenase promotes poly (ADP-ribose) polymerase-1 activation during oxidative/nitrosative stress in stroke. J. Biol. Chem. 290, 14493–14503.

Neely, J., Grotyohann, L., 1984. Role of glycolytic products in damage to ischemic myocardium: dissociation of adenosine triphosphate levels and recovery of function of reperfused ischemic hearts. Circ. Res. 55, 816–824.

Owen, P., Dennis, S., Opie, L., 1990. Glucose flux rate regulates onset of ischemic contracture in globally underperfused rat hearts. Circ. Res. 66, 344–354.

Rovetto, M., Lamberton, W., Neely, J., 1975. Mechanisms of glycolytic inhibition is ischemic hearts. Circ. Res. 37, 742–751.

Sirover, M., 1999. New insights into an old protein: the functional diversity of mammalian glyceraldehyde-3-phosphate dehydrogenase. Biochim. Biophys. Acta 1432, 159–184.

Sirover, M., 2011. On the functional diversity of glyceraldehyde-3-phosphate dehydrogenase: biochemical mechanisms and regulatory control. Biochim. Biophys. Acta 1810, 741–751.

Sirover, M., 2014. Structural analysis of glyceraldehyde-3-phosphate dehydrogenase functional diversity. Int. J. Biochem. Cell Biol. 57, 20–26.

Tanaka, R., Mochizuki, H., Suzuki, A., et al., 2002. Induction of gyceraldehyde-3-phosphate dehydrogenase (GAPDH) expression in rat brain after focal ischemia/reperfusion. J. Cereb. Blood Flow Metab. 22, 280–288.

Wang, M., Li, S., Zhang, H., et al., 2011. Direct interaction between GluR2 and GAPDH regulates AMPAR-mediated excitotoxicity. Mol. Brain 5, 13.

Yogalingam, G., Hwang, S., Ferreira, J., Mochly-Rosen, D., 2013. Glyceraldehyde-3-phosphate dehydrogenase (GAPDH) phosphorylation by protein kinase Cδ (PKCδ) inhibits mitochondria elimination by lysosomal-like structures following ischemia and reoxygenation-induced injury. J. Biol. Chem. 288, 18947–18960.

Zhai, D., Chin, K., Wang, M., Liu, F., 2014. Disruption of the nuclear p53-GAPDH complex protects against ischemia-induced neuronal damage. Mol. Brain 7, 20.

Zhai, D., Li, S., Wang, M., Chin, K., Liu, F., 2013. Disruption of the GluR2/GAPDH complex protects against ischemia-induced neuronal damage. Neurobiol. Dis. 54, 392–403.

# Section III

# The Pathology of GAPDH Functional Diversity

Chapter 11

# GAPDH and Tumorigenesis: Molecular Mechanisms of Cancer Development and Survival

*There are no coincidences*

An American Proverb

Although cancer is described classically as uncontrolled cell growth, its organ specific pathology may exhibit distinctive histological characteristics as well as defined patterns of gene and protein expression. Each is required for the well-described stages of cancer cell growth. As indicated in Fig. 11.1, moonlighting GAPDH may play a significant role in different stages of tumorigenesis, i.e., its tumor-specific expression may not be coincidental.

Accordingly, the goal of this chapter is to consider the diverse mechanisms through which cancer cells utilize the functional diversity of GAPDH as part of their pathology. In particular, this includes its classical role in energy generation, in cancer gene regulation and mRNA function, in their angiogenesis, and, perhaps most notably, its role in the prevention of apoptosis and caspase-induced cell death, thereby facilitating cancer cell survival. Conversely, GAPDH may also play a role in the prevention of cancer development by inhibiting tumor cell proliferation and the induction of cancer cell senescence.

## 1. GAPDH AND THE WARBURG EFFECT

As with its historical role as a metabolic "housekeeping" protein of little interest in normal cells, increased expression of GAPDH has long been considered as a "marker" of cancer cells. This was in accord with the well-recognized Warburg effect (Warburg et al., 1927; Warburg, 1956). The latter is characterized by increased rates of glycolysis in tumor cells as compared with that observed in normal cells. It is considered that such increases in the rates of anaerobic energy production vis a vis aerobic ATP generation is required due to the hypoxic conditions under which cancer cells may need to grow. Hence, the increased activity "across the board" of those proteins that participate in anaerobic metabolism is required to fulfill those energy requirements (Altenberg and Greulich, 2004).

Glyceraldehyde-3-Phosphate Dehydrogenase (GAPDH). http://dx.doi.org/10.1016/B978-0-12-809852-3.00011-X

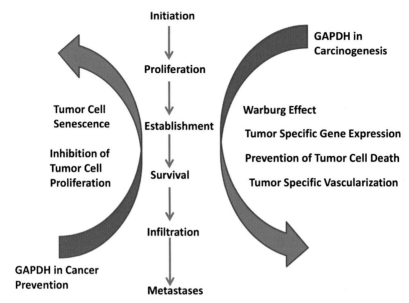

**FIGURE 11.1**   Role of GAPDH in tumorigenesis.

The findings in Table 11.1 provide a detailed analysis of GAPDH expression in each of the human tumors which were examined, establishing GAPDH overexpression as a general characteristic of human cancer. Analysis of these findings indicates that they may be grouped into four separate categories based on the experimental approaches, which were utilized. For example, Category I studies involve the in situ quantitation of GAPDH expression in tumor as compared with adjacent normal tissue. These immunohistochemical studies in colorectal cancer (Tang et al., 2012), in lung (Hao et al., 2015), and in esophagus (Hao et al., 2015) are visually striking. In the former, qualitative analysis was defined by statistical analysis. In the latter two, direct examination of GAPDH protein by gel analysis was used to quantitate GAPDH overexpression.

Category II represents studies in which GAPDH expression was determined in cancer patient tissue samples as compared with that observed in normal individuals. For these investigations GAPDH mRNA and/or protein were quantitated in lung cancer (Tokunaga et al., 1987), pancreatic adenocarcinoma (Schek et al., 1988), renal cell cancer (Vilà et al., 2000), cervical cancer (Hansen et al., 2009), and in ovarian cancer (Hansen et al., 2009). In some cases, comparative analysis was performed for other glycolytic genes/proteins (Schek et al., 1988; Vilà et al., 2000), which demonstrated increases in those genes/proteins as well. The latter would be in accord with the Warburg effect. In the renal cancer studies, gene amplification could be eliminated as a cause of the respective increase in GAPDH (Vilà et al., 2000). Comparisons were also performed in relevant cell lines as well as analysis of tumor production in the nude mouse model. In toto, for Category II studies, each parameter utilized reflected tumor-specific increases in GAPDH.

**TABLE 11.1 Regulation of GAPDH in Human Neoplasia**

| Organ | Tumor | Increased Tumor GAPDH Expression | "Unique" Finding | References |
|-------|-------|----------------------------------|------------------|------------|
| Breast | Primary | Correlated with histopathogenic grading | Expression inverse to estrogen and progesterone receptor levels | Révillion et al. (2000) |
| Cervical | Unspecified | Increase in tumor mRNA versus normal | Increased in proliferating cells | Kim et al. (1998) |
| Cervical | Unspecified | Increase in tumor mRNA versus normal | Increase independent of human papillomavirus (HPV) status | Hansen et al. (2009) |
| Colon | Carcinoma | Tumor RNA versus surrounding normal tissue | Increased expression in metastatic liver | Tang et al. (2012) |
| Esophagus | Squamous cell carcinoma | Increase in tumor protein versus normal using microarray analysis | GAPDH depletion inhibits tumorigenesis | Hao et al. (2015) |
| Lung | Unspecified | Increase in tumor mRNA and protein versus normal tissue | No erythrocyte protein contamination | Tokunaga et al. (1987) |
| Lung | Non–small cell | Tumor GAPDH RNA vs. B2M[a] and GUSB[b] RNA control | Increased GAPDH levels correlate with poor prognosis | Puzone et al. (2013) |
| Lung | Non–small cell | Tumor RNA versus noncancer controls | GACC[c] gene levels correlate with tumor stage and prognosis | Wang et al. (2013) |
| Lung | Squamous cell carcinoma | Increase in tumor protein versus normal using microarray analysis | GAPDH depletion inhibits tumorigenesis | Hao et al. (2015) |
| Kidney | Carcinoma | Increased tumor GAPDH versus normal tissue | Increased levels correlates with tumor metastases | Vilà et al. (2000) |

*Continued*

**TABLE 11.1 Regulation of GAPDH in Human Neoplasia—cont'd**

| Organ | Tumor | Increased Tumor GAPDH Expression | "Unique" Finding | References |
|-------|-------|----------------------------------|------------------|-----------|
| Ovary | Unspecified | Increased tumor GAPDH versus normal tissue | Increase independent of HPV status | Hansen et al. (2009) |
| Ovary | Epithelial | Increased GAPDH mRNA versus HPRT1[d] and β-2M[a] mRNA | Early disease progression | Hjerpe et al. (2013) |
| Pancreas | Adenocarcinoma | Increase in tumor cell mRNA versus normal pancreas | Increased tumor GAPDH protein versus normal pancreas | Schek et al. (1988) |
| Prostate | Unspecified | Increased tumor GAPDH versus normal tissue | Increased levels correlate with tumor stage | Rondinelli et al. (1997) |

[a]β-2-microglobulin.
[b]β-glucuronidase.
[c]GAPDH Associated Cell Cycle.
[d]Hypoxanthine guanine phosphoribosyltransferase 1.

Category III presents an alternative method for examining GAPDH expression in human cancer. Instead of comparisons between tumor tissue and its normal counterpart, these studies quantitated GAPDH expression strictly in tumor tissue as a function of human cancer development. In these studies, GAPDH expression was correlated with tumor progression in prostate cancer (stages B–D, Rondinelli et al., 1997). In breast cancer, GAPDH expression was correlated with histopathogenic grading (Révillion et al., 2000). In contrast, although GAPDH was increased in human cervical cancer, its aberrant regulation was independent of tumor grade (Kim et al., 1998). The basis for the latter is unknown at present.

The studies included in Category IV present the last approach used to examine GAPDH regulation in human tumors. In these studies, GAPDH levels in the tumor tissue were compared with other genes in that single human sample or by tissue microarray analysis as compared to a normal control. In the former, as indicated in Table 11.1, comparisons with internal controls demonstrated GAPDH overexpression in non–small cell cancer (Puzone et al., 2013) and in ovarian cancer (Hjerpe et al., 2013). With respect to the latter, protein expression was compared to IMR-90, a normal human fibroblast cell strain (Wang et al., 2013).

This extensive data set also reveals several other intriguing aspects of GAPDH regulation in human cancer (termed "unique findings" in Table 11.1). In particular, they shed light not only on the consequences of GAPDH overexpression but also on its physiological significance. Notably, GAPDH regulation may be an indicator as to the success (or lack thereof) of patient therapy. These studies demonstrate that GAPDH overexpression correlates with a poor prognosis and poor survival in non–small cell lung cancer patients (Puzone et al., 2013; Wang et al., 2013) as well as early disease progression in ovarian cancer (Hjerpe et al., 2013). GAPDH expression is associated with increased tumor cell proliferation in lung and esophageal cancer (Wang et al., 2013; Hao et al., 2015). Ominously, GAPDH overexpression is associated with metastasis in renal cancer (Vilà et al., 2000) and in colorectal cancer (Tang et al., 2012). Individual "unique findings" indicate a potential inverse relationship between GAPDH overexpression and the level of estrogen and progesterone receptors in breast cancer (Révillion et al., 2000). This may be of particular interest, given the aggressiveness and poor prognosis, which characterize triple negative breast cancer. Lastly, in cervical cancer, GAPDH expression appeared to be independent of human papillomavirus status. This is intriguing (and puzzling) given the causal nature of such infection with cervical cancer development.

In toto, the "take home" message from the studies presented in Table 11.1 is that, by whatever protocol that was used, GAPDH was overexpressed in all of the human cancers that were examined. This establishes GAPDH dysregulation as a common characteristic of human tumors. In consideration of the Warburg effect, this may be presumed as a given, i.e., GAPDH overexpression is expected. Thus, it may not be worthy of any further investigation. In the ensuing discussion, evidence will be presented to disprove that assumption, describing significant GAPDH functions, which are not only independent of the Warburg effect but also required for tumor development. As such, they, along with the "unique findings" summarized in Table 11.1, provide the requisite evidence, indicating the physiological significance of GAPDH moonlighting activities in human tumorigenesis.

## 2. GAPDH AND CANCER CELL SURVIVAL

Apoptosis and caspase-independent cell death (CICD) are essential functions in normal cells providing the organism with a genetic program to remove damaged cells. In contrast, their activation in tumor cells presents a challenge to the cancer cell specifically and to tumor development in general. Accordingly, to ensure their survival, cancer cells need to develop a defense mechanism to prevent apoptosis as well as CICD. Recent evidence suggests not only that cancer cells have developed such mechanisms but also that GAPDH plays a central role in these evasive maneuvers. Surprisingly, at its core, this appears to be based on a GAPDH—Akt, protein–protein interaction, which may require the posttranslational modification of each protein.

Akt is a serine/threonine kinase thought to function in the regulation of cell proliferation through its participation in several signaling pathways (Chan et al., 1999). As Akt binding to downstream targets may be directly related to its signaling function, a search was initiated to identify and characterize those cellular macromolecules. The model system chosen was Akt regulation in cardiac muscle cells as a function of changes in blood glucose levels (Baba et al., 2010).

To determine potential binding partners, coimmunoprecipitation analysis was performed using an antiphospho Akt antibody as the modified protein appears to be its active form. These studies identified a major p40 phosphoprotein which was purified and determined by protein and peptide analysis to be GAPDH. Although these studies would seem to be typical of those which identified GAPDH as a binding partner, it should be noted that both GAPDH and Akt had undergone posttranslational modification. Further, kinetic analysis indicated their coordinate phosphorylation and dephosphorylation as a function of blood glucose concentration.

Although these studies provided not only the first indication of a GAPDH[P]–Akt[P] protein complex but also its potential significance, they shed no light on its role in cancer cell survival or in tumorigenesis. For that demonstration, the interaction of GAPDH[P] and Akt[P] was examined in ovarian cancer cells using $H_2O_2$-induced oxidative stress activation of Akt2 (Huang et al., 2011).

These studies revealed several salient findings with respect to GAPDH[P]–Akt[P] binding, its physiological relevance to cancer cell survival and the specific Akt isozyme, which was responsible. Protein binding to activated Akt2[P] was identified by coimmunoprecipitation and gel analysis. The 36-kDa bound protein, which selectively bound to Akt2[P], was identified as GAPDH[P] by standard protein protocols. The formation of the complex reduced apoptosis enhancing cell survival as defined by terminal deoxynucleotidyl transferase dUTP nick end and labeling (TUNEL) and DNA ladder assay. Control studies indicated that siRNA knockdown by Akt1 had no effect on apoptosis while Akt3 levels were undetectable, thereby demonstrating the specific effect of $Akt2^p$.

Analysis of the mechanism underlying this effect of $Akt2^p$ revealed a surprising (and exciting) finding. As defined by both immunocytochemical and by subcellular localization/immunological criteria, $GAPDH^p$ remained in the cytoplasm. As noted in earlier chapter, GAPDH participates in apoptosis through the transfer of an SNO–GAPDH–Siah1 protein complex from the cytoplasm into the nucleus, thereby initiating both Siah1 activity and GAPDH-mediated gene expression. In cells which contained an activated $Akt2^p$ thereby forming the $GAPDH^p$– $Akt^p$ complex, this did not occur, which emphasizes the physiological significance of $GAPDH^p$–$Akt2^p$ complex formation.

How does $Akt^p$ binding to cytoplasmic GAPDH prevent its nuclear transfer? Previous studies identified a consensus $Akt^p$ phosphorylating sequence, which, nevertheless, is lacking in GAPDH. However, it does contain a "nontraditional" sequence, which includes $Thr^{237}$. To determine the role of that amino acid, a $GAPDH^{T237A}$ was constructed. Little GAPDH phosphorylation was observed. As previous studies indicated that nuclear translocation of the SNO–GAPDH–Siah1 complex requires $GAPDH^{lys225}$ (Hara et al., 2005), it was suggested that $Akt2^p$ phosphorylation of $GAPDH^{thr237}$ results in stearic hindrance of that complex formation due to its production of a conformational change in the $GAPDH^{thr237p}$. There is an old saying that, "a small change in structure may have a great change in activity." These findings suggest that this small alteration in GAPDH structure may provide the basis for tumor cell survival in the face of apoptotic challenges.

The studies presented so far describe downstream consequences of Akt activation. Further analysis sheds light on the upstream mechanisms, which regulate Akt function (Leisner et al., 2013). In these studies, gene deletion of CIB1 in neuroblastoma and breast cancer cells was used to determine its relationship to the established regulatory sequence: $Akt \rightarrow Akt^p \rightarrow Akt^p$–$GAPDH \rightarrow Akt^p$–$GAPDH^p \rightarrow$ inhibition of apoptosis $\rightarrow$ tumor cell survival. CIB1 is a 22 kDa protein also thought to be involved in cell proliferation and survival.

This investigation also yielded several salient findings with respect to the mechanism through which Akt–GAPDH interactions prevented tumor cell death ensuring cancer cell survival. CIB1 depletion diminished cancer cell viability and growth. Cell analysis indicated that tumor cell death did not occur due to caspase activity as no activation was observed. Further, there did not appear to be any effect on mitochondrial function. As such, it was concluded that CIB1 depletion resulted in the initiation of CICD.

Analysis of GAPDH subcellular localization as a function of CIB1 depletion revealed that reduction in CIB1 resulted in both the nuclear accumulation of GAPDH and an increase in GAPDH-mediated cell death. The latter conclusion is based on studies using deprenyl, which inhibits nuclear GAPDH translocation.

In those studies using CIB1-depleted cells, GAPDH nuclear accumulation and cell death were prevented.

The involvement of Akt was indicated by its reduction in CIB1-depleted cells, which exhibited GAPDH nuclear translocation; Akt inhibitor studies demonstrating that diminution of Akt enzymatic activity resulted in nuclear GAPDH accumulation; and, lastly, decreased Akt phosphorylation in the inhibitor studies. As previous studies indicate a role for GAPDH in CICD (Colell et al., 2007), these results identify a second, distinct mechanism which indicate that cancer cells utilize Akt–GAPDH interactions to enhance their survival by mitigating CICD. This would appear to be separate and distinct from the Akt→GAPDH→prevention of apoptosis→cancer cell survival pathway itemized above.

These findings clearly and definitively link GAPDH and Akt in the prevention of tumor cell CICD. Other studies defined the nature of that interaction and the subsequent downstream effects on gene function (Jacquin et al., 2013). Initial findings in that study include the demonstration that overexpression of GAPDH stabilizes the active form of Akt while other glycolytic enzymes (phosphoglycerate kinase or enolase) do not; that this stabilization protected cells from CICD; that GAPDH and Akt formed a protein complex detected by coimmunoprecipitation; and that the presence of GAPDH and Akt inhibitors, koningic acid and Akti, respectively, affected Akt stabilization.

Although these results are of interest, it may be argued that the truly salient findings are those which define the downstream effects of GAPDH–Akt binding on the function of genes involved in the implementation or prevention of CICD. These investigations revealed that GAPDH overexpression increased Bcl-xL levels in an Akt-dependent manner and that this increase was required for protecting cells from CICD. Bcl-xL functions to prevent mitochondrial permeabilization, a process which results in cell death. Further analysis indicated that GAPDH-mediated Bcl-xL expression was Akt dependent, i.e., its stabilization prevented Bcl-6 induction by FoxO. The latter is an Akt target (Zhang et al., 2011), while the former represses Bcl-xL expression. These findings indicate that GAPDH–Akt interaction results in a downstream cascade affecting the expression, or lack thereof, of genes critical for either the induction or prevention of CICD in tumor cells.

Given these series of studies, it is possible to construct a model that defines the mechanisms through which GAPDH facilitates cancer cell survival through the inhibition of both apoptosis and CICD (Fig. 11.2). In particular, the model highlights three steps (Fig. 11.2, 1–3) at which tumor cells utilize GAPDH as a "blocking agent" to prevent each of these cell destruction pathways, thereby facilitating cancer cell survival.

As indicated by step 1 (Fig. 11.2, left), formation of the GAPDH$^P$–Akt$^P$ complex prevents the nuclear translocation of the SNO–GAPDH–Siah1 complex. This precludes all of the downstream events, which that complex initiates. That being said, the mechanism through which this inhibition occurs is

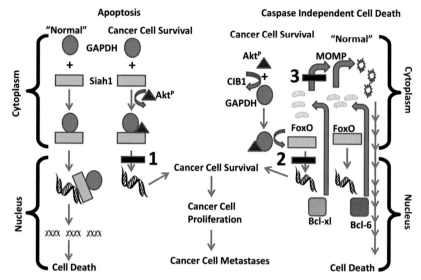

**FIGURE 11.2**   Mechanisms of GAPDH-mediated cancer cell survival.

unknown. It is unclear whether Akt$^P$ prevents the formation of the complex or whether it binds to the SNO–GAPDH–Siah1 complex forming a ternary intermediate, which is followed by GAPDH phosphorylation and Siah1 dissociation. As discussed in an earlier chapter, with respect to the competition between GOSPEL and Siah1 for SNO–GAPDH, it was hypothesized that Siah1 bound to the GAPDH–GOSPEL protein complex forming such a ternary intermediate followed by GOSPEL dissociation (Sirover, 2011).

In contrast to Step 1, which, although elegant, is at its base a "simple" steric hindrance mechanism, Steps 2 and 3 require a complex series set of gene regulatory events to effect GAPDH inhibition of CICD in tumor cells (Fig. 11.2, right). As indicated, during "normal" CICD, FoxO expression induces Bcl-6 synthesis, which induces mitochondrial disruption. In tumor cells, the temporal sequence appears to be CIB1→formation of Akt$^P$–GAPDH$^P$ complex→inhibition of FoxO→Bcl-xl synthesis→Bcl-xl cytoplasmic translocation→mitochondrial preservation→prevention of CICD. This is a much more complex process but still is based on GAPDH–Akt protein–protein interactions. In toto, the findings described in Fig. 11.2 illustrate the physiological significance of GAPDH in tumorigenesis.

## 3. GAPDH AND CANCER GENE REGULATION

Cancer is clonal in origin, i.e., it commences with an initiation event within a single cell. Paradoxically, tumors are chromosomally unstable with significant levels of chromosomal rearrangements. The latter gives rise to cancer-specific proteins resulting from the fusion of portions of two separate genes. Perhaps the most notable is the Bcr–Abl tyrosine kinase identified in chronic myelog-enous leukemia (Konopka et al., 1984). Ironically, the formation of the former in the latter is the basis for the successful treatment of this tumor by imatinib (Gleevac, O'Dwyer et al., 2002).

Similarly, the cancer transcription factor hTAF$_{II}$68-TEC arises from chro-mosomal translocation between chromosomes 9 and 17 (Labelle et al., 1999). This results in a chimeric protein containing the N-terminal portion of the TATA-binding protein associated factor II68 with a nuclear receptor protein. Analysis of hTAF$_{II}$68-TEC binding partners by affinity chromatography using hTAF$_{II}$68 (N-terminal domain) as a probe identified GAPDH as such a binding protein. The functional significance of GAPDH binding was demonstrated by the colocaliza-tion of hTAF$_{II}$68-TEC and GAPDH in the nucleus; the observation that GAPDH modulates hTAF$_{II}$68-TEC-dependent transcription; and the requirement for phys-ical association of both proteins into a complex as a requirement for GAPDH modulation of hTAF$_{II}$68-TEC-mediated transcription (Kim et al., 2007).

Intriguingly, these studies suggest that hTAF$_{II}$68-TEC promotes GAPDH nuclear localization such that nuclear GAPDH levels are equivalent to that observed in the cytoplasm. This may be unusual in that some studies demonstrat-ing nuclear GAPDH–protein interactions reveal that only a small percentage of GAPDH is involved (Hara and Snyder, 2006). Accordingly, the significance of this finding is unknown at the present time.

## 4. GAPDH IN OVARIAN CANCER: EFFECT ON mRNA STABILITY AND CANCER DEVELOPMENT

Ovarian cancer, a.k.a. "the silent killer," does not present any unique symptom-ology nor are there any novel biomarkers for its identification at an early stage. Accordingly, diagnosis is usually at a later stage of cancer development, which presents difficulties with respect to both treatment and prognosis.

Macrophage colony stimulating factor-1 (CSF-1) has normal functions in phagocytes and in bone marrow cells (Bradley et al., 1971; Stanley et al., 1983). CSF-1 may be overexpressed in ovarian cancer as well as other tumors con-tributing to cancer development by facilitating tumor invasiveness (Chambers et al., 1995). The extent of CSF-1 dysregulation in ovarian cancer patients may be considered as an indicator of poor prognosis (Baiocchi et al., 1991).

Recent studies indicate not only that GAPDH binding to CSF-1 mRNA in ovarian cancer cells may be fundamental to the role of CSF-1 in ovarian can-cer development but also reveal an intriguing molecular mechanism that may underlie GAPDH function (Bonafé et al., 2005; Zhou et al., 2008). As noted in an earlier chapter, GAPDH may affect mRNA stability by its binding to 5′- or 3′-UTR sequences. Analysis of CSF-1 mRNA revealed a novel decay element at its 3′ terminus. Comprising a 144-nucleotide sequence derived from CSF-1 exon 10, it is AU rich in nature and downregulates gene expression using a het-erologous chloramphenicol acetyltransferase vector.

Accordingly, as such sequences may be utilized for modulation of mRNA function in vivo, an ovarian cancer model system was used to identify potential CSF-1 mRNA–binding factors (Bonafé et al., 2005). Two ovarian epithelial cell lines, one of which was highly aggressive and invasive in vivo and one which was not, were utilized. The former overexpressed CSF-1 mRNA, while the latter did not. GAPDH was identified as a unique protein, which bound to CSF-1 mRNA in the highly aggressive cell line, while binding was not observed in the normal counterpart. Notably, in agreement with studies using human tissue (Table 11.1; Hansen et al., 2009), GAPDH expression was increased in the highly aggressive cell line. As with other studies of this kind, commercially available human erythrocyte GAPDH bound equally well to the ovarian cancer CSF-1 mRNA. The physiological significance of the ovarian GAPDH–CSF-1 mRNA interaction was examined using GAPDH siRNA (Zhou et al., 2008). This analysis demonstrated that transduction of the GAPDH siRNA not only reduced GAPDH levels in the ovarian cancer cells but also diminished CSF-1 mRNA and protein half-life. Mutational analysis of CSF-1 mRNA in its AU-rich 3′-UTR decay element demonstrated that those regions were essential for GAPDH binding. As discussed previously with other studies demonstrating GAPDH binding to mRNA 3′-UTR sequences, molecular modeling indicated that there appears to be a high degree of secondary structure in the CSF-1 mRNA AU-rich regions. This would be consistent with other GAPDH–3′-UTR mRNA binding studies.

These findings identify GAPDH as an important regulator of CSF-1 function in ovarian cancer. In particular, its ability to prolong CSF-1 mRNA stability underlies the 10-fold increase in secreted CSF protein observed in highly aggressive ovarian cancer cells (Bonafé et al., 2005). In vivo, such an effect would have significant effects on ovarian cancer progression. Further, the tumor-specific increases in both CSF-1 and GAPDH mRNA reflect the specificity of gene expression required for cancer development. That being said, the ability of human erythrocyte GAPDH to enhance the stability of CSF-1 mRNA suggests that there is no specific change in the pattern of GAPDH posttranslational modification in ovarian cancer cells.

## 5. GAPDH AND TUMOR ANGIOGENESIS

Tumor angiogenesis is one of a series of sequential steps necessary for cancer development. In a sense, it may be considered as a pathological recapitulation of a normal developmental process, i.e., the need of organisms to provide for the transport of necessary nutrients, factors, and minerals required for cell function. Thus, it is no surprise that cancer cells hijack this process for their development. That being said, the recent finding that GAPDH is involved intimately in tumor angiogenesis may be considered both as expected and as unexpected; the former being based on the role of GAPDH in hypoxia gene regulation and the latter based still on the conception of GAPDH as a classical housekeeping gene and protein.

This study, using both a mouse model and human lymphoma tissue, presents not only several seminal findings with respect to the role of GAPDH in tumor angiogenesis but also the specificity of this new GAPDH function in relation to the role of other glycolytic proteins in cancer development (Chiche et al., 2015). In particular, they report that GAPDH may be linked not only to morphological formation of cancer vasculature structure but also may play a fundamental role in the molecular mechanisms of signal transduction, which underlie that change in structure. Included among these are morphological changes in mouse lymphomas as a consequence of GAPDH expression in inoculated tumor cells; GAPDH-induced increases in VEGF-A mRNA expression; and HIF-1α mRNA and protein. The latter appears to be mediated by NF-κB. Each increase is based on GAPDH binding to tumor necrosis factor receptor-associated factor-2 (TRAF2) as defined by coimmunoprecipitation analysis. Incubation in cell culture with koningic acid, a specific GAPDH inhibitor, which binds to its active site cysteine, prevented GAPDH–TRAF2 complex formation. Mutational analysis of its active site cysteine indicated that the GAPDH active site was required for binding. Lastly, the finding that human lymphoma tissue samples exhibit similar increases in GAPDH expression as well as increases in the levels of HIF-1α, VEGF-A and the NF-κB targeted gene nfkbia underscores the physiological relevance of these studies to human neoplasia.

As other studies indicated a general increase in glycolytic flux and the enzymes involved in anaerobic respiration, it may be of interest to note the specificity with which GAPDH affects lymphoma vascularization. Parallel analysis of other glycolytic genes and enzymes, including 6-phosphofucto-2-kinase/fructose-2,6-bisphosphatase 3, hexokinase II, enolase, and pyruvate kinase M2 (PKM2) demonstrated no increase in either enzyme activity or mRNA levels during mouse lymphogenesis. In particular, the PKM2 studies may be especially notable, considering its role in tumorigenesis (Yang and Lu, 2015).

A surprise observation was that, in the Eμ–Myc mouse lymphoma model, two categories of mice were detected in the GAPDH-transduced experiment. The first was mice with highly aggressive tumors; the second with low aggressive tumors. The animals in the former group succumbed quickly, those in the latter group survived longer. GAPDH expression was noted in the aggressive tumor-bearing animals, while its expression was lacking in the lesser aggressive mice. Transduction of shGAPDH into the aggressive animals increased survival. These findings are in accord with the studies described in Table 11.1 relating GAPDH to a poor prognosis in patients.

This study not only identifies GAPDH as a factor in tumor angiogenesis but also suggests the signaling pathway through which it affects cancer vasculature development. In particular, the sequence appears to be GAPDH–TRAF2 binding, which stimulates NF-κB resulting in HIF-1α stimulation of VEGF-A leading to VEGF secretion. It will be of interest to determine whether other tumors employ a similar strategy for the development of their vascular structure.

## 6. GAPDH AND THE PREVENTION OF TUMOR DEVELOPMENT

As a moonlighting protein, GAPDH may be involved paradoxically in what may seem to be opposite activities which may seem to be contradictory in nature, i.e., a cellular version of the "pushmi–pullyu" animal in the 1967 movie, "Dr. Dolittle." In this section, recent evidence will be presented, suggesting that GAPDH may have a role in cancer prevention.

## 6.1 Modulation of Cancer Cell Growth-I

Tumor–host cell interactions may play a pivotal role in cancer development. In particular, cell-signaling mechanisms may provide a means for tumor cells to influence adjacent cells, thereby facilitating cancer development. As discussed above, tumor cell–induced host angiogenesis is not only an a priori requirement for cancer development but also an example of such mechanisms of tumor cell signaling.

Recent evidence suggests that tumor cell interactions with adjacent stromal cells may facilitate cancer cell growth. Components of this cell–cell interaction involve the secretion by tumor cells of factors that not only stimulate stromal cell differentiation but also their secretion of factors, which facilitate cancer cell growth. Conversely, stromal cells may secrete factors, which inhibit tumor cell growth, thereby preventing cancer development.

With respect to the latter, a recent study indicated a potential role for stromal GAPDH in the prevention of tumor cell growth (Kawada et al., 2015). The model system used gastric stromal cells cocultivated with a series of gastric cell lines characterized by differences in differentiation (well differentiated or undifferentiated), rates of tumorigenesis (high and low), and their response to stromal cell modulation (responsive or nonresponsive).

Analysis of stromal cell–conditioned media not only identified the presence of extracellular GAPDH in the media but also determined that, it by itself could inhibit gastric cancer cell growth. As with many such studies of moonlighting GAPDH, commercially available rabbit and erythrocyte GAPDH exhibited similar activity. Binding studies indicated that GAPDH bound E-cadherin within the cancer cell membrane and downregulated the mTOR-p70S6k pathway. Surface localization was consistent with previous studies in human breast cancer cell lines (Correa et al., 2010). Using truncated plasmids, the GAPDH inhibitory site was identified within the N-terminal GAPDH domain (residues 4–81), demonstrating that GAPDH enzyme activity was unnecessary for this new GAPDH function. The latter studies are in contrast to a previous finding indicating that the catalytic domain was essential for the effect of extracellular GAPDH on cell spreading (Yamaji et al., 2005).

This presents an interesting dynamic with respect to gastric cancer cell/stromal cell interactions specifically and with respect to the question of cancer cell/ normal cell in general. In particular, it would appear that these are two competing processes in tumorigenesis, i.e., the ability of cancer cells to control their microenvironment to facilitate cancer development versus the ability of normal cells adjacent to the tumor to influence negatively tumor progression. Stromal cell secretion of extracellular GAPDH may facilitate the latter.

## 6.2 Modulation of Cancer Cell Growth-II

Telomerase is a ribonucleoprotein whose function is to replicate teleomeric structure during DNA replication. Recent evidence indicates that GAPDH inhibits telomerase function in breast cancer cells resulting in their senescence (Nicholls et al., 2012). In particular, GAPDH overexpression in MCF-7 cells alters cell morphology into a "senescent-like" phenotype. In those cells, examination of teleomeric structure reveals significant telomere shortening as well as a reduction in telomerase activity quantitated in vitro. As with other studies of GAPDH as a moonlighting protein, commercially available GAPDH inhibited telomerase in vitro.

GAPDH not only interacts with the breast cancer telomerase RNA portion but also binds to the breast cancer telomerase protein itself, thereby inhibiting its activity. With respect to the former, RNA-binding assays demonstrated not only GAPDH binding but also, using competition analysis, that this interaction was specific to GAPDH as it was not observed with unrelated RNA. Using truncated GAPDH constructs, the telomerase RNA-binding site was identified within the GAPDH N-terminal region. Single base mutation studies indicated the GAPDH Rossman fold was required for binding.

With respect to the latter, further construct analysis indicated that the GAPDH C-terminal region was responsible for its inhibition of telomerase activity. Modification of the active site cysteine with S-nitrosoglutathione (GSNO) diminished telomerase inhibition as did competition with glyceraldehyde3-phosphate. In contrast, $NAD^+$ did not affect telomerase activity. Mutational analysis of $lys^{259}$ (K259N) but not that of $lys^{260}$ (K260A) eliminated GAPDH-induced telomere shortening and reduction of telomerase activity. This is noteworthy as previous studies indicated the significance of $lys^{259}$ in GAPDH functional diversity (Brown et al., 2004).

These results are consistent with the proposed model that, for the most part, the $NAD^+$-binding site is responsible for GAPDH–macromolecular binding resulting in the orientation of the GAPDH catalytic site for the respective activity (Sirover, 2014). That being said, these studies present an apparent paradox with respect to GAPDH function. Maintenance of telomere structure is required for continued cell growth and for cell proliferation. Conversely, telomere shortening results in cell cycle arrest and the induction of cell senescence. Thus, cancer cells require the conservation of telomere structure as a prerequisite for tumor development. Any mechanism that results in telomere shortening in cancer cells would preclude tumorigenesis. Accordingly, as it would seem that this GAPDH activity would be detrimental to cancer development, it would be of great interest to understand the rationale for this new cancer GAPDH function.

## 7. SUMMARY

It is hoped that, from this discussion, the reader will gain insight into the complexity of GAPDH function in tumorigenesis. As described, it is a highly adaptable protein, which is utilized in a series of important steps required for cancer development. The physiological relevance of these investigations is reflected, regrettably, in the inverse correlation between GAPDH expression and patient prognosis. In contrast, evidence is presented indicating a potential role for GAPDH in cancer prevention, indicating again the utility of this moonlighting protein.

## REFERENCES

Altenberg, B., Greulich, K.O., 2004. Genes of glycolysis are ubiquitously overexpressed in 24 cancer classes. Genomics 84, 1014–1020.

Baba, T., Kobayashi, H., Kawasaki, H., Mineki, R., Naito, H., Ohmori, D., 2010. Glyceraldehyde3-phosphate dehydrogenase interacts with phosphorylated Akt resulting from increased blood glucose in rat cardiac muscle. FEBS Lett. 584, 2796–2800.

Baiocchi, G., et al., 1991. Expression of the macrophage colony-stimulating factor and its receptor in gynecologic malignancies. Cancer 67, 990–996.

Bradley, T.R., Stanley, E., Sumner, M.A., 1971. Factors from mouse tissues stimulating colony growth of mouse bone marrow cells in vitro. Aust. J. Exp. Biol. Med. Sci. 49, 595–603.

Bonafé, N., Gilmore-Hebert, M., Folk, N.L., Azodi, M., Zhou, Y., Chambers, S.K., 2005. Glyceraldehyde3-phosphate dehydrogenase binds to the AU-rich 3′ untranslated region of colony-stimulating factor-1 (CSF-1) messenger RNA in human ovarian cancer cells: possible role in CSF-1 posttranscriptional regulation and tumor phenotype. Cancer Res. 65, 3762–3771.

Brown, V., Krynetski, E., Krynetskaia, N., et al., 2004. A novel CRM1-mediated nuclear export signal governs nuclear accumulation of glyceraldehyde-3-phosphate dehydrogenase following genotoxic stress. J. Biol. Chem. 279, 5984–5992.

Chambers, S.K., Wang, Y., Gertz, R.E., Kacinski, B.M., 1995. Macrophage colony-stimulating factor mediates invasion of ovarian cancer cells through urokinase. Cancer Res. 55, 1578–1585.

Chan, T.O., Rittenhouse, S.E., Tsichlis, P.N., 1999. AKT/PKB and other D3 phosphoinositide-regulated kinases: kinase activation by phosphoinositide-dependent phosphorylation. Annu. Rev. Biochem. 68, 965–1014.

Chiche, J., Pommier, S., Beneteau, M., et al., 2015. GAPDH enhances the aggressiveness and the vascularization of non-Hodgkin's B lymphomas via NF-κB-dependent induction of HIF-1α. Leukemia 29, 1163–1176.

Colell, A., Ricci, J.-E., Tait, S., et al., 2007. GAPDH and autophagy preserve survival after cytochrome c release in the absence of caspase activation. Cell 129, 983–997.

Correa, C.R., Bertollo, C.M., Zouain, C.S., Goes, A.M., 2010. Glyceraldehyde-3-phosphate dehydrogenase as a surface associated antigen on human breast cancer cell lines MACL-1 and MGSO-3. Oncol. Rep. 24, 677–685.

Hao, L., Zhou, X., Shuqing, L., et al., 2015. Elevated GAPDH expression is associated with the proliferation and invasion of lung and esophageal squamous cell carcinomas. Proteomics 15, 3087–3100.

Hansen, C.N., Ketabi, Z., et al., 2009. Expression of CPEB, GAPDH and U6snRNA in cervical and ovarian tissue during cancer development. APMIS 117, 53–59.

Hara, M.R., Agrawal, N., Kim, S.F., et al., 2005. S-nitrosylated GAPDH initiates apoptotic cell death by nuclear translocation following Siah1 binding. Nat. Cell Biol. 7, 665–674.

Hara, M.R., Snyder, S.H., 2006. Nitric oxide-GAPDH-Siah: a novel cell death cascade. Cell. Mol. Neurobiol. 26, 527–538.

Hjerpe, E., Brage, S.E., Carlson, J., et al., 2013. Metabolic markers GAPDH, PKM2, ATP5B and BEC-Index in advanced serous ovarian cancer. BMC Clin. Pathol. 13, 30.

Huang, Q., Lan, F., Zheng, Z., et al., 2011. Akt2 kinase suppresses glyceraldehyde-3-phosphate dehydrogenase (GAPDH)-mediated apoptosis in ovarian cancer cells via phosphorylating GAPDH at threonine 237 and decreasing its nuclear translocation. J. Biol. Chem. 286, 42211–42220.

Jacquin, M., Chiche, J., Zunino, B., et al., 2013. GAPDH binds to active Akt, leading to Bcl-xL increase and escape from caspase-independent cell death. Cell Death Differ. 20, 1043–1054.

Kawada, M., Inoue, H., Ohba, S., et al., 2015. Stromal cells positively and negatively modulate the growth of cancer cells: stimulation via the PGE2-TNFα-IL-6 pathway and inhibition via secreted GAPDH-E-cadherin interaction. PLoS One. http://dx.doi.org/10.1371/journal.pone.0119415.

Kim, J.W., Kim, S.J., Han, S.M., et al., 1998. Increased glyceraldehyde-3-phosphate dehydrogenase gene expression in human cervical cancers. Gynecol. Oncol. 71, 266–269.

Kim, S., Lee, J., Kim, J., 2007. Regulation of oncogenic transcription factor hTAFll68-TEC activity by human glyceraldehyde-3-phosphate dehydrogenase (GAPDH). Biochem. J. 404, 197–206.

Konopka, J.B., Watanabe, S.M., Witte, O.N., 1984. An alteration of the human c-abl protein in K562 leukemia cells unmasks associated tyrosine kinase activity. Cell 37, 1035–1042.

Labelle, Y., Bussieres, J., Courjal, F., Goldring, M.B., 1999. The EWS/TEC fusion protein encoded by the t(9,22) chromosomal translocation in human chondrosarcomas is a highly potent transcriptional activator. Oncogene 18, 3303.

Leisner, T., Moran, C., Holly, S., Parse, L., 2013. CIB1 prevents nuclear GAPDH accumulation and non-apoptotic tumor cell death via AKT and ERK signaling. Oncogene 32, 4017–4027.

Nicholls, C., Pinto, A.R., Li, H., Ling, L., Wang, L., Simpson, R., Liu, J.-P., 2012. Glyceraldehyde-3-phosphate dehydrogenase (GAPDH) induces cancer cell senescence by interacting with telomerase RNA component. Proc. Natl. Acad. Sci. U.S.A. 109, 13308–13313.

O'Dwyer, M.E., Mauro, M.J., Druker, B.J., 2002. Recent advancements in the treatment of chronic myelogenous leukemia. Annu. Rev. Med. 53, 369–381.

Puzone, R., Savarino, G., Salvi, S., et al., 2013. Glyceraldehyde3-phosphate dehydrogenase gene over expression correlates with poor prognosis in non small cell lung cancer patients. Mol. Cancer 12, 97–105.

Révillion, F., Pawlowski, V., Hornez, L., Peyrat, J.-P., 2000. Glyceraldehyde3-phosphate dehydrogenase gene expression in human breast cancer. Eur. J. Cancer 36, 1038–1042.

Rondinelli, R.H., Epner, D.E., Tricoli, J.V., 1997. Increased glyceraldehyde3-phosphate dehydrogenase gene expression in late pathological stage human prostate cancer. Prostate Cancer Prostatic Dis. 1, 66–72.

Schek, N., Hall, B.L., Finn, O.J., 1988. Increased glyceraldehyde3-phosphate dehydrogenase gene expression in human pancreatic adenocarcinoma. Cancer Res. 48, 6354–6359.

Sirover, M.A., 2011. On the functional diversity of glyceraldehyde-3-phosphate dehydrogenase: biochemical mechanisms and regulatory control. Biochim. Biophys. Acta 1810, 741–751.

Sirover, M.A., 2014. Structural analysis of glyceraldehyde-3-phosphate dehydrogenase functional diversity. Int. J. Biochem. Cell Biol. 57, 20–26.

Stanley, E.R., Guilbert, L.J., Tushinski, R.J., Bartelmiz, S.H., 1983. CSF-1- a mononuclear phagocyte lineage-specific hemopoietic growth factor. J. Cell. Biochem. 21, 151–159.

Tang, Z., Yuan, S., Hu, Y., et al., 2012. Over-expression of GAPDH in human colorectal carcinoma as a preferred target of 3-bromopyruvate propyl ester. J. Bioenerg. Biomembr. 44, 117–125.

Tokunaga, K., Nadamura, Y., Sakata, K., et al., 1987. Enhanced expression of a glyceraldehyde-3-phosphate dehydrogenase gene in human lung cancers. Cancer Res. 47, 5616–5619.

Vilà, M.R., Nicolás, A., Morote, J., de Torres, I., Meseguer, A., 2000. Increased glyceraldehyde-3-phosphate dehydrogenase expression in renal cell carcinoma identified by RNA-based, arbitrarily primed polymerase chain reaction. Cancer 89, 152–164.

Wang, D., Moothart, D.R., Lowy, D.R., Gian, X., 2013. The expression of glyceraldehyde-3-phosphate dehydrogenase associated cell cycle (GACC) genes correlates with cancer stage and poor survival in patients with solid tumors. PLoS One 8, e61262.

Warburg, O., Wind, F., Negelein, K., 1927. The metabolism of tumors in the body. J. Gen. Physiol. 8, 519–530.

Warburg, O., 1956. On the origin of cancer cells. Science 123, 309–314.

Yamaji, R., Chatani, E., Harada, N., Sugimoto, K., Inui, H., Nakano, Y., 2005. Glyceraldehyde-3-phosphate dehydrogenase in the extracellular space inhibits cell spreading. Biochim. Biophys. Acta 1728, 261–271.

Yang, W., Lu, Z., 2015. Pyruvate kinase M2 at a glance. J. Cell Sci. 128, 1655–1660.

Zhang, X., Tang, N., Hadden, T.J., Rishi, A.K., 2011. Akt, FoxO and regulation of apoptosis. Biochim. Biophys. Acta 1813, 1978–1986.

Zhou, Y., Yi, X., et al., 2008. The multifunctional protein glyceraldehyde-3-phosphate dehydrogenase is both regulated and controls colony-stimulating factor-1 messenger RNA stability in ovarian cancer. Mol. Cancer Res. 6, 1375–1384.

# Chapter 12

# Moonlighting GAPDH and Age-Related Neurodegenerative Disease: Diversity of Protein Interactions and Complexity of Function

*A single stone can cause an avalanche.*

Edmund A. Walsh

Age-related neurodegenerative diseases may pose one of the greatest medical challenges of the 21st century. In spite of the considerable progress in the understanding of their pathogenesis, their etiology and their treatment remain stubborn foes. Perhaps the greatest challenge is to identify and characterize that initial change in cell function, which then results in the progression and clinical presentation of Alzheimer's disease (AD), Parkinson's disease (PD), as well as Huntington's disease (HD) and other triplet repeat disorders, i.e., the movement of that single stone that produces the devastating avalanche. That being said, many of the pieces of each of the pathological puzzles have been identified and investigated.

Accordingly, the goal of this chapter is to consider the interrelationship between moonlighting GAPDH and the genotype/phenotype of these neurodegenerative diseases. In particular, this analysis will focus on the significance of GAPDH protein–protein interactions that may provide the basis for its role in each disease considered as causative for each disorder (AD: β-amyloid protein; PD: α-synuclein; HD: huntingtin and other triplet repeat proteins); changes in subcellular localization and perhaps function of GAPDH or these neuronal proteins; the association of GAPDH with the pathological structures identified in AD (amyloid plaques) and in PD (Lewy bodies); and, lastly, the role of GAPDH aggregation as a determining factor in these protein–protein interactions. In toto, it is hoped that this discussion will define this novel protein as one of the interlocking pieces that is required to construct the puzzle of each human disease.[1]

---

1. This is not intended to diminish the role of GAPDH genetic changes (Li et al., 2004; Lin et al., 2006; Allen et al., 2012) or oxidative stress in neurodegenerative disease etiology (Praticò and Mecocci, 2013). Rather, it is intended to focus specifically on GAPDH protein interactions as causative factors.

Glyceraldehyde-3-Phosphate Dehydrogenase (GAPDH). http://dx.doi.org/10.1016/B978-0-12-809852-3.00012-1

**199**

## 1. GAPDH PROTEIN–PROTEIN INTERACTIONS IN AGE-RELATED NEURODEGENERATIVE DISEASE-I: IDENTIFICATION AND CHARACTERIZATION OF NEUROPROTEIN BINDING

As indicated in Table 12.1, GAPDH binds specifically to individual proteins implicated as causative factors in age-related neurodegenerative disorders. With respect to the β-amyloid protein, an initial study (Schulze et al., 1993) probed its interactions using fusion constructs containing the 42 residue β-amyloid protein (Aβ 1–42) or a 98 residue peptide containing its C-terminal domain, termed Amy and AmyC, respectively. The former is thought to be the critical portion of the β-amyloid precursor protein involved in AD cytotoxicity, while the latter represents the cytosolic "free" C-terminal tail of this transmembrane protein.

To determine brain proteins that selectively bind to either construct, each was coupled to Affi-Gel for affinity chromatography analysis. As compared to a "mock" column, identical brain homogenate proteins were retained by the Amy containing Affi-Gel. In contrast, using the AmyC Affi-Gel column, a unique 35 kDa was retained as defined by its elution with 3 M NaSCN. Immunoblot analysis and protein sequencing identified the protein as GAPDH. As confirmation of these findings, each β-APP fragment was incubated with commercially available GAPDH. Subsequently, binding was analyzed by gel filtration analysis. These results demonstrated that only the AmyC peptide formed a complex with GAPDH as defined by their coelution using Sephacryl G-200 chromatography. No GAPDH binding was observed as defined by immunoreactivity with an antibody directed against the β-amyloid protein.

Finally, the effect of Amy or AmyC binding on GAPDH catalysis was examined. Quantitation of $V_{max}$ failed to reveal any significant differences in catalysis due to the presence of either β-APP fragment (<10% diminution of $V_{max}$ with either peptide). In toto, these results provided evidence that GAPDH did not bind to the cytotoxic Aβ 1–42 peptide but did bind to the cytoplasmic region of the β-APP. However, the binding of the latter did not appear to affect catalysis to a significant degree.

A second study indicated that GAPDH did recognize and bind to the β-amyloid protein (1–40) (Oyama et al., 2000). In these investigations, a synthetic β-amyloid protein (1–40) was synthesized and then immobilized on Affi-Gel. Binding to proteins in a rat brain supernatant was identified by gel permeation chromatography following elution with 7 M guanidine hydrochloride. N-terminal sequencing subsequent to SDS-PAGE identified a number of proteins including GAPDH. The latter was verified by immunoblot analysis. It was suggested that the discrepancy between this study and that discussed above may relate to differences in the immobilization protocols.

A more recent investigation suggested that GAPDH does bind to Aβ 1–42 but only in its fibrillary form, termed fAβ 1–42 (Verdier et al., 2008). In these studies, fAβ 1–42 or fAβ 1–40 was prepared by its synthesis followed by incubation in

**TABLE 12.1 GAPDH Protein–Protein Interactions in Age-Related Neurodegenerative Disease**

| Neurodegenerative Disease Protein | Pathology | Detection of GAPDH: Neuroprotein Binding | "Unique" Finding | References |
|---|---|---|---|---|
| β-Amyloid precursor protein (β-APP) | Alzheimer's disease(AD) | Affinity chromatography; gel filtration analysis | GAPDH binds to β-APP cytoplasmic domain; no decrease in GAPDH catalysis | Schulze et al. (1993) |
| β-Amyloid protein (1–40) | AD | Affinity chromatography, immunoblot analysis | GAPDH binds to the β-amyloid protein | Oyama et al. (2000) |
| β-Amyloid protein (1–42) | AD | Transmission electron microscopy; coimmunoprecipitation | Differential GAPDH binding to fibrillary and nonfibrillary Aβ 1–42; residues 41–42 required for binding | Verdier et al. (2008) |
| Huntingtin (htt) | Huntington's disease | Affinity chromatography-20 or 60 glutamine polypeptide, GAPDH | GAPDH binding to N-terminal htt protein | Burke et al. (1996) |
| DRLPA[a] | DRLPA | Affinity chromatography-20 or 60 glutamine polypeptide, GAPDH | GAPDH recognition of 190-kDa and 100-kDa DRLPA protein | Burke et al. (1996) |
| Ataxin-1 | SCA1[b] | Yeast two-hybrid screen; GAPDH affinity chromatography | N-terminal ataxin-1 binds to GAPDH NAD+ binding site | Koshy et al. (1996) |
| Androgen receptor (AR) | Spinobulbar muscular atrophy | Yeast two-hybrid screen; Affi-Gel immobilization | N-terminal AR binding to GAPDH NAD+-binding site | Koshy et al. (1996) |

[a]Dentatorubral-pallidoluysian.
[b]Spinocerebellar ataxia type 1.

aqueous solution for 48 h. The nonfibrillar form of Aβ 1–42 was prepared in aqueous solution immediately prior to use. Formation of fAβ 1–42 or fAβ 1–40 structures was confirmed by transmission electron microscopy (TEM). Commercially available GAPDH or rat brain subcellular fractions were used to examine GAPDH binding. The latter was monitored by TEM or by coprecipitation analysis.

As defined by visualization by TEM using anti-GAPDH gold-labeled antibody, incubation of commercially available GAPDH with fAβ 1–42 results in the formation of a GAPDH–fAβ 1–42 complex. In the control (fAβ 1–42 plus antibodies), sparse, nonspecific staining was observed. Coimmunoprecipitation studies confirmed the TEM results.

The specificity of these findings was then examined. Substitution of fAβ 1–42 with fAβ 1–40 followed by coimmunoprecipitation demonstrated a substantial reduction of bound GAPDH.

Substitution of subcellular GAPDH fractions showed equivalent binding to fAβ 1–42. These findings demonstrate not only that fAβ 1–42 binds GAPDH but also that its 41–42 residues are critical for this protein–protein interaction. In toto, with respect to each study, the reasons for the discrepancies between these separate findings are unknown but could relate either to the affinity protocols used in the former or the specific configuration of fAβ 1–42 used in the latter. That being said, these studies identify a specific protein–protein GAPDH–β-APP structural interaction.

Further studies indicated that GAPDH binding was a common characteristic of what have been termed "gain of function" CAG triplet repeat neurodegenerative disorders. The latter are defined by increases in CAG repeat gene sequences (usually in the N-terminal regions) resulting in additional polyglutamine domains in a specific protein. The latter may result in new, albeit deleterious, properties of that protein that is thought to give rise to the specific clinical presentations of each disorder.

As indicated in Table 12.1, the indicated causative proteins from four such "gain of function" neurodegenerative disorders have been identified as GAPDH binding proteins. Again, the basic experimental protocols involved used affinity chromatography or yeast two-hybrid analysis to identify proteins of interest. With respect to the former, a selective strategy was employed, i.e., the use of two different-sized glutamine (gln) repeats as the affinity probes (Burke et al., 1996). The rationale was that a 20 gln repeat was characteristic of the normal protein, while a 60 gln repeat was characteristic of the mutant protein identified in afflicted individuals.

Using that strategy, in brain homogenates, a small number of proteins were retained using the 20 gln repeat probe. In contrast, using the larger, 60 gln repeat probe an unknown protein was retained that was subsequently identified as GAPDH by microsequencing. Subsequently, a reverse affinity probe was used, i.e., GAPDH was bound, and proteins binding to it were isolated and identified. The first protein so identified was DRPLA that is present in two forms, a 190 kDa and a 100 kDa protein (Yazawa et al., 1995). Both were detected as

GAPDH-binding proteins. The second protein so identified using the GAPDH affinity protocol was Huntingtin (Htt). In this instance, several Htt fragments were observed although full length Htt was not identified. With respect to the latter, it was postulated that either the C-terminal Htt domain may hinder binding or the binding of GAPDH to the agarose matrix might preclude that interaction (as might have happened with Amy in the β-APP studies).

The yeast two-hybrid protocol as utilized to examine whether ataxin-1 or the androgen receptor bound GAPDH (Koshy et al., 1996). In this manner, GAPDH was identified as an ataxin-1-binding protein. This binding was deemed specific as the use of a series of control heterologous bait fusion proteins did not result in GAPDH identification. To identify binding domains, ataxin-1 deletion mutants were used in the hybrid analysis. This study demonstrated that the ataxin-1 N-terminal region was essential for GAPDH binding. Further, analysis of the GAPDH clone used as bait, it was noted that it contained amino acids 1–170 that included the GAPDH NAD⁺-binding site and its catalytic cysteine 149.

In a second study, the in vitro association of GAPDH and ataxin-1 was determined using Affi-Gel-bound rabbit muscle GAPDH. Wild-type (wt) ataxin-1 (32 repeats) and mutant ataxin-1 (82 repeats) were tested. Each bound tightly to the affinity column being eluted with a 1 M NaCl wash. The latter indicated a strong affinity of each protein for GAPDH, reminiscent of the resistance of GAPDH–DNA binding to 3–5 M NaCl (Sawa et al., 1997). Similarly, using protein extracts from patient-derived lymphoblastoid cell lines, GAPDH binding was detected irrespective of CAG repeat length (30, 43, and 60 gln, respectively).

Similar results were obtained for the androgen receptor (AR). In this instance, AR amino terminal portions (24 gln, amino acids 13–134; 45 gln, amino acids 13–155; and 66 gln, 13–176) were tested against the abovementioned GAPDH clone. All three of the AR clones bound GAPDH. Further, three full-length ARs with the identical CAG repeats were introduced into COS-7 cells. Protein extracts from those cells were tested with the GAPDH–Affi-Gel resin. Each bound to the resin. In toto, these two reports identify GAPDH as a specific binding partner for the proteins indicated as causal for each disorder. For that reason, in accord with the β-amyloid findings, it was postulated that GAPDH–neuronal protein interactions are a common characteristic of age-related neurodegenerative disorders (Mazzola and Sirover, 2002b).

## 2. GAPDH PROTEIN–PROTEIN INTERACTIONS IN AGE-RELATED NEURODEGENERATIVE DISEASE-II: AMYLOID PLAQUES, NEUROFIBRILLARY TANGLES, AND LEWY BODIES

The studies described above defined specific GAPDH–neuroprotein binding in vitro and in vivo. However, they did not indicate either the cellular consequences of such protein–protein interactions or their physiological significance. In part, those were defined by investigations that demonstrated the presence of

GAPDH in amyloid plaques, in neurofibrillary tangles (NFTs), both clinical characteristics of AD, as well as its localization in Lewy bodies, which present in PD.

## 2.1 Role of Moonlighting GAPDH in Alzheimer's Disease: Amyloid Plaques and Neurofibrillary Tangles

In an initial study, monoclonal antibodies were prepared against proteins present in amyloid plaques (Sunaga et al., 1995). The first, Am-3, was raised against plaque proteins from the brain of an Alzheimer's patient. The second, AmT-1, was produced against a synthetic β-amyloid protein. During an examination of neuronal cell apoptosis, each antibody was tested against a 38-kDa protein, which was overexpressed during that program of cell death and which was identified as GAPDH. The Am-3 antibody, but not the AmT-1 antibody, recognized that protein. It also was immunoreactive with commercially available GAPDH. This study represented the first investigation that GAPDH may be present in amyloid plaques.

Subsequently, in an investigation of proteins in NFTs, a hallmark of AD, it was determined that not only was GAPDH present in NFTs but also that it bound to paired helical filaments (PHF-tau) in vivo (Wang et al., 2005). These conclusions were based on immunohistochemical analysis and coimmunoprecipitation. Further, it was noted that GAPDH accumulates in the detergent (sarkosyl)-insoluble fraction from AD brain samples but not in samples from age-matched controls.

Although these studies demonstrated the presence of GAPDH in both amyloid plaques and NFTs, they did not indicate either the subcellular origin of that GAPDH protein or the mechanism through which it may be recruited for plaque or NFT formation. One possibility is that GAPDH (and perhaps other proteins as well) may be of membrane origin. This was suggested by an investigation, which reported that a series of synaptic plasma membrane proteins were coprecipitated with the fibrillary form of the β-amyloid protein (fAβ, Verdier et al., 2005). In that study, triton X-100 soluble brain plasma proteins were tested for binding to a synthesized Aβ 1–42 peptide by coprecipitation followed by LC-MS/MS identification.

Six proteins were identified including GAPDH. Although all six were said to be associated with AD, the identification of synaptic plasma membrane GAPDH may be of special physiological significance. Synaptic GAPDH is involved as a nucleotide transporter (Schläfer et al., 1994); for ATP synthesis (Wu et al., 1997; Ikemoto et al., 2003); is complexed with F-actin (Rogalski-Wilk and Cohen, 1997); and acts as a $GABA_A$ receptor kinase (Laschet et al., 2004). Accordingly, diminution of synaptic plasma membrane GAPDH through amyloid plaque or NFT formation could have a serious effect on normal neuronal function.

The question then arises as to the mechanism through which fibril formation proceeds. A study by Zhao et al. (2004) suggests that the latter may arise through the action of phosphatidylserine-containing membranes. The latter is an acidic phospholipid. In these studies, using a liposome paradigm, a number of proteins, including GAPDH, could form fibers as detected by phase contrast microscopy. However, membranes containing phosphatidylserine were required. It was hypothesized that those membranes created an acidic environment that facilitated fibrillary formation. A further hypothesis was that such an environment may be required for amyloid formation by β-AP.

The interrelationship between this study and the role of GAPDH–β-AP protein–protein interactions in the etiology of AD relates to the demonstration that GAPDH contains a phosphatidylserine-binding site at amino acid residues 70–94 located within its NAD+-binding site (Kaneda et al., 1997; Nakagawa et al., 2003). The site contains several basic amino acids that would be likely candidates for binding to the phosphatidylserine-containing acidic membrane. As such, these combined studies suggest a mechanism through which GAPDH binding to the β-AP could facilitate amyloid fibril formation.

Further analysis indicated that moonlighting GAPDH could undergo significant structural changes as a consequence of its binding to phosphatidylserine-containing acidic membranes (Cortez et al., 2010). In this investigation, infrared spectroscopic analysis was used to monitor both GAPDH structure and the kinetics of fibril formation. With respect to the latter, following a lag of 10 min, there was a consistent increase in protein aggregation. With respect to the former, it appeared that there was a coordinate increase in GAPDH tetramer dissociation due to protein unfolding that was consistent with the progressive loss of GAPDH enzymatic activity.

Conformational changes in GAPDH itself may facilitate its association into amyloid structures. In particular, recent evidence indicates not only that GAPDH may exist in native and nonnative states but also that antibodies may be raised, which are directed against the nonnative protein (Grigorieva et al., 1999). The latter were used to probe differences in their interactions with the β-amyloid protein (Naletova et al., 2008). In those studies, the latter antibodies were reacted with sensor chips coated with nonnative or native GAPDH. Immunoreactivity was observed only with the nonnative GAPDH bound chip. Subsequently, binding of β-amyloid protein to the nonnative and native GAPDH bound chips was examined. It also bound only to the nonnative GAPDH chip. These studies suggest that GAPDH denaturation may be a contributory factor in the formation of GAPDH–β-amyloid protein complexes.

GAPDH aggregates may also influence the formation of β-amyloid structures. The former, produced as a consequence of oxidative stress, may be involved in GAPDH-mediated apoptosis (Nakajima et al., 2007, 2009a,b). Analysis in vitro and in vivo suggests their role in the kinetics of β-amyloid amyloidogenesis (Itakura et al., 2015). Changes in fluorescence analysis, in

Congo red binding, and in atomic force microscopy imaging were used as indicators of amyloidogenesis when Aβ40 was incubated in vitro with GAPDH aggregates.

The physiological significance of GAPDH aggregate–induced β-amyloid amyloidogenesis was demonstrated both in PC-12 cells in culture and in mice in vivo. PC-12 cells treated with $50\,\mu M$ Aβ exhibited a decrease in cell viability of 28%. Coincubation with GADPH aggregates increased this to 58% and to 78% (1% or 10% GAPDH aggregates, $P<.05$ or $P<.01$, respectively). Treatment with the GAPDH aggregates by themselves had no effect. In addition, PC-12 cells incubated with both Aβ and with the 10% GAPDH aggregate exhibited disruption of mitochondrial membrane potential as defined by increased rhodamine fluorescence.

Similarly, GAPDH aggregates enhanced Aβ40-induced neuronal toxicity in vivo. In these experiments, mice were injected with Aβ40 as well as a solution of Aβ40 + 10% GAPDH aggregates. In the coinjected mice, the number of hippocampal cells with pyknotic nuclei increased as did astrocyte accumulation. Further, immunocytochemical analysis indicated the accumulation of cytoplasmic apoptotic inducing factor and of cytochrome c, indicating mitochondrial dysfunction. Cumulatively, these findings suggest that GAPDH aggregation as a consequence of oxidative stress may provide a pathway for the latter to influence β-amyloid amyloidogenesis (Butterfield et al., 2010).

## 2.2 Role of Moonlighting GAPDH in Parkinson's Disease: Lewy Bodies and α-Synuclein

As with AD, recent studies demonstrate the association of GAPDH with Lewy bodies, the pathological structure characteristic of PD. Initially, immunocytochemical analysis of patient brain samples, as compared to age-matched controls, localized GAPDH within Lewy bodies (Tatton, 2000). However, this did not appear to a singular alteration in GAPDH distribution. In PD patients, there was an increase in caspase 3 and Bax immunoreactivity, along with GAPDH. Further, there was an increase in GAPDH nuclear localization in PD patients that corresponded to the frequency of apoptotic nuclei as compared to controls.

In a second study, the interrelationship between GAPDH and α-synuclein expression (and the consequences thereof) was examined in COS-7 cells in culture and in PD patient samples (Tsuchiya et al., 2005). COS-7 cells were transfected with a series on constructs including full- and partial-length GAPDH (C66 that contains the C-terminal amino acids 268–333), wt α-synuclein and β-synuclein.

Immunocytochemical analysis demonstrated the expression of wt α-synuclein, the α-synuclein mutant, or β-synuclein resulted solely in their cytoplasmic localization. Similarly, cotransfection wt α-synuclein with full-length GAPDH also resulted in a predominantly cytoplasmic distribution albeit with some perinuclear localization. However, in the cotransfected cells, the presence of Lewy body–like structures (termed LB-like) was observed. No such inclusions were observed on cotransfection of α-synuclein with the C66 GAPDH construct. Similarly, no such inclusions were detected on cotransfection of full-length GAPDH and β-synuclein.

Subsequent coimmunoprecipitation studies demonstrated the physical association of GAPDH and α-synuclein. In these studies, antibodies to both proteins were used in the coimmunoprecipitation protocol followed by immunoblot analysis with the other antibody. The potential physiological significance of the cell culture studies was indicated by immunohistochemical analysis of PD brain tissue. Colocalization of both α-synuclein and GAPDH was detected in Lewy bodies.

In a third study, the interrelationship between the binding of tubulin polymerization promoting protein (TPPP/p25) and GAPDH in PD was examined (Oláh et al., 2006). As with many moonlighting GAPDH studies, the major focus was on the determination of TPPP/p25 binding partners. The interest on PD relates to a previous study demonstrating the possibility that TPPP/p25 may be a biomarker for α-synuclein pathologies (Kovács et al., 2004).

Initially, two standard protocols were used to determine TPPP/p25 interacting proteins, i.e., coimmunoprecipitation analysis and affinity chromatography. Bovine brain extracts were used as the experimental paradigm. In the former, a 35-kDa protein was detected following SDS-PAGE of the immunoprecipitate. Mass spectrum analysis identified it as GAPDH. With respect to tubulin, prior to affinity chromatography did not alter GAPDH binding. However, binding could be inhibited in a concentration-dependent manner by $NAD^+$, indicating that the TPPP/p25 GAPDH-binding site was located in its N-terminal region. This finding is in accord with the observation that most moonlighting GAPDH recognition sites are located in its $NAD^+$ domain, thereby freeing its catalytic site for the new activity (Sirover, 2014).

Subsequently, cell culture and patient studies were performed to test the potential physiological significance of these in vitro analyses. With respect to the former, HeLa cells were transfected with an EGFP-TPPP/p25 construct. Its subcellular localization was determined by fluorescence. GAPDH was defined by immunostaining. A dual effect was observed. At low EGFP-TPPP/p25 construct expression, a microtubular localization was noted, consistent with its tubulin binding properties. No GAPDH colocalization was detected. In contrast, at high construct expression resulted in an alteration of microtubular structure, the appearance of a protein aggregate and colocalization with GAPDH within the protein aggregate.

With respect to the latter, colocalization of TPPP/p25, GAPDH, and α-synuclein was examined in human brain tissue. With respect to TPPP/p25 and α-synuclein, consistent colocalization was observed in human brain tissue while GAPDH was restricted to the rim region. The latter finding appears similar to the finding of Tsuchiya et al. (2005) that GAPDH was located in the halo zone of Lewy bodies.

## 3. FUNCTIONAL CONSEQUENCES OF GAPDH PROTEIN–PROTEIN INTERACTIONS IN AGE-RELATED NEURODEGENERATIVE DISEASE-I: DETERMINATION OF GAPDH GLYCOLYTIC ACTIVITY IN VIVO

The studies described above provide the requisite evidence indicating the binding of GAPDH to neurodegenerative proteins. Accordingly, further studies are required to demonstrate their physiological significance, i.e., in vivo, are there changes in GAPDH activity that are characteristic of these age-related neurodegenerative disorders?

In an initial study, brain GAPDH activity was determined in patients with CAG repeat disorders as well as in AD patients as compared to controls (Kish et al., 1998). Crude cell homogenates were used. These studies indicated that there was no difference in catalysis in samples from spinocerebellar ataxia-1 (SCA-1), SCA-2, or SCA-3 patients. In contrast, there appeared to be slight reductions in GAPDH activity (c.12%, $P < .001$) in the caudate nucleus from HD patients and in the temporal cortex from patients with AD(c.19%, $P < .02$). These results were the average of an $n = 10$ for both HD and AD as compared to an $n = 10$ for control.

In contrast, in a second study, no difference was observed in GAPDH activity in any brain area that was examined (Tabrizi et al., 1999). Four brain areas were examined (caudate, putamen, cortex, and cerebellum). The results were the average of an $n = 5$ to an $n = 15$. However, differences were detected in aconitase and in enzymes that comprised the mitochondrial respiratory chain. The latter two activities were also examined in patient fibroblasts. Presumably due to the lack of brain GAPDH activity alteration, it was not apparently examined.

In a third study, cerebral GAPDH activity was determined in four different animal models of AD (Shalova et al., 2007). These included a transgenic mouse strain that develops amyloid plaques at 10–12 months of age as well as rats injected intracerebrally with β-amyloid 1–42, treated with DL-thirophan that reduces Aβ catabolism; and rats treated with inflammatory lipopolysaccharides, which results in brain Aβ accumulation.

In the transgenic mouse model, GAPDH activity was reduced by 25%–30% even though there was no decrease in immunoreactive protein. Diminution of GAPDH activity was observed in both the cortex and in the cerebellum ($P < .05$). Decreased GAPDH activity was also detected in the cortex and hippocampus of animals treated with either thirophan or with LPS ($P < .05$, respectively). No decrease in GAPDH immunoreactivity was observed. However, it was noted that the enzymatic results were not always statistically significant. In addition, no decrease in catalysis was observed in rats that were injected with β-amyloid 1–42.

## 4. FUNCTIONAL CONSEQUENCES OF GAPDH PROTEIN–PROTEIN INTERACTIONS IN AGE-RELATED NEURODEGENERATIVE DISEASE-II: FORMATION IN VIVO OF GAPDH–NEURODEGENERATIVE PROTEIN COMPLEXES

The results described above indicate that there may, or may not, be a considerable decrease in GAPDH activity in AD and in HD in spite of the demonstrative binding of GAPDH to either the β-amyloid protein (β-APP or β-AP) or huntingtin. That being said, one of the demonstrative features of moonlighting GAPDH is the interrelationship between its subcellular localization and its functional diversity (reviewed in Sirover, 2005, 2012; Tristan et al., 2011).

Accordingly, intracellular GAPDH was determined in AD and in HD fibroblasts as the experimental paradigm. Twelve different cell strains from separate individuals were used to provide an $n = 4$ for AD, HD, and for age-matched controls. Three samples were examined in 12 different cell strains: whole-cell homogenates; the nuclear and postnuclear fractions. The latter two were obtained by low-speed centrifugation of the whole-cell homogenate to pellet the nuclei while the supernatant was designated as the postnuclear sample (supernatant) that contained the cytoplasm, cytosolic organelles, and membrane.

As indicated in Table 12.2, analysis of whole-cell homogenates demonstrated that no difference was observed in those cell extracts from AD and HD cell strains as compared to the age-matched controls (Mazzola and Sirover, 2001). In contrast, different findings were observed for subcellular GAPDH in both AD and HD cells. In AD cells, postnuclear and nuclear GAPDH activities were reduced as compared with those observed in the age-matched controls ($P = .002$ and $.045$, respectively). Immunoblot analysis demonstrated no difference in immunoreactive GAPDH. In contrast, in HD cells, no difference was observed in HD postnuclear GAPDH activity as compared with the activity observed in the postnuclear age-matched control. However, a statistical difference ($P = .025$) was observed in the HD postnuclear fraction. Again, immunoblot analysis demonstrated no difference in subcellular HD GAPDH protein levels as compared with the age-matched controls.

Although these findings demonstrate a reduction in GAPDH activity, there was no indication as to the mechanism that resulted in the diminution of catalytic activity. As previous studies indicated the physical interaction between GAPDH and neurodegenerative proteins, it seemed reasonable to suggest that the subcellular reduction in catalysis may be due to the formation of a high molecular weight (HMW) protein complex.

For that reason, the formation of such an HMW protein complex was investigated in HD nuclei using a glycerol gradient sedimentation protocol previously developed to analyze GAPDH–anti-GAPDH monoclonal antibody complex formation (Arenaz and Sirover, 1983). Using that protocol,

**TABLE 12.2** Subcellular GAPDH Activity in Alzheimer's Disease and in Huntington's Disease Cells[a]

| Pathology | GAPDH Activity (Whole-Cell Sonicate) | Subcellular GAPDH Activity | Determination of High Molecular Weight Species | References |
|---|---|---|---|---|
| Alzheimer's disease(AD), Huntington's disease(HD) | No difference in GAPDH activity in AD or HD cells | Reduced activity in AD postnuclear and nuclear fractions; reduced activity in HD nuclei only | Not determined | Mazzola and Sirover (2001) |
| HD | No difference in GAPDH activity in HD cells | Reduced activity in HD nuclei | High molecular weight (HMW) nuclear GAPDH species | Mazzola and Sirover (2002) |
| AD | No difference in GAPDH activity in AD cells | Reduced activity in AD postnuclear fraction | HMW postnuclear GAPDH species; HMW species absent in whole cell sonicate | Mazzola and Sirover (2003) |
| AD | No difference in GAPDH activity in AD cells | Not determined | HMW GAPDH and β-APP postnuclear species | Mazzola and Sirover (2004) |

[a]As compared to age-matched controls.

the distribution of GAPDH in HD and age-matched control nuclei was determined by enzyme activity and by immunoblot analysis (Mazzola and Sirover, 2002a,b). In the latter, there was coordinate sedimentation of catalytic and immunoreactive GAPDH. In contrast, in HD nuclei, immunoreactive GAPDH was detected sedimenting at a higher density. In those fractions, no GAPDH enzymatic activity was observed. This was the first in vivo demonstration of a potential GAPDH–huntingtin complex. The amount of HMW immunoreactive GAPDH appeared to be approximately equal to the reduction of enzyme activity observed in HD nuclei.

As GAPDH activity was decreased similarly in the AD postnuclear fraction, a similar examination was performed to determine whether an HMW GAPDH species could be detected in that subcellular fraction (Mazzola and Sirover, 2003). As with the age-matched control in the HD study, no HMW species was observed when the postnuclear control was analyzed by glycerol gradient sedimentation, i.e., GAPDH glycolytic activity and GAPDH immunoreactivity cosedimented. In contrast, analysis of AD postnuclear GAPDH demonstrated the formation of an immunoreactive HMW GAPDH species that did not exhibit any GAPDH enzyme activity. This suggested the formation of an HMW GAPDH protein complex.

As a further control, the sedimentation of GAPDH using AD whole-cell preparations was examined. In this gradient, cosedimentation of GAPDH enzyme activity and its immunoreactivity were observed. These findings suggest that the HMW GAPDH protein complex observed in the AD postnuclear fraction was dissociated by the extensive sonication protocols intrinsic to the whole-cell preparation (except when it was used for nuclear and postnuclear sample preparation).

As the observation of an HMW GAPDH species in the AD postnuclear sample suggested the formation of a GAPDH protein complex, the sedimentation of the β-APP was examined (Mazzola and Sirover, 2004). In the age-matched postnuclear control, immunoreactive β-APP was readily detected sedimenting at its individual molecular weight position. In contrast, in the AD postnuclear sample, an HMW β-APP species was observed. Using the identical age-matched and AD postnuclear samples, the position of immunoreactive GAPDH was determined. In the former, no HMW GAPDH species was detected. In contrast, in the AD postnuclear sample, HMW GAPDH was observed, sedimenting at the identical position at which the HMW β-APP was detected. Accordingly, these cumulative studies in AD and HD cells not only demonstrate specific subcellular reductions in GAPDH activity but also indicate those reductions in catalysis are due to the formation of a postnuclear AD HMW GAPDH–β-APP and a nuclear HD GAPDH–huntingtin protein complex, respectively.

## 5. FUNCTIONAL CONSEQUENCES OF GAPDH PROTEIN–PROTEIN INTERACTIONS IN AGE-RELATED NEURODEGENERATIVE DISEASE-III: ROLE OF GAPDH: NEURODEGENERATIVE PROTEIN COMPLEXES IN HUNTINGTIN TOXICITY

As discussed above not only does GAPDH form a protein complex with huntingtin but also examination of GAPDH structure in HD cells indicates the formation of an HMW GAPDH–huntingtin complex. The question then arises as to whether there are any functional consequences to the formation of that complex.

### 5.1 Role of Intracellular GAPDH–Huntingtin Complexes in Huntington's Disease Neuronal Toxicity

Recent studies demonstrate the role of expanded CAG repeats in the pathology of HD. In particular, repeats of <35 do not result in overt toxicity. In contrast, repeats >35 appear to be the "single stone which causes the avalanche" (Gusella et al., 1996). The latter have been termed mutant huntingtin (mHtt).

A series of studies indicated that mHtt translocates to the nucleus and induces both apoptosis and the formation of intranuclear inclusions. Recent evidence suggests that the former rather than the latter is responsible for cell death (Saudou et al., 1998). Further, nuclear localization of N-terminal huntingtin fragments (which contain the polyglutamine repeat) increases cell loss (Peters et al., 1999). Subcellular distribution of caspase-cleaved mHtt indicates a cytoplasmic→perinuclear→nuclear transition of the N-terminal fragment (Sawa et al., 2005). GAPDH was first implicated in nuclear mHtt function in an HD transgenic mouse model (Senatorov et al., 2003). In that investigation, immunocytochemical analysis was performed in a series of brain sections. Those studies indicated that GAPDH was localized in neurons with a specific subcellular nuclear distribution.

A critical study interrelating GAPDH–huntingtin complexes to HD toxicity demonstrated not only that mHtt formed a ternary complex with GAPDH and Siah1 but also that the latter was responsible for mHtt cytotoxicity (Bae et al., 2006). In many of these studies, Siah1ΔRING and Siah1ΔNLS were used as the former domain catalyzes Siah1 self-degradation while the latter is responsible for nuclear translocation. Salient findings include the observations that GAPDH binds in vitro to an N-terminal Htt fragment containing 23 polyglutamine domain that is noncytotoxic; coimmunoprecipitation studies indicated that Htt, GAPDH, and Siah1 form a ternary complex; Siah1 levels are not only higher in cell lysates that contain mHtt (148 polygln) as compared to Htt but also coimmunoprecipitate to a greater degree with the former as compared to the latter; cells overexpressing Siah1 exhibit increased mHtt nuclear localization; cells transfected with Htt augmented with GAPDH are not cytotoxic, but cells overexpressing mHtt and Siah1 exhibit increased cell death; depletion of GAPDH

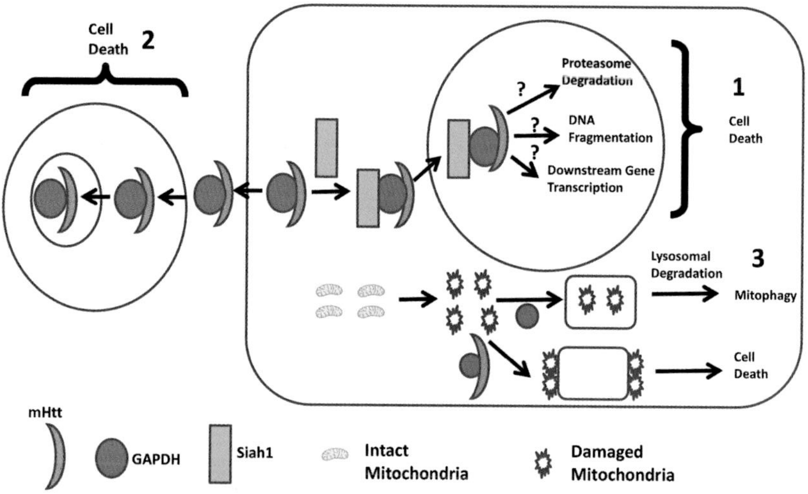

**FIGURE 12.1**    The role of GAPDH in Huntington's disease.

or Siah1 reduces levels of nuclear inclusions while increasing their perinuclear localization. In toto, these findings suggest that GAPDH and Siah1 facilitate the nuclear translocation of mHtt thereby increasing cytotoxicity by as a yet unknown mechanism. Accordingly, although previous studies indicated apoptosis as such a mechanism, this uncertainty is illustrated in Fig. 12.1 (labeled **1**).

## 5.2  Role of Extracellular GAPDH–Huntingtin Complexes in Huntington's Disease Neuronal Toxicity

The studies described above consider a role for mHtt within the cell. As described previously in Chapters 3 and 7, moonlighting GAPDH is capable of extracellular actions in both iron metabolism and the control of cell proliferation, respectively. The role of extracellular GAPDH will also be considered in Chapter 13 with respect to its role in infection and immunity. The "take-home" message is that extracellular functions of GAPDH are not unusual.

With that in mind, a recent study suggests a new function for GAPDH–mHtt complexes in the etiology of HD. In this investigation, a novel method was used to examine the cytotoxicity of such complexes to adjacent neuronal cells (Mikhaylova et al., 2016). For that purpose, mHtt (103 polygln) under the control of an ecdysone-regulated promoter was overexpressed in rat pheochromocytoma PC-12 cells (termed PC-12HttQ103). Expression was induced by ponasterone A (PonA), an ecdysone analogue.

In the transfected cells, subsequent to PonA addition, there was not only a time-dependent increase in cell death (MTT and LDH assay) but also an increase in the release of both GAPDH and mHtt poly Q into the cell medium. The latter was detected by a two-site ELISA following ultrafiltration used as a filter trap assay. Each increased in an identical manner that was coordinate with the increase in cell death, i.e., immunoreactivity for each was observed 9 h after PonA addition and continued to increase throughout the 48-h experimental period. In contrast, PC-12 control cells displayed only a slight sensitivity to PonA and did not secrete GAPDH extracellularly.

The toxicity of the extracellular GAPDH–mHtt poly Q aggregate was examined by using conditioned media from PonA-stimulated PC-12HttQ103 cells versus that of the PC-12 control. Each was added to PC-12 control cells as well as PC-12HttQ103 cells not exposed to PonA. These studies demonstrated that significant cell death was observed in both cell lines exposed to the PC-12HttQ103 conditioned media. Neither cell line displayed significant cell death when exposed to the conditioned media from the PC-12 control cells. Further, centrifugation of the PC-12HttQ103 conditioned media removed the GAPDH–mHTT poly Q aggregate. The resultant supernatant failed to increase cell death in either cell line.

The mechanism through which the aggregate is transported into the recipient cell was determined using SK-N-SH cells to which purified GAPDH or purified Q58 was added. Uptake of GAPDH by itself was observed. In contrast, Q58 required the presence of GAPDH for its transport into the SK-N-SH cells. Inhibitor analysis demonstrated that clathrin-dependent endocytosis was required to uptake. This is consistent with previous studies demonstrating GAPDH-mediated endocytosis in mammalian cells (Robbins et al., 1995). In toto, as illustrated in Fig. 12.1 (labeled **2**), these studies suggest another mechanism through which GAPDH–mHtt binding may result in HD-specific cell toxicity.

## 5.3 Role of GAPDH–Huntingtin Complexes in Mitophagy Dysregulation

As previously described in Chapter 10, GAPDH fulfills a necessary role in mitophagy, the process by which damaged mitochondria may be removed by lysosomal degradation thereby facilitating cell survival. Further, as defined in that chapter, cell death may occur as a consequence of interference with mitophagy-associated GAPDH function (Yogalingam et al., 2013).

Recent studies suggest that such cell death may be a consequence of GAPDH–mHtt binding (Hwang et al., 2015). In these investigations, three experimental paradigms were utilized: PC-12 cells expressing either an N-terminal Htt containing 23 polygln residues or expressing a construct containing 74 polygln residues; striatal cell lines isolated from HD knock-in mice containing 7 or 111 polygln repeats; and HD patient–derived fibroblasts (used with normal controls).

Morphological analysis demonstrated significant structural differences in cells containing mHtt > 35 polygln as compared to those with Htt < 35 polygln. The latter displayed normal mitochondrial structure, while, in the former, the mitochondria were small and swollen. In cells with Htt < 35, the mitochondria could be observed localized in lysosomal vacuoles, while in cells with mHtt > 35, such vacuoles were noticeably empty. In the latter instance, mitochondria were observed around the lysosomal vacuoles. Further, in cells with mHtt > 35, the level of a number of mitochondrial proteins was increased in cell extracts.

The mitophagic role of GAPDH–mHtt was analyzed by a number of criteria. In cells with mHtt > 35 polygln, GAPDH coimmunoprecipitated with mitochondrial marker proteins. Further, as compared to cells with Htt < 35 polygln, its mitochondrial associated levels increased, its mitochondrial activity decreased, and its mitochondrial level of oxidation increased. Coimmunoprecipitation analysis in cells treated with 1% formaldehyde to induce protein cross-links also demonstrated that GAPDH associated only with mHtt > 35 polygln. Submitochondrial localization analysis indicated that both GAPDH and mHtt were observed at the outer mitochondrial membrane. Overexpression of GAPDH appeared to restore mitophagy in cells with mHtt > 35 polygln. Accordingly, it was suggested that the mitochondrial formation of the GAPDH–mHtt protein complex disrupted mitophagy-associated GAPDH function (Fig. 12.1, labeled **3**). Thus, cumulatively, there appear to be three different mechanisms through which the formation of a GAPDH–mHtt protein complex can contribute to the pathology of HD. That being said, the relative contributions of each remain to be defined.

## 6. SUMMARY

The studies presented in this chapter highlight the structural and functional roles of moonlighting GAPDH protein–neurodegenerative protein complexes. As compared to Chapter 9, in which a single GAPDH posttranslational modification induced pleiotropic cellular consequences, these studies demonstrate many different protein interactions of GAPDH and relate their potential role in the etiology of age-related neurodegenerative disorders. Further, they demonstrate the contribution of several different GAPDH structures (native, denatured, and aggregate) to the respective pathologies. In particular, the latter relate to their role in β-amyloid amyloidogenesis and subsequent plaque formation. The role of dynamic changes in GAPDH–neuronal protein complex intercellular and intracellular localizations is indicated by the studies in HD. The latter elucidated three different potential mechanisms through which GAPDH–neuronal protein complexes may mediate cellular neurotoxicity. In toto, these studies provide the requisite evidence for a complex role of moonlighting GAPDH in this aspect of cell pathology.

## REFERENCES

Allen, M., Cox, C., Belbin, O., et al., 2012. Association and heterogeneity at the GAPDH locus in Alzheimer's disease. Neurobiol. Aging 33, 203e25–203e33.

Arenaz, P., Sirover, M., 1983. Isolation and characterization of monoclonal antibodies directed against the DNA repair enzyme uracil DNA glycosylase from human placenta. Proc. Natl. Acad. Sci. U.S.A. 80, 5822–5826.

Bae, B.-I., Hara, M., Cascio, M., ct al., 2006. Mutant huntingtin: nuclear translocation and cytotoxicity mediated by GAPDH. Proc. Natl. Acad. Sci. U.S.A. 103, 3405–3409.

Burke, J., Enghild, J., Martin, M., et al., 1996. Huntingtin and DRPLA proteins selectively interact with the enzyme GAPDH. Nat. Med. 2, 347–350.

Butterfield, D., Hardas, S., Bader Lange, M., 2010. Oxidatively modified glyceraldehyde-3-phosphate dehydrogenase (GAPDH) and Alzheimer's disease: many pathways to neurodegeneration. J. Alzheimers Dis. 20, 369–393.

Cortez, L., Ávila, C., Torres Bugeau, C., et al., 2010. Glyceraldehyde-3-phosphate dehydrogenase tetramer dissociation and amyloid fibril formation induced by negatively charged molecules. FEBS Lett. 584, 625–630.

Grigorieva, J., Dainiak, M., Katrukha, A., Muronetz, V., 1999. Antibodies to the nonnative forms of D-glyceraldehyde-3-phosphate dehydrogenase: identification, purification, and influence on the renaturation of the enzyme. Arch. Biochem. Biophys. 369, 252–260.

Gusella, J., McNeil, S., Perischetti, F., et al., 1996. Huntington's disease. Cold Spring Harb. Symp. Quant. Biol. 61, 615–626.

Hwang, S., Disatnik, M.-H., Mochly-Rosen, D., 2015. Impaired GAPDH-induced mitophagy contributes to the pathology of Huntington's disease. EMBO Mol. Med. 7, 1307–1326.

Ikemoto, A., Bole, D., Ueda, T., 2003. Glycolysis and glutamate accumulation into synaptic vesicles: role of glyceraldehyde-3-phosphate dehydrogenase and 3-phosphoglycerate kinase. J. Biol. Chem. 278, 5929–5940.

Itakura, M., Nakahima, H., Kubo, T., et al., 2015. Glyceraldehyde-3-phosphate dehydrogenase aggregates accelerate amyloid-β amyloidogenesis in Alzheimer disease. J. Biol. Chem. 290, 26072–26087.

Kaneda, M., Takeuchi, K., Inoue, K., Umeda, M., 1997. Localization of the phosphatidylserine-binding site of glyceraldehyde-3-phosphate dehydrogenase responsible for membrane fusion. J. Biochem. 122, 1233–1240.

Kish, Lopes-Cendes, I., Guttman, M., et al., 1998. Brain glyceraldehyde-3-phosphate dehydrogenase activity in human trinucleotide repeat disorders. Arch. Neurol. 55, 1299–1304.

Koshy, B., Matilla, T., Burright, E., et al., 1996. Spinocerebellar ataxia type-1 and spinobulbar muscular atrophy gene products interact with glyceraldehyde-3-phosphate dehydrogenase. Hum. Mol. Genet. 5, 1311–1318.

Kovács, G., Lászlo, L., Kovács, J., 2004. Natively unfolded tubulin polymerization promoting protein TPPP/p25 in a common marker of alpha-synucleinopathies. Neurobiol. Dis. 17, 155–162.

Laschet, J., Minier, F., Kurcewicz, I., et al., 2004. Glyceraldehyde-3-phosphate dehydrogenase is a GABA$_A$ receptor kinase linking glycolysis to neuronal inhibition. J. Neurosci. 24, 7614–7622.

Li, Y., Nowotny, P., Holmans, P., et al., 2004. Association of late-onset Alzheimer's disease with genetic variation in multiple members of the GAPD gene family. Proc. Natl. Acad. Sci. U.S.A. 101, 15688–15693.

Lin, P., Martin, E., Bronson, P., et al., 2006. Exploring the association of glyceraldehyde-3-phosphate dehydrogenase and Alzheimer disease. Neurology 97, 64–68.

Mazzola, J., Sirover, M., 2001. Reduction of glyceraldehyde-3-phosphate dehydrogenase activity in Alzheimer's disease and in Huntington's disease fibroblasts. J. Neurochem. 76, 442–449.

Mazzola, J., Sirover, M., 2002a. Alteration of nuclear glyceraldehyde-3-phosphate dehydrogenase activity in Huntington's disease fibroblasts. Mol. Brain Res. 100, 95–101.

Mazzola, J., Sirover, M., 2002b. Alteration of intracellular structure and function of glyceraldehyde-3-phosphate dehydrogenase: a common phenotype of neurodegenerative disorders? Neurotoxicology 23, 603–609.

Mazzola, J., Sirover, M., 2003. Subcellular alteration of glyceraldehyde-3-phosphate dehydrogenase in Alzheimer's disease fibroblasts. J. Neurosci. Res. 71, 279–285.

Mazzola, J., Sirover, M., 2004. Subcellular analysis of aberrant protein structure in age-related neurodegenerative disorders. J. Neurosci. Methods 137, 241–246.

Mikhaylova, E., Lazarev, V., Nikotina, A., Margulis, B., Guzhova, I., 2016. Glyceraldehyde-3-phosphate dehydrogenase augments the intercellular transmission and toxicity of polyglutamine aggregates in a cell model of Huntington disease. J. Neurochem. 136, 1052–1063.

Nakajima, H., Amano, W., Fujita, A., et al., 2007. The active site cysteine of the proapoptotic protein glyceraldehyde-3-phosphate dehydrogenase is essential in oxidative stress-induced aggregation and cell death. J. Biol. Chem. 282, 26562–26574.

Nakajima, H., Amano, W., Fukuhara, A., et al., 2009a. An aggregate-prone mutant of human glyceraldehyde-3-phosphate dehydrogenase augments oxidative stress-induced cell death in SH-SY5Y cells. Biochem. Biophys. Res. Commun. 390, 1066–1071.

Nakajima, H., Amano, W., Kubo, T., et al., 2009b. Glyceraldehyde-3-phosphate dehydrogenase aggregate formation participates in oxidative stress-induced cell death. J. Biol. Chem. 284, 34331–34341.

Nakagawa, T., Hirano, Y., Inomata, A., et al., 2003. Participation of a fusogenic protein glyceraldehyde-3-phosphate dehydrogenase, in nuclear membrane assembly. J. Biol. Chem. 278, 20395–20404.

Naletova, I., Schmalhausen, E., Kharitonov, A., et al., 2008. Nonnative glyceraldehyde-3-phosphate dehydrogenase can be an intrinsic component of amyloid structures. Biochim. Biophys. Acta 1784, 2052–2058.

Oláh, J., Tókési, N., Vincze, O., et al., 2006. Interaction of TPPP/p25 protein with glyceraldehyde-3 phosphate dehydrogenase and their colocalization in Lewy bodies. FEBS Lett. 580, 5807–5814.

Oyama, R., Yamamoto, H., Titani, K., 2000. Glutamine synthetase, hemoglobin α-chain, and macrophage migration inhibition factor binding to amyloid β-protein: their identification in rat brain by a novel affinity chromatography and in Alzheimer's disease brain by immunoprecipitation. Biochim. Biophys. Acta 1479, 91–102.

Peters, M., Nucifora Jr., F., Kushi, J., et al., 1999. Nuclear targeting of mutant Huntington increases toxicity. Mol. Cell. Neurosci. 14, 121–128.

Praticò, D., Mecocci, P. (Eds.), 2013. Studies on Alzheimer's Disease. Humana Press, New York.

Robbins, A., Ward, R., Oliver, C., 1995. A mutation in glyceraldehyde-3-phosphate dehydrogenase alters endocytosis in CHO cells. J. Cell Biol. 130, 1093–1104.

Rogalski-Wilk, A., Cohen, R., 1997. Glyceraldehyde-3-phosphate dehydrogenase activity and F-actin associations in synaptosomes and postsynaptic densities of porcine cerebral cortex. Cell. Mol. Neurobiol. 17, 51–70.

Saudou, F., Finkbeiner, S., Devys, D., Greenberg, M., 1998. Huntingtin acts in the nucleus to induce apoptosis but death does not correlate with the formation of intranuclear inclusions. Cell 95, 55–66.

Sawa, A., Khan, A., Hester, L., Snyder, S., 1997. Glyceraldehyde-3-phosphate dehydrogenase: nuclear translocation participates in neuronal and nonneural cell death. Proc. Natl. Acad. Sci. U.S.A. 94, 11669–11674.

Sawa, A., Nagata, E., Sutcliffe, S., et al., 2005. Huntingtin is cleaved by caspases in the cytoplasm and translocated to the nucleus via perinuclear sites in Huntingtin's disease patient lymphoblasts. Neurobiol. Dis. 20, 267–274.

Schläfer, M., Volknandt, W., Zimmermann, H., 1994. Putative synaptic vesicle nucleotide transporter identified as glyceraldehyde-3-phosphate dehydrogenase. J. Neurochem. 63, 1924–1931.

Schulze, H., Schuyler, A., Stüber, D., et al., 1993. Rat brain glyceraldehyde-3-phosphate dehydrogenase interacts with the recombinant cytoplasmic domain of Alzheimer's β-amyloid precursor protein. J. Neurochem. 60, 1915–1922.

Senatorov, V., Charles, V., Reddy, P., Tagle, D., Chuang, D.-M., 2003. Overexpression and nuclear accumulation of brain glyceraldehyde-3-phosphate dehydrogenase in a transgenic mouse model of Huntington's disease. Mol. Cell. Neurosci. 22, 285–297.

Shalova, I., Cechalova, K., Rehakova, Z., et al., 2007. Decrease of dehydrogenase activity of cerebral glyceraldehyde-3-phosphate dehydrogenase in different animal models of Alzheimer's disease. Biochim. Biophys. Acta 1770, 826–832.

Sirover, M., 2005. New nuclear functions of the glycolytic protein, glyceraldehyde-3-phosphate dehydrogenase, in mammalian cells. J. Cell. Biochem. 95, 45–52.

Sirover, M., 2012. Subcellular dynamics of multifunctional protein regulation: mechanisms of GAPDH intracellular translocation. J. Cell. Biochem. 113, 2193–2200.

Sirover, M., 2014. Structural analysis of glyceraldehyde-3-phosphate dehydrogenase functional diversity. Int. J. Biochem. Cell Biol. 57, 20–26.

Sunaga, K., Takahashi, H., Chuang, D.-M., Ishitani, R., 1995. Glyceraldehyde-3-phosphate dehydrogenase is over-expressed during apoptotic death of neuronal cultures and is recognized by a monoclonal antibody against amyloid plaques from Alzheimer's brain. Neurosci. Lett. 200, 133–136.

Tabrizi, S., Cleeter, M., Suereb, J., et al., 1999. Biochemical abnormalities and excitotoxicity in Huntington's brain. Ann. Neurol. 45, 25–32.

Tatton, N., 2000. Increased caspase 3 and Bax immunoreactivity accompany nuclear GAPDH translocation and neuronal apoptosis in Parkinson's disease. Exp. Neurol. 166, 29–43.

Tristan, C., Shahani, N., Sedlak, T., Sawa, A., 2011. The diverse functions of GAPDH: views from different subcellular compartments. Cell. Signal. 23, 317–323.

Tsuchiya, K., Tajima, H., Kuwae, T., et al., 2005. Pro-apoptotic protein glyceraldehyde-3-phosphate dehydrogenase promotes the formation of Lewy body-like inclusions. Eur. J. Neurosci. 21, 317–326.

Verdier, Y., Földi, I., Sergeant, N., et al., 2008. Characterization of the interaction between Aβ 1-42 and glyceraldehyde-3-phosphate dehydrogenase. J. Pept. Sci. 14, 755–762.

Verdier, Y., Huszár, E., Penke, B., et al., 2005. Identification of synaptic plasma membrane proteins co-precipitated with fibrillary β-amyloid peptide. J. Neurochem. 94, 617–628.

Wang, Q., Woltjer, R., Uhinuo, F., et al., 2005. Proteomic analysis of neurofibrillary tangles in Alzheimer disease identifies GAPDH as a detergent-insoluble paired helical filament tau binding protein. FASEB J. 19, 869–871.

Wu, K., Aoki, C., Elste, A., et al., 1997. The synthesis of ATP by glycolytic enzymes in the postsynaptic density and the effect of endogenously generated nitric oxide. Proc. Natl. Acad. Sci. U.S.A. 94, 13273–13278.

Yazawa, I., Nukina, N., Hashida, H., et al., 1995. Abnormal gene product identified in hereditary dentatorubral-pallidoluysian atrophy. Nat. Genet. 10 (99), 103.

Yogalingam, G., Hwang, S., Ferreira, J., Mochly-Rosen, D., 2013. Glyceraldehyde-3-phosphate dehydrogenase (GAPDH) phosphorylation by protein kinase Cδ (PKCδ) inhibits mitochondria elimination by lysosomal-like structures following ischemia and reoxygenation-induced injury. J. Biol. Chem. 288, 18947–18960.

Zhao, J., Tuominen, E., Kinnunen, P., 2004. Formation of amyloid fibers triggered by phosphatidylserine-containing membranes. Biochemistry 43, 10302–10307.

## FURTHER READING

Chuang, D.-M., Ishitani, R., 1996. A role for GAPDH in apoptosis and neurodegeneration. Nat. Med. 2, 609–610 (reply by Ross et al., 1996. Nat Med 2, 610).

Cooper, A., Rex, K.-F., Burke, J., et al., 1997. Transglutaminase-catalyzed inactivation of glyceraldehyde-3-phosphate dehydrogenase and a-ketoglutarate dehydrogenase complex by polyglutamine domains of pathological length. Proc. Natl. Acad. Sci. U.S.A. 94, 12604–12609.

Cooper, A., Sheu, K.-F.R., Burke, J., Strittmatter, W., Blass, J., 1998. Glyceraldehyde-3-phosphate dehydrogenase abnormality in metabolically stressed Huntington's disease fibroblasts. Dev. Neurosci. 20, 462–468.

Huang, J., Xiong, N., Chen, C., et al., 2011. Glyceraldehyde-3-phosphate dehydrogenase: activity inhibition and protein overexpression in rotenone models for Parkinson's disease. Neuroscience 192, 598–608.

Ruoppolo, M., Orrù, S., Francese, S., Caputo, I., Esposito, C., 2003. Structural characterization of transglutaminase-catalyzed cross-linking between glyceraldehyde-3-phosphate dehydrogenase and polyglutamine repeats. Protein Sci. 12, 170–179.

Shiozawa, M., Fukutani, Y., Arai, N., et al., 2003. Glyceraldehyde-3-phosphate dehydrogenase and endothelin-1 immunoreactivity is associated with cerebral white matter in dentatorubral-pallidoluysian atrophy. Neuropathology 23, 36–43.

# Chapter 13

# Functional Diversity of GAPDH in Infection and Immunity: The Complexity of "Simple" Organisms

*An ugly duckling can become a swan*

Hans Christian Anderson

The pathology of infectious disease may be divided into two main categories. The first category comprises those mechanisms through which an organism is able to invade successfully its target host and to propagate within that target. To accomplish these goals, infectious agents need to attach to target cells and need to obtain required nutrients or cofactors from their host environment. The second category comprises those mechanisms through which the invading organism seeks to evade successfully the defense mechanisms used by the host target to thwart this threat to its existence. Termed the immune response, such defense mechanisms function to target and to destroy these threats to cell viability. Accordingly, to survive, infectious agents need to develop their own countermeasures to vitiate those defense mechanisms.

The goal of this chapter is to define the significant role of moonlighting GAPDH not only in the progression of infectious disease but also in its role in the evasion of host defense mechanisms, which seek to thwart infection. With respect to the former, this includes the role of membrane GAPDH in the attachment of the infectious agent to the host cell; the intriguing use of secreted GAPDH to obtain, sequester, and transport host $Fe^{++}$ into the infectious agent, the novel use of parasitic GAPDH in different organs as part of their life cycle and the innovative role of bacterial GAPDH in the formation of intracellular protein replicative complexes. With respect to the latter, this includes GAPDH induction of apoptosis in host macrophages (with a potential interaction with bacterial- or host-derived nitric oxide) as well as its ability to affect the expression of genes involved in the immune response. In toto, these studies indicate not only the complex role of moonlighting GAPDH in infectious disease but also the multiplicity of infectious organisms, which utilize GAPDH functional diversity in their respective pathologies.

Glyceraldehyde-3-Phosphate Dehydrogenase (GAPDH). http://dx.doi.org/10.1016/B978-0-12-809852-3.00013-3

**221**

# 1. THE ROLE OF MOONLIGHTING GAPDH IN INFECTION

As indicated in the chapter title, in comparison to higher species, bacteria, and other comparable organisms may be thought of as "simple creatures" possessing only the basic properties needed to survive during their limited lifespan. Accordingly, it would be considered highly unlikely that GAPDH would be present in any locale other than the cytosol, given its role in energy production. Thus, the thought that they may exhibit a moonlighting GAPDH protein with a complex role in the pathology of infectious disease would be both unexpected and surprising. Yet, as indicated in Table 13.1, GAPDH from a variety of "simple" organisms exhibits unique functions, which qualify it as a moonlighting protein.

## 1.1 Role of Moonlighting GAPDH in Infectious Agent Attachment to, and Invasion of, Host Target Cells

The first indications that bacterial GAPDH may exhibit moonlighting activity were studies, which defined concurrently the homology of the plasminogen receptor to GAPDH (Lottenberg et al., 1992) and its unexpected subcellular localization (Pancholi and Fischetti, 1992). The goal of these latter investigations was to identify cell surface proteins, which may be involved in infection by group A streptococci. SDS-PAGE analysis following lysin digestion of the streptococcal cell was used as the experimental paradigm. As with many such studies, a 39 kDa protein was observed, purified, and its N-terminal amino acid sequence determined. The latter identified the protein as GAPDH and was termed streptococcal surface dehydrogenase (SDH). Subsequently, the enzyme activity of the 39 kDa protein was determined and shown to be characteristic of GAPDH.

The cell surface location of the 39 kDa protein was determined by immunoassay. Intriguingly, it could not be removed by a 2M NaCl wash or by 2% ODU suggesting its tight, nonserendipity, binding on the cell surface. This is reminiscent of the tight binding of nuclear GAPDH to DNA seen in apoptosis (Chapter 8; Sawa et al., 1997). Furthermore, analysis of other streptococcal types indicated that this SDH/GAPDH protein was present in groups B, C, E, G, and L. The lack of recognition in other groups was thought to be due the efficiency of lysin cell wall digestion. Immunoanalysis was then performed to determine whether streptococcal SDH could bind to cell proteins. These findings indicated binding to myosin, actin, and fibronectin. In toto, these initial results established the cell wall location of a bacterial GAPDH species and indicated its ability to bind specifically to host proteins.

A further investigation demonstrated the significance of cell wall–associated streptococcal SDH with respect to host cell phosphorylation (Pancholi and Fischetti, 1997). These studies demonstrated not only the phosphorylation of a 17 kDa pharyngeal cell protein by whole streptococci or by SDH but also that this posttranslational modification could be abolished by genistein, a tyrosine

**TABLE 13.1 Required Functions of Moonlighting GAPDH in Infectious Disease Pathogenesis**

| Function | Organism[a] | Binding Protein | "Unique" Findings | References[b] |
|---|---|---|---|---|
| Attachment and infection | Candida albicans | Fibronectin, Plasminogen, Collagen | Escherichia coli | Lottenberg et al. (1992), Pancholi and Fischetti (1992, 1997), Gil-Navarro et al. (1997), Seifert et al. (2003); Bergmann et al. (2004), Egea et al. (2007), Lama et al. (2009), Maeda et al. (2013) and Terrasse et al. (2012, 2015) |
| | E. coli | | GAPDH isoforms-only one secreted Group A Streptococci | |
| | Streptococci Group A, B | | Receptor-mediated cell signaling-control of host gene expression C. albicans | |
| | Streptococcus pneumoniae | | Tight binding to surface T. vaginalis Fe++ control of GAPDH expression | |
| | Trichomonas vaginalis | | | |
| Cocolonization | Porphyromonas gingivalis | Streptococcus oralis GAPDH binding to P. gingivalis | P. gingivalis binding to human epithelial GAPDH | Maeda et al. (2004a,b), Sojar and Genco (2005) and Nagata et al. (2009) |
| Apical complex formation | Plasmodium falciparum | Rab 2 | Membrane fusion, intracellular trafficking | Daubenberger et al. (2003) |
| Formation of Brucella-containing vacuoles | A. abortus | Rab 2 | Host GAPDH required for agent pathology | Fugier et al. (2009) |
| Iron metabolism | M. tuberculosis | Transferrin | Iron acquisition, extracellular transport, internalization | Modun and Williams (1999), Modun et al. (2000), Boradia et al. (2014) and Vázquez-Zamorano et al. (2014) |
| | S. pneumoniae | Hemoglobin[c] | | |

[a]In alphabetical order.
[b]Chronological order.
[c]Includes heme binding.

kinase inhibitor. N-terminal sequence analysis identified the 17 kDa protein as histone H3. The significance of this finding was indicated by further inhibitor studies demonstrating that abolishment of host cell protein tyrosine and serine phosphorylation did not affect streptococcal attachment to pharyngeal cells but did reduce significantly their invasive capacity.

Morphological analysis of SDH-treated pharyngeal cells demonstrated pronounced changes in their ultrastructure. This included changes in nucleus-to-cytoplasmic size ratios and chromatin condensation. Surprisingly, no drastic changes were observed in cell viability. To determine the presence of pharyngeal SDH receptor proteins, immunoblot ligand–binding assays were performed. These studies indicated the presence of two SDH receptors, a 30 and a 32 kDa protein. From these cumulative studies, it was suggested that binding of streptococcal SDH to their pharyngeal target cells initiates a receptor-mediated signal transduction pathway resulting in the phosphorylation of histone H3 subsequently affecting host cell chromatin structure and nucleosome assembly.

Subsequent studies focused on the identification of the putative GAPDH/SDH receptor present on the pharyngeal cell membrane (Jin et al., 2005). In these investigations, pharyngeal cell membrane–associated proteins were resolved on SDS-PAGE then probed with $^{125}$I-labeled SDH protein. Several proteins were identified, among them moesin, enolase, GAPDH, and 14-3-3ε as well as an unknown 55 kDa protein. The latter was identified by MALDI–TOF and LC–MS/MS spectroscopy as the urokinase plasminogen activator (uPAR). As controls, it was determined that $^{125}$I-labeled SDH bound to recombinant uPAR; preincubation of uPAR with an anti-uPAR antibody abolished binding; and, lastly, SDH bound to soluble uPAR from enzymatically digested pharyngeal cells.

The SDH-binding domain of uPAR was examined using purified recombinant uPAR as well as its three domains (D1, D2, and D3) in an immunoblot-based ligand-binding assay. These studies demonstrated that SDH bound only to the D1 domain. Conversely, a series of recombinant SDH constructs were used to determine its uPAR-D1 binding site. Of interest, these studies revealed two uPAR-D1 binding domains, a C-terminal and an N-terminal site. However, analysis of binding site intensity suggests that the former was a strong binding site while the latter was weak, perhaps marginal in nature.

The functional significance of SDH in group A streptococcal adherence and infection was then examined by mutation analysis (Boël et al., 2005). SDH does not contain either an N-terminal signal sequence or a C-terminal hydrophobic tail, each of which is required for their transport to the cell surface. Accordingly, it was hypothesized that the addition of a C-terminal hydrophobic tail may result in the inability of the pathogen to export the mutant protein to the cell surface, i.e., it would accumulate in the cytoplasm. Should that be the case, it would be possible then to demonstrate the significance of SDH as a cell surface protein by observation of defects in cell function due to the absence of the SDH surface protein.

Following mutant construction, its properties were compared to those of the parent wt group A streptococci. This analysis demonstrated that both the parent and mutant pathogen displayed similar growth rates; comparable mRNA and protein expression; and similar enzyme kinetics using crude cell extracts. In contrast, quantitation of cell wall GAPDH activity demonstrated a 5.5-fold reduction in the mutant as compared to its parent as well as a threefold reduction in activity in intact mutant cells.

Immunofluorescence analysis demonstrated SDH-specific fluorescence in the wt strain, which was lacking in the mutant. This was corroborated by immunoblot analysis following subcellular fractionation, i.e., there was c. fourfold less SDH in the mutant cell wall as compared to that in the parent strain. In contrast, the mutant cytosol contained a greater SDH concentration than that observed in the parent strain cytosol.

The physiological significance of the deficiency of cell wall SDH was then determined. As previously described, SDH binds to plasminogen, having been characterized as a plasminogen-binding protein (Plr) (Lottenberg et al., 1992). Quantitation of that binding using equal concentration of mutant and wt SDH protein indicated no difference in Plr binding. In contrast, as defined by a ligand-binding assay Plr acquisition was reduced by 60% in the mutant cell strain, which was confirmed by immunofluorescence analysis. Furthermore, the mutant strain exhibited a 99% decrease in antiphagocytic activity as well as a threefold reduction in its adherence to pharyngeal cells. In toto, these mutant studies demonstrate that changes in SDH structure affecting its surface localization have significant functional consequences for the infectivity of group A streptococci.

Other studies indicated the surface localization of GAPDH from group B *Streptococcus agalactiae* and from *Streptococcus pneumoniae* (Seifert et al., 2003; Bergmann et al., 2004, respectively). In the *S. agalactiae* investigations, surface proteins were analyzed by SDS-PAGE, which resulted in the identification of a distinctive 173.5 kDa protein. Sequence analysis identified it as the plasminogen-binding protein Plr subsequently shown to be GAPDH. ELISA analysis demonstrated its recognition of a number of cytoskeletal proteins including plasminogen, fibrinogen, and actin.

In *S. pneumoniae*, immunoblot analysis identified GAPDH not only as a cell wall constituent but also as a plasminogen-binding protein. Immunoelectron microscopy indicated the presence of cell surface GAPDH in both unencapsulated and encapsulated *S. pneumoniae*. In a subsequent study, it was demonstrated that *S. pneumoniae* GAPDH was released by cell lysis and interacted with peptidoglycan but not teichoic acid both of which are cell wall components (Terrasse et al., 2015). Furthermore, *S. pneumoniae* GAPDH bound to the cell wall in intact cells. In contrast, in an autolysin LytA mutant, GAPDH cell surface localization was reduced by 70%. For that reason, it was suggested that cell lysis may provide a mechanism used by this pathogen to localize GAPDH on the cell surface.

Further analyses demonstrated the diversity of moonlighting GAPDH function in "simple" organisms. In an immunoscreening study of *Candida albicans* proteins from infected patients using a λgt11 cDNA screening protocol, a 0.9-kb cDNA was isolated encoding a protein homologous to the *Saccharomyces cerevisiae* GAPDH genes *TDH1 and TDH3* (Gil-Navarro et al., 1997). Immunoblot analysis identified the surface localization of the *C. albicans* GAPDH protein and enzymatic assay quantitated its GAPDH glycolytic activity. Of note, this surface localized *C. albicans* GAPDH protein was resistant to dissociation by treatment with 2M NaCl or with 2% SDS, in agreement with the findings described above for the streptococcal SDH protein (Pancholi and Fischetti, 1992). Subsequent analysis demonstrated that, in accord with bacterial surface GAPDH, the comparable *C. albicans* GAPDH surface protein bound to fibronectin. However, it also was a laminin-binding partner that differentiated it from the bacterial SDH (Gozalbo et al., 1998).

The role of surface GAPDH in enterohemorrhagic (EHEC) and enteropathic (EPHC) *Escherichia coli* revealed several, new and intriguing findings apart from the identification of GAPDH surface localization and its binding to plasminogen and fibrinogen (Egea et al., 2007). These included the presence of two GAPDH isoforms-only one of which, the more basic, is secreted into the culture medium; the mechanism of secretion does not depend on vesicular formation; the secreted GAPDH species is enzymatically active; surface, or secreted GAPDH bind to intestinal epithelial cells; and nonpathogenic *E. coli* do not secrete GAPDH.

As distinct from higher organisms that contain a single somatic cell GAPDH gene, bacteria contain three genes that encode a GAPDH species (Gap A, B, and C). As the presence of two *E. coli* GAPDH isoforms resulted solely from the expression of the gapA gene, not only would posttranslational modification provide presumrably the mechanism for the second isoform but also it would be required for secretion of the basic species.

As contrasted with pathogenic bacteria, *Lactobacillus reuteri* interactions with intestinal cells provide a positive host effect. Recent studies identified the role of surface localized GAPDH in the ability of *L. reuteri* to adhere to its target cells (Zhang et al., 2015). This study analyzed surface layer proteins in seven different *L. reuteri* cell strains. In each cell strain, a 37 kDa protein was identified as a blotting partner. Mass spectroscopy analysis identified it as GAPDH.

The *L. reuteri* GAPDH gene and protein was expressed in *E. coli*; the latter was identified using an anti-His-tagged IgG and used to prepare an anti-*L. reuteri* polyclonal antibody. Using that antibody, the role of GAPDH in *L. reuteri* adherence to intestinal cells was determined. Treatment with the antibody reduced binding. Indirect immunofluorescence analysis demonstrated the surface localization of the *L. reuteri* GAPDH protein.

The studies described so far focus primarily on the functional diversity of bacterial GAPDH. As reported, protozoan GAPDH exhibits both a similar surface localization and a comparable ability to bind fibronectin (Lama et al., 2009).

The goal of these studies was to identify *Trichomonas vaginalis* proteins, which were responsible for its binding to target urogenital cells. Using a monoclonal antibody (Mab ws1), which inhibited *T. vaginalis* binding to fibronectin, a 39 kDa protein was identified by immunoblot analysis of whole cell lysates. Subsequently, a *T. vaginalis* cDNA library with Mab ws1, which identified a full-length GAPDH cDNA, which was used to produce a recombinant *T. vaginalis* GAPDH protein.

Subsequent studies demonstrated that the recombinant protein bound to fibronectin, plasminogen, and collagen in a dose-dependent manner. No binding was observed to laminin. Fluorescence analysis detected GAPDH on the surface of *T. vaginalis* cells. Intriguingly, it was noted that iron regulated the expression of GAPDH mRNA, i.e., low levels of iron correlated with low levels of GAPDH mRNA/protein and high levels of the former was associated with high levels of the latter. This will be considered later.

## 1.2 The Role of GAPDH As a Coadhesion Factor Promoting the Infection of a Heterologous Infectious Agent

The studies described in Section 1.1 detail the mechanisms through which surface or secreted GAPDH facilitates the actions of that individual infectious agent. As described in this section, this moonlighting protein possesses the ability to facilitate the pathology of a second, heterologous infectious agent.

Fimbriae are extracellular structures that enable bacteria to adhere to target cells. As such, they represent the mechanism through which those infectious agents are able to initiate their respective pathologies. Recent evidence suggests that this GAPDH may function as a coattachment factor facilitating the ability of other pathogens to adhere to and to infect their target cells.

*Porphyromonas gingivalis* is an oral pathogen that utilizes fimbriae to attach to oral cavity cells. This attachment is stimulated by *Streptococcus. oralis*. To identify those factors in the latter, which promote the attachment of the former, *S. oralis* extracts were incubated with *P. gingivalis* recombinant fimbriae (rFimA), coimmunoprecipitated with anti-rFimA antibody and subjected to SDS-PAGE. These studies revealed the coimmunoprecipitation of a 40 kDa *S. oralis* protein with the *P. gingivalis* rFimA (Maeda et al., 2004a). The N-terminal of the 40 kDa protein was determined and identified as GAPDH. The corresponding cDNA was prepared, amplified, and the resulting *S. oralis* recombinant protein prepared. This protein interacted with *P. gingivalis* rFimA.

*S. oralis* coaggregation capacity was observed with 12 different *S. oralis* cell strains although there were differences in the strength of that coaggregation capacity (Maeda et al., 2004b). Analysis of cell surface GAPDH enzyme activity suggested that there was a correlation between coaggregation capacity and the extent of enzyme function. Surface GAPDH was purified from five *S. oralis* strains demonstrating the strongest coaggregation activity and highest surface GAPDH activity. Each bound to *P. gingivalis* rFimA.

Subsequently, studies were performed to determine the identity of the epithelial cell protein, which bound to *P. gingivalis* fimbriae (Sojar and Genco, 2005). Affinity purification using immobilized fimbriae identified two proteins of $M_r = 50$ and $40\,kDa$. Cyanogen bromide analysis identified the $50\,kDa$ protein as keratin I while N-terminal sequence analysis identified the $40\,kDa$ protein as GAPDH. Rabbit muscle GAPDH could substitute for its epithelial cell counterpart. Incubation with antifimbriae peptide antibodies. The identity of the GAPDH *P. gingivalis* fimbriae–binding site was identified in binding assays using GAPDH peptides derived from lysyl endopeptidase digestion of *S. oralis* rGAPDH (Nagata et al., 2009). Those studies indicated that amino acids 163–216 in the G-3-P binding domain represented that binding site. In toto, these studies interrelating the role of *S. oralis* rGAPDH in the pathology of *P. gingivalis* demonstrates further not only the diversity of moonlighting GAPDH function but also the complexity of those activities.

## 1.3 Apical Complex Formation in Parasitic Infection

Apical complexes are specific vesicular structures formed by parasitic organisms, which function not only in the adhesion and invasion phases of infection but also in later phases of parasitic disease progression. Recent studies in *Plasmodium falciparum*, the organism that causes malaria, suggest that moonlighting GAPDH fulfills a number of different function in its infection cycle both in the liver and in erythrocytes (Daubenberger et al., 2003).

For this study, an anti-*P. falciparum* Mab ws1 was prepared, which recognized specifically the *P. falciparum* protein as defined by immunoblot analyses in infected and uninfected erythrocytes. It was then used to determine the intracellular localization of *P. falciparum* GAPDH, first in the liver then in erythrocytes. This experimental paradigm was used as the temporal sequence of *P. falciparum* infection involves the liver first and erythrocytes second.

In early liver infection *P. falciparum* GAPDH is localized in the cytoplasm. As infection proceeds, its intracellular distribution changes to the periphery, the subcellular region that contains the *P. falciparum* apical complex. This is in contrast to the *P. falciparum* aldolase that remains in the cytoplasm and was used as a negative control.

Immunocytochemical localization analyses were then performed in infected erythrocytes, which represent the latter phases of the *P. falciparum* infectious cycle. Subcellular fractionation followed by immunoblot analysis indicated that *P. falciparum* GAPDH was present both in the cytosol and in the membrane fraction. In contrast the *P. falciparum* aldolase was identified only in the cytosol.

As previously described, moonlighting GAPDH functions in Rab2-mediated intracellular vesicular trafficking (Chapter 4). Accordingly, a microsomal membrane-binding assay was used to quantitate the effect of Rab2 on *P. falciparum* GAPDH membrane binding. To ensure consistency, a Rab2 *P. falciparum* homolog was used as there may be some sequence dissimilarity between it and the host protein.

These studies demonstrated that *P. falciparum* Rab2 increased *P. falciparum* GAPDH membrane binding. Further, studies with N-terminal (1–133) and C-terminal (133–337) *P. falciparum* GAPDH domains indicated that only the N-terminal fragment competed with *P. falciparum* Rab2-mediated membrane binding of the full-length protein. This suggests that, as is the case for many moonlighting GAPDH functions, the *P. falciparum* Rab2 binding site resides in the GAPDH NAD$^+$ domain (Sirover, 2014).

In toto, these findings suggested that *P. falciparum* GAPDH may be associated with both *P. falciparum* vesicular protein structures and with apical complexes. For that reason, it was hypothesized that during infection, parasitic-derived moonlighting GAPDH may exhibit a membrane fusion function required for apical complex formation and/or may exhibit a cytoskeletal structural activity required for apical complex biogenesis. In either instance, these studies illuminate the significant role of GAPDH in malarial infection.

## 1.4 The Role of Intracellular Membrane Trafficking–Associated Moonlighting GAPDH in *Brucella abortus* Pathology

As indicated in Chapter 4 and in Section 1.3 above, moonlighting GAPDH fulfills an important function in normal cell vesicular trafficking and in the pathology of *P. falciparum*, respectively. In these studies, its role in the formation of "*Brucella*-containing vacuoles" (BCV), a required step in the pathology of *Brucella abortus*, will be discussed, illuminating again the diversity and complexity of GAPDH function in infectious disease.

*B. abortus*, a human pathogen, exhibits a rather unique mechanism for its successful invasion and replication in both phagocytic and nonphagocytic cells. In particular, it forms a unique intracellular structure, termed as "BCV," which is associated with the endoplasmic reticulum. In this endosomal-like structure not only is it able to replicate but also it is able to evade host defense mechanisms.

In a recent investigation, the protein composition of the BCV was examined by two-dimensional gel electrophoresis (Fugier et al., 2009). Although the BCV is comprised of some 1000 proteins, one, host GAPDH, identified by mass spectroscopy, seemed of particular interest given its association with the endoplasmic reticulum and its role in intraorganelle vesicular trafficking. Its presence on the BCV membrane was confirmed by immunoblotting as was Rab 2.

The significance of host GAPDH in *B. abortus* was tested using small interfering GAPDH RNA. As expected, the latter significantly reduced GAPDH expression. It also resulted in a tenfold decrease in *B. abortus* replication. Of further significance, changes were observed in the subcellular distribution of *B. abortus*, i.e., it was not able to reach the endoplasmic reticulum, remaining localized with the lysosomal-associated membrane protein 1 (LAMP-1). The latter is reflective of the lysosomal localization of *B. abortus*, which is an early step in its infective pathology prior to BCV formation. Inhibition of GAPDH expression also resulted in the inability of *B. abortus* to recruit Rab 2 as well.

## 1.5 The Role of GAPDH and Iron Metabolism in the Pathology of Infectious Disease

As most, if not all, living organisms require $Fe^{++}$ as a necessary nutrient/cofactor, it may not be surprising to note that infectious agents have developed mechanisms through which they obtain the requisite $Fe^{++}$ necessary for viability. That being said, what may be intriguing is that they, along with mammalian cells (Chapter 3), use moonlighting GAPDH for this purpose. In particular, studies from a variety of infectious agents demonstrate a sequential pattern of GAPDH involvement in iron acquisition, transport, and uptake (Fig. 13.1). Of interest, this may involve both host and agent GAPDH.

With respect to the former, Sato et al. (2012) demonstrated that erythrocyte GAPDH was released preferentially by *Prevotella oris* hemolysin. As with many GAPDH studies, they identified an unknown 38 kDa protein, which was released into the cell supernatant by incubation with *P. oris* hemolysin. N-terminal sequence analysis identified the unknown protein as GAPDH. Of significance, it appeared that this was a specific action of *P. oris* hemolysin and did not represent a general bacterial induced release of erythrocyte membrane proteins.

By itself, this study, although intriguing, does not indicate the importance of erythrocyte membrane GAPDH release. The latter may be suggested by studies using *S. pneumoniae*, which demonstrated the binding of secreted GAPDH

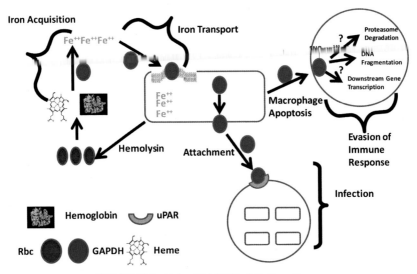

**FIGURE 13.1** Role of GAPDH in infectious disease.

to hemoglobin (Hb) and to heme (Vázquez-Zamorano et al., 2014). Initially, it was determined that Hb and heme supported *S. pneumoniae* growth while transferrin did not. Accordingly, affinity chromatography was used as a screening mechanism to identify those secreted proteins, which could bind to either Hb or heme. Those studies identified a 38 kDa heme-binding protein. Analysis of the protein–heme complex indicated the presence of $Fe^{++}$. It was then suggested that this unknown protein was secreted by *S. pneumoniae* as a bacterial scavenging protein to acquire $Fe^{++}$ from host Hb or heme.

Peptide sequence analysis following trypsin digestion identified the secreted protein as GAPDH. Further amino acid sequence studies demonstrated that the *S. pneumoniae* GAPDH protein contained the requisite heme-binding domain (KVAFDH). Molecular modeling (3D analysis) indicated that this domain was exposed on the GAPDH protein thereby facilitating its heme interaction.

Although the studies described above indicate the mechanisms through which infectious agents may acquire $Fe^{++}$, it remains necessary for them to possess a means to import that essential nutrient into the cell. In that regard, recent studies indicate that the bacterial cell surface GAPDH may function as a transferrin-binding protein (Tpn) (Modun and Williams, 1999; Modun et al., 2000).

In these investigations, a 42 kDa cell wall Tpn was identified in both *Staphylococcus aureus* and the *Staphylococcus epidermidis*. Characterization of the protein indicated that its 20 N-terminal domain was comparable to that of bacterial GAPDH. Characterization of the cell surface in intact cells and in cell membrane fractions demonstrated that each exhibited GAPDH glycolytic activity. Intriguingly, the extent of that activity appeared to be $Fe^{++}$ dependent, i.e., cell wall fractions from $Fe^{++}$-depleted cells contained greater GAPDH enzyme activity as compared to that observed in equivalent preparations from $Fe^{++}$-replete cells. It was stated that this finding was comparable to that observed for Tpn.

Subsequently, the latter was purified by $NAD^+$ affinity chromatography. Native Tpn exhibited an $M_r = 172$ kDa on nondenaturing gel electrophoresis comparable to that observed for tetrameric GAPDH. Further analysis demonstrated not only that it bound plasmin but also that plasmin binding inhibited its transferrin-binding activity. In toto, as illustrated in Fig. 13.1, these findings suggest that the mechanism through which acquired $Fe^{++}$ is transported into infectious agents resides in the transferrin-binding activity of cell surface GAPDH.

The final step in $Fe^{++}$ acquisition would be its internalization into the infectious agent. Examination of *Mycobacterium tuberculosis* indicated that its cell surface GAPDH functioned both as an $Fe^{++}$-binding protein but also may be responsible for its internalization (Boradia et al., 2014). Coimmunoprecipitation studies identified *M. tuberculosis* GAPDH as a Tpn. Subcellular localization analysis indicated *M. tuberculosis* GAPDH was a cell wall-associated protein with enzyme activity.

Fluorescence analysis demonstrated the binding of GAPDH and transferrin on the *M. tuberculosis* cell surface. Intriguingly, iron depletion resulted in a twofold increase in cell wall GAPDH expression and a threefold increase in transferrin binding. In vitro analysis demonstrated the binding of recombinant *M. tuberculosis* GAPDH to transferrin.

The internalization of both transferrin and surface GAPDH was examined using transferrin-labeled gold particles, which were detected by transmission electron microscopy. After incubation of cells for 1 h at 37°, cytoplasmic particles were detected. Immunoblot analysis confirmed the presence of transferrin. To confirm that internalization occurred in intracellular bacilli, analysis was performed in *M. tuberculosis* infected macrophages. Transferrin colocalized with cytoplasmic GAPDH as defined by coimmunoprecipitation. Accordingly, this study, indicating internalization of $Fe^{++}$ by surface GAPDH, fulfills the final step of the model illustrated in Fig. 13.1.

## 2. THE ROLE OF GAPDH IN THE EVASION OF HOST DEFENSIVE MEASURES

As indicated, the studies described in Section 1, above, belie the designation of those organisms as "simple creatures." Instead, those investigations document the complexity of those infectious agents specifically with respect to the diversity to which they utilize moonlighting GAPDH as part of their pathology. Similarly, in this section, as indicated in Table 13.2, evidence will be presented indicating the complexity with which moonlighting GAPDH is used as part of their mechanisms to evade the defensive means through which their respective targets seek to destroy the invading pathogen.

### 2.1 GAPDH-Mediated Initiation of Macrophage Apoptosis

Macrophages fulfill an important role in the immune response to infectious agents. This includes their production of nitric oxide through an increase in iNOS gene expression thereby facilitating the induction of oxidative damage and infectious agent cell death. Accordingly, the latter would need to devise appropriate countermeasures to ensure their survival. Recent evidence suggests that this may involve both GAPDH and, ironically, nitric oxide as inducers of macrophage apoptosis.

As expected, in studies of pneumococcal infection of human macrophages, nitric oxide levels were increased as a consequence of iNOS induction resulting in pneumococcal cell death (Marriott et al., 2004). Unexpectedly, the production of nitric oxide was also associated with the induction of macrophage apoptosis as defined by terminal deoxynucleotidyl transferase dUTP nick end and labeling (TUNEL) analysis. This finding may be of special significance as it indicates the nuclear transmission of an apoptotic signal. In particular, as discussed in Chapter 8, programmed cell death is a complex pathway involving discrete subcellular steps in a defined temporal sequence.

**TABLE 13.2 Host Evasion Mechanisms of Moonlighting GAPDH in Infectious Disease Pathogenesis**

| GAPDH Effect | Organisms | Host Protein/Effect | GAPDH Function | References |
|---|---|---|---|---|
| Apoptosis | Group B Streptococcus | DNA fragmentation | N.D. | Oliveira et al. (2012) |
| | S. pyogenes, S. aureus | | | |
| Innate immunity | S. pyogenes | Complement 5a | Inhibition of $H_2O_2$ production | Terao et al. (2006) |
| | S. pneumoniae | Complement C1q | Activation of C1q protein; complement binding protein | Terrasse et al. (2012) and Vedamurthy et al. (2015) |
| | Haemonchus controtus | | | |
| | H. controtus | Complement 3 | Complement binding protein | Vedamurthy et al. (2015) |
| Gene expression | Escherichia coli | TRAF2 | Binding to TRAF2 inhibits Nf-κB activation | Gao et al. (2013) |
| Immunomodulation | S. agalactiae | IL-10 | Facilitates host colonization | Madureira et al. (2007) |

In a further detailed analysis, the role of nitric oxide was investigated in relation to group B *streptococcus* (GBS)-induced murine macrophage apoptosis (Ulett and Adderson, 2005). DNA fragmentation analysis was used to quantitate programmed cell death. The regulation of gene expression was performed by microarray and northern blot determinations. In cultured murine macrophages cells, apoptosis related increases in expression were observed for tumor necrosis factor alpha, interleukin-1 and in iNOS. However, only inhibition of iNOS activity diminished apoptosis. Analysis of macrophages from wt versus iNOS$^{-/-}$ mice demonstrated that macrophages from the latter exhibited less nitric oxide production as well as a lower extent of programmed cell death. These findings suggest that nitric oxide represents a major factor in GBS-induced murine macrophage apoptosis.

Recently, a third study examined the role of GAPDH in GBS-induced murine macrophage apoptosis (Oliveira et al., 2012). An immortalized murine macrophage cell line and primary bone marrow-derived murine macrophages were exposed either to a recombinant bacterial GAPDH (rGAPDH) or to culture supernatants from a GBS wt strain. The latter contained secreted GAPDH. Apoptosis was determined again by a fluorometric TUNEL protocol.

Addition of either rGAPDH or the GBS culture supernatant resulted in significant apoptosis as defined as percentage of TUNEL positive cells (60%–80% as compared to control). Similar results were observed using culture supernatants from either *Streptococcus. pyogenes* or *S. aureus* cultures. Further, immunodepletion of GAPDH from the cell supernatant reduced the percentage of TUNEL positive cells to 10%–20%.

In toto, these studies suggest that both nitric oxide and bacterial GAPDH provide a means for infectious agents to evade the host immune response. That being said, these investigations do not interrelate the effects of nitric oxide and of GAPDH. Each would appear to be independent of the other. However, considering the role of SNO-GAPDH$^{cys149}$ in apoptosis (Chapter 8), it is tempting to speculate that it is the active agent in each study. This is suggested in Fig. 13.1. Further analysis is required to ascertain the validity of this suggestion.

## 2.2 The Role of Moonlighting GAPDH in the Evasion of Innate Immunity

In contrast to what may be termed adaptive immunity, innate immunity rests, in part, with the ability of complement proteins to recognize and to bind to infectious agents. This results in neutrophil and macrophage recruitment ultimately resulting in infectious agent destruction. Accordingly, to maintain their viability and integrity, it would be necessary for such agents to devise countermeasures to mitigate these host protective effects. Recent evidence suggests that bacterial GAPDH is involved intimately in the defenses against complement protein function mounted by infectious agents.

In studies of *S. pyogenes* infection, a protease was isolated, which degraded the complement C5a protein. Initially compared to the streptococcal plasmin receptor (Plr) and to the streptococcal surface dehydrogenase, it was recognized as bacterial GAPDH and termed the Plr/SDH/GAPDH protein (Terao et al., 2006). This analysis demonstrated that not only was this multifunctional protein located on the cell surface but also that it was secreted into the supernatant where it formed a soluble complex with the complement C5a protein. Further analysis indicated that a recombinant Pir/SDH/GAPDH protein inhibited neutrophil chemotactic activity in a dose-dependent manner. As neutrophils may induce bacterial cell death by production of $H_2O_2$, the ability of the rPir/.SDH/GAPDH protein to inhibit C5a production of $H_2O_2$ was determined. Those studies demonstrated a 34% reduction in $H_2O_2$ generation.

In a subsequent study, the recognition and binding properties of the human C1q protein were examined (Terrasse et al., 2012, 2015). These studies were performed using both *S. pneumoniae* and human cells. Intriguingly, the human C1q protein bound to human plasma membrane proteins as well as to the *S. pneumoniae* bacterial cell surface. With respect to the former, there appeared to be preferential binding at the cell surface of early apoptotic cells. With respect to the latter, the human complement protein bound to the cell surface as well. Mutational studies indicated that reduction in the amount of GAPDH at the bacterial cell surface resulted in a demonstrable decrease in the colocalization of human C1q. In an in vitro assay using recombinant human and *S. pneumoniae* GAPDH, the latter activated the C1q protein preferentially. From these combined human and bacterial studies, it was hypothesized that the binding activity of surface *S. pneumoniae* GAPDH to activated C1q protein might represent an attempt by the pathogen to mimic a human cell, i.e., a "cloaking" mechanism to confuse the host complement defense mechanism.

Studies with the parasite *Haemonchus controtus* examined the structural basis for GAPDH binding to the complement proteins C1q and C3 (Vedamurthy et al., 2015). These studies used full-length and partial GAPDH recombinant proteins in microtitre plate, coimmunoprecipitation and affinity chromatography to define not only the ability of *H. controtus* GAPDH to bind complement proteins but also that N-terminal GAPDH region was responsible for that interaction. This is in accord with the suggest that the N-terminal GAPDH $NAD^+$ domain contains the binding sites for it moonlighting functions thereby freeing its catalytic site for the performance of those moonlighting activities (Sirover, 2014). Further, competition experiments indicated that C1q and C3 bind to different N-terminal GAPDH sequences.

## 2.3 Role of Moonlighting GAPDH in the Subversion of Host Cell Gene Regulation

The selective regulation of gene expression is an integral part of host defense mechanisms. Conversely, the ability of infectious agents to disrupt those gene regulatory pathways provides a means to evade the immune response thereby facilitating their respective pathologies. Recent evidence suggests that host GAPDH represents a critical focal point through which infectious agents interfere with host gene regulation (Gao et al., 2013).

As described earlier, GAPDH fulfills an important role in the regulation of gene expression by the transcription factor, NF-κB (Chapter 2). Conversely, in EHEC and EPHC, NleB, a bacterial effector, disrupts NF-κB activation. Initially, the latter was thought to involved its inhibition of TNF receptor-associated factor 2 (TRAF2), an E3 ubiquitin ligase. The latter is required for NF-κB activation but also that its activity requires its own polyubiquitination.

It was determined that NleB inhibits TRAF2 polyubiquitination. However, the mechanism through which that occurs was unclear. In particular, NleB did not coimmunoprecipitate with TRAF2 indicating that a direct interaction was not the causative factor. Instead, using affinity chromatography, it was determined that a 37 kDa protein was the NleB interacting partner. Mass spectroscopy identified the 37 kDa protein as GAPDH.

As GAPDH–TRAF2 interactions are required for NF-κB activation, disruption of that protein–protein interaction could provide the foundation for the effect of NleB as a bacterial effector. Accordingly, as NleB contains an O-linked N-acetyl glucosamine (O-GlcNac) transferase activity, its ability to so modify GAPDH, and the effect thereof, was determined. Those studies revealed that wt NleB blocked GAPDH–TRAF2 binding while an O-GLcNac NleB mutant did not. Accordingly, these findings suggest that NleB functions, at least in part, as a means to evade host defense mechanisms through its disruption of GAPDH–TRAF2 mediation of NF-κB activation.

## 2.4 Role of GAPDH As a Virulence-Associated Immunomodulatory Protein

Although counterintuitive, recent evidence suggests not only that infectious agents may activate B and T lymphocytes but also that such immunomodulation may facilitate infection. In *S. agalactiae*, a 45 kDa protein was isolated from culture supernatants, which stimulated B and T cells as defined by measurement of CD69 levels on lymphocyte surfaces (Madureira et al., 2007). It was noted that the effect was greater in B cells versus T cells. As defined by N-terminal sequencing, the 45 kDa protein was identified as GAPDH.

Subsequently, a recombinant *S. agalactiae* GAPDH (rGAPDH) protein was prepared and tested for its ability to enhance *S. agalactiae* colonization. The recombinant protein stimulated spleen mononuclear cell proliferation in culture; its injection into L57BL/10 mice resulted in B-cell differentiation into Ig secreting cells as well as increased host IL-10 production, facilitating host colonization. The exact mechanism through which *S. agalactiae* GAPDH activates B and T cells are unknown. However, it was noted that infections by a number of diverse agents other than bacteria (virus, fungi, or protozoa) results in B-cell activation.

## 3. SUMMARY

The studies presented in this chapter highlight the structural and functional roles of moonlighting GAPDH in the pathology of infectious disease. In particular, these studies highlight the complex role of GAPDH in the mechanisms through which such organisms invade their target cells as well as those through which they seek to evade host defense mechanisms. In the former, moonlighting GAPDH is used as an adherence protein to form an infectious agent–host cell complex, in the formation of structures required for successful propagation and as an $Fe^{++}$ acquisition, transport, and internalization protein. In the latter, moonlighting GAPDH is used as an apoptosis inducing protein resulting in macrophage programmed cell death, as a means to counter innate immunity measures by complement binding, to alter host gene expression and as a modulator to modify the immune response. These varied functions are exhibited in diverse organisms indicating the generality of its activities. In toto, these studies provide the requisite evidence for a complex role of moonlighting GAPDH in infection and immunity.

## REFERENCES

Bergmann, S., Rohde, M., Sammerschmidt, S., 2004. Glyceraldehyde-3-phosphate dehydrogenase of *Streptococcus pneumoniae* is a surface-displayed plasminogen-binding protein. Infect. Immun. 72, 2416–2419.

Boël, G., Jin, H., Pancholi, V., 2005. Inhibition of cell surface export of group A streptococcal anchorless surface dehydrogenase affects bacterial adherence and antiphagocytic properties. Infect. Immun. 73, 6237–6248.

Boradia, V., Malhotra, H., Thakkar, J., et al., 2014. *Mycobacterium tuberculosis* acquires iron by cell-surface sequestration and internalization by human holo-transferrin. Nat. Commun. 5, 4730.

Daubenberger, C., Tisdale, E., Curcic, M., et al., 2003. The N′-terminal domain of glyceraldehyde-3-phosphate dehydrogenase of the apicomplexan *Plasmodium falciparum* mediates GTPase Rab-2 dependent recruitment to membranes. Biol. Chem. 384, 1227–1237.

Egea, L., Aguilera, L., Giménez, R., et al., 2007. Role of secreted glyceraldehyde-3-phosphate dehydrogenase in the infection mechanism of enterohemorrhagic and enteropathic *Escherichia coli*: interaction of the extracellular enzyme with human plasminogen and fibrinogen. Int. J. Biochem. Cell Biol. 39, 1190–1203.

Fugier, E., Salcedo, S., de Chastellier, C., et al., 2009. The glyceraldehyde-3-phosphate dehydrogenase and the small GTPase Rab 2 are crucial for *Brucella* replication. PLoS Pathog. 5, e1000487.

Gao, X., Wang, X., Pham, T., et al., 2013. NleB, a bacterial effector with glycosyltransferase activity, targets GAPDH function to inhibit NF-κB activation. Cell Host Microbe 13, 87–99.

Gil-Navarro, I., Gil, M., Casonova, M., et al., 1997. The glycolytic enzyme glyceraldehyde-3-phosphate dehydrogenase of *Candida albicans* is a surface antigen. J. Bacteriol. 179, 4992–4999.

Gozalbo, D., Gil-Navarro, I., Azorin, I., et al., 1998. The cell wall-associated glyceraldehyde-3-phosphate dehydrogenase of *Candida albicans* is also a fibronectin and laminin binding protein. Infect. Immun. 68, 2052–2059.

Jin, H., Song, Y., Boel, G., Kochar, J., Pancholi, V., 2005. Group A streptococcal GAPDH, SDH, recognizes uPAR/CD87 as its receptor on the human pharyngeal cell and mediates bacterial adherence to host cells. J. Mol. Biol. 350, 27–41.

Lama, A., Kucknoor, A., Mundodi, V., Alderete, J., 2009. Glyceraldehyde-3-phosphate dehydrogenase is a surface-associated fibronectin-binding protein of *Trichomonas vaginalis*. Infect. Immun. 77, 2703–2711.

Lottenberg, R., Broder, C., Boyle, M., et al., 1992. Cloning, sequence analysis, and expression in *Escherichia coli* of a streptococcal plasmin receptor. J. Bacteriol. 174, 5204–5210.

Madureira, P., Baptista, M., Vieira, M., et al., 2007. *Streptococcus agalactiae* GAPDH is a virulence-associated immunomodulatory protein. J. Immunol. 178, 1379–1387.

Maeda, K., Nagata, H., Yamamoto, Y., et al., 2004a. Glyceraldehyde-3-phosphate dehydrogenase of *Streptococcus oralis* functions as a coadhesion for *Porphyromonas gingivalis* major fimbriae. Infect. Immun. 72, 1341–1348.

Maeda, K., Nagata, H., Nonaka, A., et al., 2004b. Oral streptococcal glyceraldehyde-3-phosphate dehydrogenase mediates interaction with *Porphyromonas gingivalis* fimbriae. Microbes Infect. 6, 1163–1170.

Maeda, K., Nagata, H., Kuboniwa, M., et al., 2013. Identification and characterization of *Porphyomonas gingivalis* client proteins that bind to *Streptococcus oralis* glyceraldehyde-3-phosphate dehydrogenase. Infect. Immun. 81, 753–763.

Marriott, H., Ali, F., Read, R., et al., 2004. Nitric oxide levels regulate macrophage commitment to apoptosis or necrosis during pneumococcal infection. FASEB J. 18, 1126–1128.

Modun, B., Williams, P., 1999. The staphylococcal transferrin-binding protein is a cell wall glyceraldehyde-3-phosphate dehydrogenase. Infect. Immun. 67, 1086–1092.

Modun, B., Morrissey, J., Williams, P., 2000. The staphylococcal transferrin receptor: a glycolytic enzyme with novel functions. Trends Microbiol. 8, 231–237.

Nagata, H., Iwasaki, M., Maeda, K., et al., 2009. Identification of the binding domain of *Streptococcus oralis* glyceraldehyde-3-phosphate dehydrogenase for *Porphyromonas gingivalis* major fimbriae. Infect. Immun. 77, 5130–5138.

Oliveira, L., Madureira, P., Andrade, E., et al., 2012. Group B *streptococcus* GAPDH is released upon cell lysis, associates with bacterial surface, and induces apoptosis in murine macrophages. PLoS One 7, e29963.

Pancholi, V., Fischetti, V., 1992. A major surface protein on group A Streptococci is a glyceraldehyde-3-phosphate dehydrogenase with multiple binding activity. J. Exp. Med. 176, 415–426.

Pancholi, V., Fischetti, V., 1997. Regulation of the phosphorylation of human pharyngeal cell proteins by a Group A streptococcal surface dehydrogenase: signal transduction between streptococci and pharyngeal cells. J. Exp. Med. 186, 1633–1043.

Sato, T., Kamaguchi, A., Nakazawa, F., 2012. The release of glyceraldehyde-3-phosphate dehydrogenase (GAPDH) from human erythrocyte membranes lysed by hemolysin of *Prevotella oris*. Anaerobe 18, 553–555.

Sawa, A., Khan, A., Hester, L., Snyder, S., 1997. Glyceraldehyde-3-phosphate dehydrogenase: nuclear translocation participates in neuronal and nonneuronal cell death. Proc. Natl. Acad. Sci. U.S.A. 94, 11669–11674.

Seifert, K., McArthur, W., Bleiweis, A., Brady, L., 2003. Characterization of group B streptococcal glyceraldehyde-3-phosphate dehydrogenase: surface localization, enzymatic activity, and protein-protein interactions. Can. J. Microbiol. 49, 350–356.

Sirover, M., 2014. Structural analysis of glyceraldehyde-3-phosphate dehydrogenase functional diversity. Int. J. Biochem. Cell Biol. 57, 20–26.

Sojar, H., Genco, R., 2005. Identification of glyceraldehyde-3-phosphate dehydrogenase of epithelial cells as a second molecule that binds to *Porphyromonas gingivalis* fimbriae. FEMS Immunol. Med. Microbiol. 45, 25–30.

Terao, Y., Yamaguchi, M., Hamada, S., Kawabata, S., 2006. Multifunctional glyceraldehyde-3-phosphate dehydrogenase of *Streptococcus pyogenes* is essential for evasion from neutrophils. J. Biol. Chem. 281, 14215–14223.

Terrasse, R., Tacnet-Delorme, P., Moriscot, C., et al., 2012. Human and pneumococcal cell surface glyceraldehyde-3-phosphate dehydrogenase (GAPDH) proteins are both ligands of human C1q protein. J. Biol. Chem. 287, 42620–42633.

Terrasse, R., Amoroso, A., Vernet, T., Di Guilmi, A., 2015. *Streptococcus pneumoniae* GADPH is released by cell lysis and interacts with peptidoglycan. PLoS One 10, e0125377.

Ulett, G., Adderson, E., 2005. Nitric oxide is a key determinant of group B *streptococcus*-induced murine macrophage apoptosis. J. Infect. Dis. 191, 1761–1770.

Vázquez-Zamorano, Z., González-López, M., Romero-Espejel, M., et al., 2014. *Streptococcus pneumoniae* secretes a glyceraldehyde-3-phosphate dehydrogenase, which binds haemoglobin and haem. Biometals 27, 683–693.

Vedamurthy, G., Sahoo, S., Devi, I., Murugavel, S., Joshi, P., 2015. The N-terminal segment of glyceraldehyde-3-phosphate dehydrogenase of *Haemonchus contortus* interacts with complements C1q and C3. Parasite Immunol. 37, 568–578.

Zhang, W.-M., Wang, H.-F., Gao, K., et al., 2015. *Lactobacillus reuteri* glyceraldehyde-3-phosphate dehydrogenase functions in adhesion to intestinal epithelial cells. Can. J. Microbiol. 61, 373–380.

# Chapter 14

# Moonlighting GAPDH and Diabetes: Pleiotropic Effects of Perturbations in GAPDH Structure and Function

---

*You don't know what you have until it's gone*

An old proverb

Hormonal regulation of cell function represents one of the primary mechanisms through which cell homeostasis is achieved. Conversely, dysregulation of hormonal control may underlie the initiation or progression of human pathologies. The best example of the latter may be the onset of diabetes as a consequence of insulin dysfunction. Perturbations in insulin regulation of cell activity may be due to a deficiency in its synthesis or cellular structural alterations, which prevent its function. In either instance, the hyperglycemia-induced diabetic pathologies that result may be both debilitating and life-threatening.

Accordingly, the goal of this chapter is to consider the interrelationship between moonlighting GAPDH and the etiology of diabetes. In particular, this analysis will focus on hyperglycemic perturbations of GAPDH structure and function as fundamental mechanisms in relation to the development of this human pathology. Apart from the normal role of insulin in GAPDH gene regulation (mediated by two insulin response elements (IREs) in the GAPDH promoter region), GAPDH may provide the focal point for a hyperglycemic cell damage pathway, which is initiated by superoxide generation, and which terminates in the induction of a series of diabetes-related cell pathologies. The latter involves oxidative stress-induced poly (ADP-ribose) polymerase-1 modification of GAPDH resulting in the production of advanced glycation end products (AGEs), the induction of protein kinase C (PKC) activity as well as the activation of the hexosamine pathway. Furthermore, GAPDH is involved in hyperglycemia-induced diabetic retinopathy as well as the initiation of apoptosis through GAPDH–Siah1 protein–protein interactions. In toto, these studies illustrate not only the cellular consequences of diabetes-related moonlighting GAPDH function but also the molecular mechanisms that underlie those new activities.

Glyceraldehyde-3-Phosphate Dehydrogenase (GAPDH). http://dx.doi.org/10.1016/B978-0-12-809852-3.00014-5

# 1. INSULIN REGULATION OF GAPDH GENE EXPRESSION: ROLE OF INSULIN RESPONSE ELEMENTS IN THE GAPDH PROMOTER REGION

Early investigations indicated the interrelationship between insulin and GAPDH gene expression. These studies included the demonstration that GAPDH mRNA expression and protein synthesis were insulin dependent in adipocytes (Alexander et al., 1985); the isolation of a human GAPDH gene (Ercolani et al., 1988); the role of the GAPDH promoter region in insulin-mediated GAPDH gene expression (Alexander et al., 1988); and the identification of two (IREs) within the GAPDH promoter each of which increased independently the synthesis of GAPDH mRNA (Nasrin et al., 1990; Alexander-Bridges et al., 1992).

The role of cis-acting sequences in insulin regulation of GAPDH gene expression was examined using the expression of the human gene in rat hepatoma cells and in a mouse adipocyte cell line (Alexander et al., 1988). Stable transfection resulted in the expression of the human GAPDH construct, which not only produced a full length mRNA but also was induced tenfold by insulin treatment. With respect to the latter, the insulin-dependent increase in human GAPDH mRNA would be observed within 4 h after treatment with nanomolar insulin concentrations. The latter demonstrated the physiological significance of the insulin effect of GAPDH gene expression.

For analysis of the mechanisms through which insulin increased GAPDH mRNA, the experimental paradigm was to use chloramphenicol transferase activity (CAT) as a marker to determine the sequence requirements for insulin function. The recombinant CAT construct produced contained 5'-upstream GAPDH sequences (−487 to +20), which was transfected into both the rat hepatoma cells and the mouse 3T3 adipocyte cell line. Treatment with insulin increased CAT activity 5- to 6-fold. Of note, the time and dose dependency for induction of CAT activity were equivalent to those observed in each cell type for the respective endogenous GAPDH mRNA.

Subsequently, deletion analyses were utilized to define the specific sequences within the GAPDH promoter region responsible for its insulin regulation (Nasrin et al., 1990). The experimental paradigm was to use transient transfection of the rat hepatoma cell line by a CAT reporter plasmid. These studies indicated that elimination of the −488 to −270 GAPDH sequence abolished the effect of insulin on CAT activity. Intriguingly, elimination of the −488 to −409 region diminished insulin stimulation of CAT activity by 50%, while no effect was detected when a 100-base sequence (−408 to −309) was eliminated. In contrast, elimination of nucleotides from −308 to −269 reduced the insulin effect by twofold. Cumulatively, it was suggested that these studies identified two DNA sequences, which not only were physically separated but also which independently contributed to the aggregate insulin effect on GAPDH gene expression. The two sequences were termed insulin response element A (IRE-A) and insulin response element B (IRE-B).

An electrophoretic mobility shift assay (EMSA) was used to determine whether IRE-A and/or IRE-B could bind to nuclear proteins. For these studies, a [$^{32}$P]-labeled probe containing both IREs was utilized (−488 to −269). EMSA revealed a series of bound proteins, the level of which was increased by insulin exposure. Sequential nucleotide deletion studies demonstrated that elimination of nucleotides −488 to −423 (which contained the IRE-A motif) diminished protein binding. The physiological significance of these studies was assessed by using hepatic nuclear protein extracts from animals shifted from a low-insulin to a high-insulin condition, a standard protocol known to increase metabolic enzymes. In this experimental paradigm, GAPDH mRNA was increased by tenfold. Furthermore, using IRE-A DNA, the DNA protein complex level was increased by eightfold. Cumulatively, these studies identified the basic mechanism through which insulin regulates GAPDH gene expression. That being said, as indicated by the role of GAPDH in hypoxia (Chapter 9), the GAPDH promoter contains other response elements as well. It also includes a TPA response element (gTRE) motif that binds to the *c-fos/c-jun* heterodimer (Alexander-Bridges et al., 1992) and a myeloid zinc finger 1 (MZF-1) transcription factor-binding site (Piszczatowski et al., 2014). Accordingly, the complexity of the GAPDH 5′-upstream region is in agreement with its multiple roles as a moonlighting protein.

## 2. PLEIOTROPIC EFFECTS OF HYPERGLYCEMIC GAPDH MODIFICATION

The reduction in cell viability and ensuring clinical diabetic complications due to hyperglycemic stress may be due to the formation of AGEs, which induces cell damage through protein glycation (Brownlee, 1995); increased activity of the hexosamine pathway, which results in Sp1-induced gene dysregulation (Kolm-Litty et al., 1998; Chen et al., 1998); and the induction of PKC activity, which results in vascular dysfunction (Koya and King, 1998) as well as increased glucose metabolism through the aldose reductase pathway (Lee et al., 1995).

Recent evidence suggests that alterations in GAPDH structure resulting in its glycolytic inhibition may provide the common mechanism underlying each hyperglycemic-induced reduction in cell viability. This is illustrated in Fig. 14.1. Inhibition of GAPDH activity would result in the following: the accumulation of both dihydroxyacetone phosphate (DHAP) and glyceraldehyde-3-phosphate (G3P), the precursors to methylglyoxal (MG) formation; increased concentrations of fructose-6-phosphate, the precursor to glucosamine formation stimulating the hexosamine pathway; and the formation of diacylglycerol resulting in PKC activation. Although not illustrated, GAPDH inhibition would also decrease glycolytic glucose utilization facilitating its metabolism through the aldose reductase pathway resulting in polyol accumulation (Lee et al., 1995). The latter would result in depletion of cellular glutathione levels thereby increasing the toxicity of reactive oxygen species (ROS).

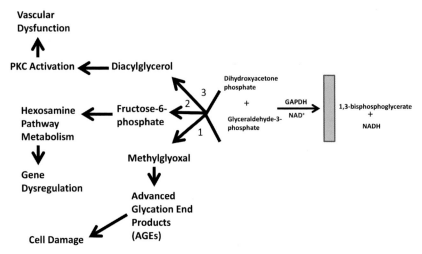

FIGURE 14.1    Role of GAPDH inhibition in the cellular pathology of diabetes.

## 2.1 The Role of Oxidative Stress-Induced Poly (ADP-Ribose) Polymerase 1 Modification of GAPDH: A Unifying Mechanism Underlying Hyperglycemic Injury

As indicated above, four distinct cellular pathologies may be due to hyperglycemia-induced inhibition of GAPDH glycolytic activity. Recent evidence suggests that a single alteration in the cellular milieu, production of superoxide, may not only underlie that diminution in enzyme function but also its pleiotropic effects in relation to hyperglycemic-induced cell damage (Nishikawa et al., 2000; Du et al., 2000, 2003; commented on by Reusch, 2003).

Initially, the role of oxidative stress in hyperglycemic damage was examined using mitochondrial electron transport chain inhibitors to reduce the formation of (ROS) (Nishikawa et al., 2000). These studies demonstrated that exposure of aortic endothelial cells to aminooxyacetate had no effect on ROS production due to hyperglycemia. The inhibitor affects NADH mitochondrial transport through the malate-aspartate shuttle. In contrast, exposure to 4-hydroxycyano-cinnamic acid abolished ROS generation. That inhibitor affects mitochondrial pyruvate transport. Additional inhibitor studies demonstrated that reduction in complex II function or uncoupling oxidative phosphorylation eliminated hyperglycemia-induced oxidative stress. As overexpression of manganese superoxide dismutase prevented generation of ROS, superoxide was identified as the ROS produced by hyperglycemia exposure.

The functional consequences of hyperglycemic induction of superoxide were then examined using inhibitor analysis and gene expression studies as the experimental paradigm. Those studies indicated that the formation of AGEs, sorbitol formation (aldose reductase pathway) and PKC-induced activation of the inflammatory protein NFκB were reduced in inhibitor-treated cells. It was noted that inhibition of GAPDH would increase levels of MG, an AGE precursor. Overall, it was suggested that the hyperglycemic induction of superoxide may provide a unifying mechanism underlying hyperglycemic cell damage.

Subsequently, the role of GAPDH and the hexosamine pathway in hyperglycemia were examined (Du et al., 2000). Using aortic endothelial cells as the experimental paradigm, a considerable inhibition of GAPDH activity (66%) was observed as a function of hyperglycemic-induced superoxide generation. Inhibitor studies performed as described above demonstrated that elimination of ROS generation reversed the effect on GAPDH catalysis. No effect was observed using azaserine, an inhibitor of the hexosamine pathway. Of note, there was no effect of hyperglycemia or any of the inhibitors used on GAPDH protein levels as defined by immunoblot analysis.

As illustrated in Fig. 14.1, inhibition of GAPDH glycolytic activity would result in the elevation of fructose-6-phosphate concentration, thereby increasing the initial substrate for the hexosamine pathway. As measured by levels of uridine diphosphate N-acetylglucosamine (UDP-GlcNAc), there was a 2.5 fold increase in that pathway as a function of hyperglycemia. That increase was abolished by inhibition of hyperglycemic mitochondrial superoxide generation or by the use of azaserine.

As indicated in Fig. 14.1, hyperglycemic activation of the hexosamine pathway may affect gene expression. This may be mediated by posttranslational modification of the transcription factor specificity protein 1 (Sp1). Sp1 immunoprecipitation studies as a function of hyperglycemia revealed a 1.7 fold increase in its modification by O-linked GlcNac as well as 80% and 70% decrease in its phosphoserine and phosphothreonine content, respectively. These changes in protein modification were reversed by inhibitors of hyperglycemic superoxide generation.

Recombinant TGF$\beta_1$ promoter luciferase constructs were utilized to examine downstream effects of modified Sp1 on gene regulation. A twofold increase in promoter activity was observed as a consequence of hyperglycemic superoxide generation, which was reduced by addition of the mitochondrial superoxide generating inhibitors. Similarly, examination of transcriptional expression of the plasminogen activator-1 (PAI-1) promoter demonstrated a comparable hyperglycemia induction (threefold) as well as an identical inhibitor reduction. Mutation of the Sp1 binding sites eliminated the hyperglycemic effect. In toto, these studies established the link between diminution of GAPDH activity and the deleterious effect of hexosamine pathway activation on gene expression.

In a third study, molecular analyses were used to define not only the significance of GAPDH inhibition as a rate-limiting step in diabetic pathogenesis but also the role of poly (ADP-ribose) polymerase 1 modification of GAPDH as a rate-determining factor as well (Du et al., 2003; commented on by Reusch, 2003). Antisense GAPDH protocols were used initially to determine whether ablation of GAPDH affected one more of the distinctive cellular changes observed in hyperglycemia. As expected cells exposed to high glucose (30 mM) exhibited increases in PKC, UDP-GlcNAc, AGEs, and NF-κB activity or formation. Intriguingly, cells exposed to antisense GAPDH in normal glucose (5 mM) exhibited the same increases. As a control, cells in normal glucose were exposed to scrambled GAPDH sequences. No changes in any of the cell parameters were observed. As such, these results demonstrate the role of GAPDH expression in hyperglycemia.

Subsequently, the role of poly (ADP-ribose) polymerase 1 (PARP-1) in GAPDH structure and function was examined. In hyperglycemic cells (30 mM glucose), immunoprecipitation followed by immunoblot analysis using an anti-poly (ADP-ribose) antibody revealed a 2.2-fold increase in the covalent modification of GAPDH. Again, inhibition of mitochondrial superoxide formation prevented this increase in ADP-ribosylation. Analysis of wt versus *PARP*-1 KO mice indicated that hyperglycemia-induced PARP-1 GAPDH modification was observed in the former but not in the latter.

As PARP-1 is a nuclear enzyme, the subcellular localization of modified GAPDH was determined. The analysis demonstrated that there was an equivalent increase in modified GAPDH in both the nuclear and cytoplasmic regions. This would suggest that GAPDH is translocated into the nucleus where it is modified then the modified protein is transported into the cytoplasm. A transport system, as yet unknown, would seem to be required for this effect.

To determine whether PARP-1 GAPDH modification is required for the four cellular consequences of hyperglycemia, cells were exposed to high glucose (30 mM) in the presence or absence of a specific PARP-1 inhibitor. Each of the four parameters was increased in the absence of the PARP-1 inhibitor. In contrast, incubation with the PARP-1 inhibitor prevented the respective cellular response to hyperglycemia. As such, these studies, along with the antisense GAPDH findings, indicate a temporal structure–function sequence in which GAPDH ADP-ribosylation is followed by diminution of GAPDH activity during hyperglycemia.

In summary, these three studies demonstrate a sequential hyperglycemic pathway that is initiated by superoxide generation and, through a series of defined steps, results in the four characteristic cellular changes associated with diabetic pathogenesis. As such, these studies indicate GAPDH inhibition of activity as perhaps the focal point relating cause and effect in cellular changes as a function of hyperglycemia.

## 2.2 Role of GAPDH Inhibition in Methylglyoxal Formation and Action

As described above, AGEs formation requires increases in MG levels. For that reason, the relationship between inhibition of GAPDH glycolytic activity and diabetic increases in MG levels were examined (Beisswenger et al., 2003). Initial studies were performed in normal erythrocytes incubated with increasing concentrations of koningic acid (KA), a specific GAPDH inhibitor. These studies demonstrated not only KA characteristic inhibition of GAPDH glycolytic activity but also a commensurate increase in MG levels.

Subsequently, studies were performed using red blood cells obtained from type 2 diabetic patients who were exposed in culture to 30 mM glucose for 32 h. This study indicated that not only was there a significant positive correlation between levels of dihydroxyacetone phosphate and MG but also there was a negative correlation between GAPDH activity and MG levels in diabetic red blood cells following their hyperglycemic exposure. Furthermore, the identical negative correlation was observed in diabetic red blood cells examined prior to the incubation with 30 mM glucose. The latter finding may be considered as highly significant as it indicated that this inverse relationship between red blood cell MG levels and GAPDH glycolytic activity was an intrinsic characteristic of type 2 diabetic patients. Similarly, red blood cells from type 1 diabetic patients exhibited a similar negative relationship between plasma MG levels and red blood cell GAPDH activity.

As earlier studies indicated an effect of aldehydes and other small molecules on GAPDH glycolytic activity (Novotny et al., 1994; Hook and Harding, 1997), the effect of MG itself on GAPDH structure and function was examined (Lee et al., 2005). Rabbit muscle GAPDH modification was used as the experimental paradigm. Incubation for 96 h with increasing concentrations of MG (1 μM–10 mM) diminished GAPDH glycolytic activity in a dose-dependent manner (20% reduction to complete inhibition, respectively). Further analysis used physiological or pathological MG concentrations (1–100 μM) with an incubation time of 48 or 96 h. A similar dose–response relationship was observed.

Isoelectric focusing of MG-treated rabbit muscle GAPDH revealed an interesting change in GAPDH structure. A pI of 8.5 was observed for untreated GAPDH, which is in accord with previous determinations. In contrast, rabbit muscle GAPDH treated with 50 or 100 μM MG exhibited a pI of 8.0, while a pI of 7.5 was detected for the protein incubated with 1 mM MG. It was noted that MG could bind to either GAPDH[lys] or GAPDH[arg] at their free amino groups, which would account for the negative change in pI. Furthermore, mass spectroscopic analysis indicated the binding of 12.75 MG residues bound to GAPDH at 50 μM MG, which increased to 55.6 MG residues at 1 mM MG. In toto, these cumulative studies indicated that MG contained a "double whammy" in diabetes based on inhibition of GAPDH, i.e., GAPDH inhibition increases MG concentration resulting in AGE-induced cell damage, while MG itself can modify GAPDH thereby potentiating its clinical effects.

## 2.3 GAPDH Inhibition in Diabetes As a Function of Its Succination

Recent evidence indicates that succination, the formation of S-(2-succinyl)cysteine (2SC), represents a physiologically significant protein modification with potential significance in diabetic pathogenesis (Alderson et al., 2006; Blatnik et al., 2008a; Blatnik et al., 2008b; Frizzell et al., 2011). In particular, these studies demonstrated not only GAPDH modification resulted in inhibition of its glycolytic activity but also they suggested an interrelationship between 2SC GAPDH modification and the pathology of diabetes.

Initial studies detected 2SC in human plasma proteins as well as in commercially obtained human serum albumin (Alderson et al., 2006). Perhaps unexpectedly, fumarate, a mitochondrial metabolite, was identified as the source of 2SC. Reaction of fumarate but not succinate with GAPDH resulted in both 2SC modification of GAPDH structure and a time-dependent inhibition of GAPDH glycolytic activity. It was stated that this reaction resulted in the modification of 8% of GAPDH[cys] with a reduction of 25% of its activity. As noted previously (Chapter 8), modification of the GAPDH active site cysteine is a physiologically significant structural alteration. Finally, examination of diabetic rat skeletal muscle as compared to control animals indicated a c. threefold increase in 2SC formation. This was comparable to the increase observed for the lysine-derived AGE product, N[6]-(carboxymethyl)lysine (CML). The latter may be formed by glyoxal exposure.

Subsequently, the interrelationships between 2SC GAPDH modification, inhibition of its glycolytic activity, and its expression in diabetic cells were examined in detail (Blatnik et al., 2008a). Kinetic analysis in vitro using rabbit muscle GAPDH as the experimental paradigm indicated the time and dose dependency of fumarate-induced inhibition of glycolytic activity, while a similar analysis using mass spectroscopy demonstrated the formation of 2SC-GAPDH.

Streptozotocin-induced diabetes was used as the experimental paradigm to examine 2SC-GAPDH modification in diabetic rat muscle as compared to age-matched controls. Quantitation of 2-SC and fumarate levels indicated their increase in the diabetic animals. In contrast, the specific activity of GAPDH was reduced by c.25%. Statistical analysis indicated a significant inverse correlation between 2SC muscle content and GAPDH catalysis ($P < .01$).

Immunoprecipitation was used to demonstrate 2SC modification of GAPDH. The latter was immunoprecipitated from diabetic and control animals using an anti-GAPDH polyclonal antibody followed by SDS-PAGE and trypsin digestion. Peptide analysis by MALDI-TOF mass spectroscopy indicated the presence of 2SC on GAPDH peptides 17 and 26 both of which are cysteine-containing peptides. Furthermore, that analysis indicated an equal modification of GAPDH[cys149] and GAPDH[cys244]. The former is the active site cysteine known to be a critical residue required for a number of moonlighting GAPDH activities. The latter is also modified by a number of compounds with resultant loss of protein function.

Tissue specific examination of 2SC formation indicated its preferential formation in adipocytes as compared to fibroblasts (Frizzell et al., 2011). In toto, these cumulative studies identify 2SC modification of GAPDH as a contributor to AGE formation in diabetes, i.e., its formation coordinate with the loss of GAPDH activity could initiate MG formation (Fig. 14.1).

## 3. ROLE OF GAPDH IN DIABETIC RETINOPATHY-INDUCED APOPTOSIS

Diabetic retinopathy is, regrettably, one of the now classical hyperglycemic-related clinical pathologies associated with this human disease. As indicated in Table 14.1, recent studies define not only the role of GAPDH in hyperglycemic-related retinal apoptosis but also they indicate that this occurs as a general phenomenon in diverse types of retinal cells.

### 3.1 Müller Cell GAPDH Function in Hyperglycemic-Induced Apoptosis

Initial studies in a transformed rat Müller retinal cell line (rMC-1) demonstrated that exposure to high glucose concentrations resulted in the nuclear translocation of GAPDH (Kusner et al., 2004). In particular, at 48 h, as defined by immunostaining, a 41% increase in nuclear GAPDH was detected. Incubation with R-(-)deprenyl decreased GAPDH nuclear localization by 63%. As described in Section IV, R-(-)deprenyl is an anti-Parkinson's disease drug, which targets GAPDH.

Subcellular fractionation followed by immunoblot analysis demonstrated that nuclear immunoreactive GAPDH protein increased by 47% as a function of hyperglycemic cell exposure. That result was comparable to that obtained by immunohistochemical analysis. Again, concomitant incubation with R-(-)deprenyl resulted in a 53% decrease in nuclear immunoreactive GAPDH protein. Similarly, in primary human Müller cells, as defined by immunostaining, hyperglycemic exposure resulted in GAPDH nuclear translocation, which was prevented by the addition of R-(-)deprenyl.

The physiological significance of these findings was tested using Müller retinal cells isolated from diabetic and normal rats. Immunostaining revealed that an increase in GAPDH nuclear expression was observed in cells from diabetic animals as compared to that observed in Müller cells from normal animals. Furthermore, in the transformed Müller rat cell line and in primary human Müller cells exposed to high glucose, apoptosis, as defined by annexin-V and caspase-3 expression, was observed at 96 h after exposure. Cell death increased threefold as compared to cells grown in low glucose. As such, these findings not only demonstrate the physiological significance of hyperglycemic nuclear GAPDH localization but also indicate that it precedes the initiation of programmed cell death. The latter is in accord with those studies previously discussed relating GAPDH nuclear translocation and apoptosis (Chapter 8).

**TABLE 14.1** Role of GAPDH in Diabetic Retinopathy[a]

| Retinal Cell Source/Organism | Experimental Paradigm | Experimental Results | "Unique" Findings | References |
| --- | --- | --- | --- | --- |
| Müller cells (transformed rat cell line; primary human cells; diabetic rat Müller cells) | Cells cultured in low or high glucose; immunoanalysis; gene expression | GAPDH nuclear translocation; apoptosis and Siah1–GAPDH binding | Nuclear GAPDH translocation precedes appearance of apoptotic markers | Kusner et al. (2004), Yego et al. (2009) and Yego and Mohr (2010) |
| Retinal ganglion cell line | Cells exposed to hyperpressure (100 mmHg); protein analysis | GAPDH protein levels increased during hyperpressure | Hyperpressure- induced nuclear GAPDH translocation | Kim et al. (2006) |
| Retinal cells (unseparated) | Cells cultured in low or high glucose; preparation of retina cell homogenates; subcellular fractionation | GAPDH mRNA and protein levels decreased during hyperglycemia; increase in nuclear GAPDH protein | Histological analysis indicates increased formation of acellular retinal capillaries | Kanwar and Kowluru (2009) |
| Bovine endothelial | Cells cultured in low or high glucose; GAPDH gene expression; GAPDH ribosylation; determination of AGEs, PKC, hexosamine pathway | Hyperglycemia affects GAPDH activity and subcellular localization; formation of AGEs, induction of PKC, activation of hexosamine pathway | GAPDH gene transfection reverses hyperglycemic effect on GAPDH function and subcellular localization | Madsen-Bouterse et al. (2010) |
| Pericytes (human primary) | Cells cultured in low or high glucose; immunoanalysis; subcellular determination of GAPDH activity; inhibitor analysis | PARP-1 inhibitor, destruction of peroxynitrite prevent hyperglycemia-associated GAPDH enzyme inhibition and apoptosis | Acetylated N-terminal GAPDH or Siah1 peptides block GADPH–Siah1 complex formation; nuclear translocation and apoptosis | Madsen-Bouterse et al. (2010) and Suarez et al. (2015) |

*AGEs, advanced glycation end products; PKC, protein kinase C.*
*[a]Chronological order.*

As inflammation is also a characteristic clinical presentation in diabetes, the role of the inflammatory cytokine IL-1β in Müller cell hyperglycemic GAPDH nuclear translocation was examined (Yego et al., 2009). Initial studies demonstrated that hyperglycemic exposure for 24 h resulted in a twofold and a threefold increase in IL-1β concentration in the rat rMC-1 cell line and in human Müller cells, respectively. No significant difference was observed in cells kept in low glucose for that time period.

Subsequently, the effect of IL-1β itself on GAPDH was examined in rMC-1 cells. Under normal conditions (5 mM glucose), IL-1β increased the nuclear localization of GAPDH as defined by both immunofluorescence and immunoblot analysis. Similar immunofluorescence findings were observed in human Müller cells. This effect of IL-1β was concentration and time dependent (0–10 ng/mL, 6–24 h, respectively). Furthermore, the temporal sequence between IL-1β-induced GAPDH nuclear translocation and IL-1β-induced cell death was in accord with that previously described (Kusner et al., 2004), i.e., nuclear GAPDH appearance preceded IL-1β-induced caspase activation (2- to 3-fold for both rMC-1 and human Müller cells, 72–96 h)

Inhibitor studies were used to probe the effect of IL-1β on hyperglycemic-induced GAPDH nuclear translocation. In these investigations, pretreatment of rMC-1 cells with 50 ng/mL IL-1β receptor blocker reduced nuclear GAPDH localization after hyperglycemic exposure by 51.6%. Similarly, blockage of endogenous IL-1β production decreased GAPDH nuclear localization as a function of hyperglycemic stress.

As described previously (Chapter 8), GAPDH nuclear translocation in apoptosis is mediated by its binding to Siah1. That binding accomplishes two goals: GAPDH does not contain a nuclear localization signal (NLS), its formation of a protein–protein complex with Siah1, which does contain an NLS facilitates GAPDH nuclear transport; similarly, as Siah1 by itself is inherently unstable, its formation of this protein–protein complex increases its stability.

In a third study, the role of the Siah1 protein was examined in relation to hyperglycemic-induced Müller cell GAPDH nuclear translocation and the observed subsequent increase in apoptotic cell death (Yego and Mohr, 2010). Initially, Siah1 expression as a function of hyperglycemia was examined in the rat rMC-1 Müller cell line. It was noted that Siah1 mRNA and protein levels were increased c. twofold 12 h after cell transfer to hyperglycemic conditions. It was also noted that Siah1 protein levels remained elevated at 24 h, an interval at which nuclear GAPDH localization is detected. Similar results were observed using the human Müller cell paradigm. In addition, immunoblot analysis revealed a significant increase in nuclear Siah1 content.

Coimmunoprecipitation analysis after a 24 h exposure was used to determine the hyperglycemic-dependent physical association of nuclear Siah1 and GAPDH. Immunoprecipitation was performed using an anti-Siah1 antibody followed by immunoblot determination of both Siah1 and GAPDH levels, i.e., probe with anti-Siah1 followed by reprobe with anti-GAPDH antibody.

A twofold increase in GAPDH–Siah1 binding was detected. Control studies demonstrated the specificity of binding by determining the specificity of the anti-Siah1 antibody as well as the lack of GAPDH binding to lactic dehydrogenase (highly expressed in Müller cells).

Siah1 knockdown studies were performed to define its requirement for GAPDH nuclear translocation. Addition of Siah1 siRNA reduced the levels of both Siah1 mRNA and protein. Under hyperglycemic conditions, Siah1 siRNA reduced nuclear localization in rMC-1 cell by twofold demonstrating that nuclear GAPDH translocation required formation of the GAPDH–Siah1 protein complex. This was further confirmed through the use of a C-terminal truncated Siah1 construct, which lacked the GAPDH binding site (Hara et al., 2005). In those studies, hyperglycemia exposure failed to increase nuclear GAPDH levels. In toto, these cumulative three studies define both the role of GAPDH in hyperglycemic Müller cell pathology and the defined temporal sequence in which GAPDH nuclear translocation initiates apoptotic cell death.

## 3.2 Pericyte GAPDH Function in Hyperglycemic-Induced Apoptosis

Initial examination of pericyte GAPDH in hyperglycemia indicated that exposure to high glucose inactivated both bovine cytoplasmic and nuclear GAPDH activity (Madsen-Bouterse et al., 2010). Addition of the poly (ADP-ribose) polymerase 1 inhibitor PJ34 or FeTPPS (which destroys peroxynitrite) prevented hyperglycemic GAPDH inactivation. Each compound reduced hyperglycemic-induced apoptosis. As such, these studies are in accord with those indicating the roles of superoxide generation and poly (ADP-ribose) polymerase 1 modification of GAPDH as basic mechanisms underlying hyperglycemia cell damage (Nishikawa et al., 2000; Du et al., 2000, 2003).

Subsequently, the mechanisms of GAPDH-mediated pericyte apoptosis as a function of hyperglycemia were examined (Suarez et al., 2015). The experimental paradigm used was cultures of primary human retinal pericytes cultured in low (5 mM) versus high (25 mM) glucose. Incubation in the latter resulted in twofold increase in Siah1 protein ($P = .0136$). Analysis of GAPDH–Siah1 binding by coimmunoprecipitation indicated a comparable increase in the formation of the protein complex. Exposure of cells to Siah1 siRNA reduced significantly its association with GAPDH. Furthermore, immunoanalysis demonstrated that either an N-terminal acetylated GAPDH or a comparable Siah1 peptide blocked complex formation. Subcellular fractionation indicated that exposure to high glucose increased the accumulation of nuclear GAPDH. This subcellular translocation was blocked by treatment with Siah1 siRNA, the N-terminal acetylated GAPDH peptide as well as by the comparable Siah1 peptide.

The functional significance of GAPDH–Siah1 complex formation during hyperglycemic stress was examined by determining its relationship to the initiation of programmed cell death. As defined by determination of Annexin V levels and caspase-3 enzyme activity, a threefold increase in apoptosis was observed initially at 72 h and was increased at later intervals. As with the Müller cell analysis, it would appear that formation of the GAPDH–Siah1 complex and its nuclear translocation preceded apoptosis. Furthermore, treatment with Siah1 siRNA, the GAPDH peptide or the Siah1 peptide blocked hyperglycemia-induced pericyte apoptosis. In toto, these studies demonstrate that, as with Müller cells, GAPDH is involved intimately in hyperglycemia-induced cell death.

## 3.3 GAPDH and Hyperglycemia-Induced Damage in Retinal Endothelial Cells

Bovine cells with or without transfected bovine GAPDH constructs were used as the experimental paradigm to consider the role of moonlighting GAPDH in hyperglycemia-induced cell damage in retinal endothelial cells (Madsen-Bouterse et al., 2010). Initial studies examined GAPDH activity as a function of hyperglycemic stress. Hyperglycemic exposure reduced both GAPDH levels and glycolytic activity. In contrast, GAPDH gene transfected cells exhibited increased levels of protein and enzymatic activity. Subcellular fractionation indicated an increase in nuclear GAPDH in hyperglycemic cells as defined by immunoblot analysis. Transfection of the bovine GAPDH gene resulted in normal levels of nuclear GADPH in cells exposed to hyperglycemic stress. Furthermore, addition of PJ34, an inhibitor of poly (ADP-ribose) polymerase-1 or FeTPPS, which destroys peroxynitrite, reversed the effect of hyperglycemic exposure on cytosolic or nuclear GAPDH activity.

The functional consequences of hyperglycemic-induced changes in GAPDH function and localization were then determined. In particular, as measured by either the cytoplasmic appearance of histone-associated DNA fragments or by the activation of caspase-3, hyperglycemic stress-initiated programmed cell death. Addition of the bovine GAPDH construct, PJ34, or FeTPPS abrogated the former, while addition of the GAPDH gene eliminated the latter.

As indicated previously, the formation of AGEs, the induction of PKC, and the activation of the hexosamine pathway are three of the characteristic cellular and clinical hyperglycemic pathologies. Quantitation of each in retinal endothelial cells indicated that each is increased as a function of hyperglycemic exposure. Similarly, immunoprecipitation analysis indicated the hyperglycemic-dependent GAPDH ribosylation, which was abolished by the PARP-1 inhibitor PJ34. In toto, these findings in bovine endothelial cells provide the requisite evidence demonstrating not only the role of GAPDH in endothelial cell hyperglycemic changes but also the presentation of the characteristic cell changes, which result in diabetic retinopathy. As such, these findings indicate a third retinal cell type, which is affected by hyperglycemic exposure.

## 3.4 Role of GAPDH in Hyperglycemic-Induced Retinal Cell Damage

Retinal cell preparations were used also to examine the effect of hyperglycemia and GAPDH structure and function (Kanwar and Kowluru, 2009). Immunoblot analysis indicated a statistically significant decrease in GAPDH mRNA and protein levels as a function of hyperglycemic stress. Similarly, PARP-1 activity, as measured by SDS-PAGE determination of protein ribosylation, increased as well, i.e., a number of ribosylated proteins were observed but were not analyzed. However, immunoprecipitation was used to determine GAPDH ribosylation. Again, a statistically significant increase of GAPDH modification was observed. Subcellular fractionation indicated a decrease in cytoplasmic GAPDH protein but a demonstrable increase in nuclear GAPDH in hyperglycemia exposed cells.

Examination of AGEs formation and PKC induction in retinal homogenates demonstrated a hyperglycemia-related increase in total AGEs formation as well as an elevation of PKC activity. Histopathological analysis indicated that there was an increase in the number of retinal acellular capillaries at extended intervals of hyperglycemic exposure. As such, those findings may be of special interest as they demonstrate the consequence of the cellular changes, which result from hyperglycemic exposure.

## 3.5 Effect of Hyperpressure on GAPDH Structure and Function in Retinal Ganglionic Cells

Increases in ocular pressure represent a significant, subsequent effect of diabetic retinopathy. Recent studies suggest that moonlighting GAPDH structure and function may be involved in this clinical complication as well (Kim et al., 2006). The experimental paradigm utilized involved the exposure of a retinal ganglion cell line (RGC-5) to hyperpressure (0–200 mmHg).

Initial studies demonstrated that exposure to 100 mmHg inhibited cell proliferation and growth. Subsequently, 2-dimensional gel electrophoresis was performed to identify those proteins affected by hyperpressure. That analysis revealed a significant number of proteins, which were differently expressed as compared to cells grown in normal pressure. GAPDH was among those proteins, which exhibited the greatest change in expression (tenfold after a 24 h exposure to 100 mmHg). Kinetic analysis as defined by immunoreactive protein indicated a time dependent increase in protein levels (0, 8, 16, and 24 h exposure). Subcellular fractionation studies indicated the nuclear translocation of GAPDH as a function of hyperpressure. A similar finding was observed by immunocytochemical microscopy. A terminal deoxynucleotidyl transferase dUTP nick end and labeling (TUNEL) assay indicated a significant number of DNA strand breaks in exposed cells. Cumulatively, these studies indicate an interrelationship between GAPDH nuclear translocation and the initiation of programmed retinal ganglion cell death. As such, moonlighting GAPDH may not only be involved in the development of diabetic retinopathy but also in its subsequent clinical symptomology.

## 4. SUMMARY

The studies presented in this chapter highlight the role of moonlighting GAPDH structure and function in relation to hyperglycemic changes in cell structure, which underlie diabetic pathogenesis. In particular, they define specific mechanisms through which hyperglycemia results in changes in GAPDH structure or activation of its moonlighting activities. These include oxidative stress-induce GAPDH modification thereby inhibiting its glycolytic activity as well as its role in increased apoptotic cell death resulting in diabetic retinopathy. Strikingly, the latter occurs in each of the four different retinal cell types, which were examined. Each of the individual changes in GAPDH structure and function provides the bases for one or more of a series of well-defined, characteristic diabetic cellular symptomologies. In toto, they provide the requisite evidence indicating the complex role of moonlighting GAPDH in this human disease.

## REFERENCES

Alderson, N., Wang, Y., Blatnik, M., et al., 2006. S-(2-succinyl)cysteine: a novel chemical modification of tissue proteins by a Krebs cycle intermediate. Arch. Biochem. Biophys. 450, 1–8.

Alexander, M., Curtis, G., Avruch, J., Goodman, H., 1985. Insulin regulation of protein synthesis in differentiated 3T3 adipocytes: regulation of glyceraldehyde-3-phosphate dehydrogenase. J. Biol. Chem. 260, 11978–11985.

Alexander, M., Lomanto, M., Nasrin, N., Ramaika, C., 1988. Insulin stimulates glyceraldehyde-3-phosphate dehydrogenase gene expression through cis-acting DNA sequences. Proc. Natl. Acad. Sci. U.S.A. 85, 5092–5096.

Alexander-Bridges, M., Dugast, I., Ercolani, L., et al., 1992. Multiple insulin-responsive elements regulate transcription of the GAPDH gene. Adv. Enzym. Regul. 32, 149–159.

Beisswenger, P., Howell, S., Smith, K., Szwergold, B., 2003. Glyceraldehyde-3-phosphate dehydrogenase activity as an independent modifier of methylglyoxal levels in diabetes. Biochim. Biophys. Acta 1637, 98–106.

Blatnik, M., Frizzell, N., Thorpe, S., Baynes, J., 2008a. Inactivation of glyceraldehyde-3-phosphate dehydrogenase by fumarate in diabetes: formation of S-(2-succinyl)cysteine, a novel chemical modification of protein and possible biomarker of mitochondrial stress. Diabetes 57, 41–49.

Blatnik, M., Thorpe, S., Baynes, J., 2008b. Succination of proteins by fumarate: mechanism of inactivation of glyceraldehyde-3-phosphate dehydrogenase in diabetes. Ann. N.Y. Acad. Sci. 1126, 272–275.

Brownlee, M., 1995. Advanced protein glycosylation in diabetes and aging. Annu. Rev. Med. 46, 223–234.

Chen, Y., Su, M., Walia, R., et al., 1998. Sp1 sites mediate activation of the plasminogen activator inhibitor-1 promoter by glucose in vascular smooth muscle cells. J. Biol. Chem. 273, 8225–8231.

Du, X., Edelstein, D., Rossetti, L., et al., 2000. Hyperglycemia-induced mitochondrial superoxide overproduction activates the hexosamine pathway and induces plasminogen activator inhibitor-1 expression by increasing Sp1 glycosylation. Proc. Natl. Acad. Sci. U.S.A. 97, 12222–12226.

Du, X., Matsumura, T., Edelstein, D., et al., 2003. Inhibition of GAPDH activity by poly(ADP-ribose) activates three major pathways of hyperglycemic pathways in endothelial cells. J. Clin. Invest. 112, 1049–1057.

Ercolani, L., Florence, B., Denaro, M., Alexander, M., 1988. Isolation and complete sequence of a functional human glyceraldehyde-3-phosphate dehydrogenase gene. J. Biol. Chem. 263, 15335–15341.

Frizzell, N., Lima, M., Baynes, J., 2011. Succination of proteins in diabetes. Free Radic. Res. 45, 101–109.

Hara, M., Agarwal, N., Kim, S., et al., 2005. S-nitrosylated GAPDH initiates apoptotic cell death by nuclear translocation following Siah1 binding. Nat. Cell Biol. 7, 665–674.

Hook, D., Harding, J., 1997. Inactivation of glyceraldehyde-3-phosphate dehydrogenase by sugars, prednisolone-21-hemisuccinate, cyanate and other small molecules. Biochim. Biophys. Acta 1362, 232–242.

Kanwar, M., Kowluru, R., 2009. Role of glyceraldehyde-3-phosphate dehydrogenase in the development and progression of diabetic retinopathy. Diabetes 58, 227–234.

Kim, C., Lee, S., Seong, G., Kim, Y., Lee, M., 2006. Nuclear translocation and overexpression of GAPDH by the hyper-pressure in retinal ganglion cell. Biochem. Biophys. Res. Commun. 341, 1237–1243.

Kolm-Litty, V., Sauer, U., Nerlich, A., Lehmann, R., Schleicher, E., 1998. High glucose-induced transforming growth factor β1 production is mediated by the hexosamine pathway in porcine glomerular mesangial cells. J. Clin. Invest. 101, 160–169.

Koya, D., King, G., 1998. Protein kinase C activation and the development of diabetic complications. Diabetes 47, 859–866.

Kusner, L., Sarthy, V., Mohr, S., 2004. Nuclear translocation of glyceraldehyde-3-phosphate dehydrogenase: a role in high glucose-induced apoptosis in retinal Müller cells. Invest. Ophthalmol. Vis. Sci. 45, 1553–1561.

Lee, A., Chung, S., Chung, S., 1995. Demonstration that polyol accumulation is responsible for diabetic cataract by the use of transgenic mice expressing the aldose reductase gene in the lens. Proc. Natl. Acad. Sci. U.S.A. 92, 2780–2784.

Lee, H., Howell, S., Sanford, R., Beisswenger, P., 2005. Methylglyoxal can modify GAPDH activity and structure. Ann. N.Y. Acad. Sci. 1043, 135–145.

Madsen-Bouterse, S., Mohammad, G., Kowluru, R., 2010. Glyceraldehyde-3-phosphate dehydrogenase in retinal microvasculature: implications for the development and progression of diabetic retinopathy. Invest. Ophthalmol. Vis. Sci. 51, 1765–1772.

Nasrin, N., Ercolani, L., Denaro, M., et al., 1990. An insulin response element in the glyceraldehyde-3- phosphate dehydrogenase gene binds a nuclear protein induced by insulin in cultured cells and by nutritional manipulation in vivo. Proc. Natl. Acad. Sci. U.S.A. 87, 5273–5277.

Nishikawa, T., Edelstein, D., Du, X., et al., 2000. Normalizing mitochondrial superoxide production blocks three pathways of hyperglycemic damage. Nature 404, 787–790.

Novotny, M., Yancey, M., Stuart, R., Wiesler, D., Peterson, R., 1994. Inhibition of glycolytic enzymes by endogenous aldehydes: a possible relation to diabetic neuropathies. Biochim. Biophys. Acta 1226, 145–150.

Piszczatowski, R., Rafferty, B., Rozado, A., Tobak, S., Lents, N., 2014. The glyceraldehyde-3-phosphate dehydrogenase gene (GAPDH) is regulated by myeloid zinc finger 1 (MZF) and is induced by calcitriol. Biochem. Biophys. Res. Commun. 451, 137–141.

Reusch, J., 2003. Diabetes, microvascular complications, and cardiovascular complications: what is it about glucose? J. Clin. Invest. 112, 986–988.

Suarez, S., McCollum, G., Jayagopal, A., Penn, J., 2015. High glucose-induced retinal pericyte apoptosis depends of association of GAPDH and Siah1. J. Biol. Chem. 290, 28311–28320.

Yego, E., Mohr, S., 2010. Siah-1 protein is necessary for high glucose-induced glyceraldehyde-3-phosphate dehydrogenase nuclear accumulation and cell death in Müller cells. J. Biol. Chem. 285, 3181–3190.

Yego, E., Vincent, J., Sarthy, V., et al., 2009. Differential regulation of high glucose-induced glyceraldehyde-3-phosphate dehydrogenase nuclear accumulation in Müller cells by IL-1β and UK-6. Invest. Ophthalmol. Vis. Sci. 50, 1920–1928.

## FURTHER READING

Li, T., Liu, M., Feng, X., et al., 2014. Glyceraldehyde-3-phosphate dehydrogenase is activated by lysine 254 acetylation in response to glucose signal. J. Biol. Chem. 289, 3775–3785.

# Section IV

# The Pharmacology of Moonlighting GAPDH

*Better Things for Better Living Through Chemistry*

Dupont Industries Slogan (1935)

The above statement, used as an advertising slogan by Dupont, proved prophetic for innumerable phases of human life. In particular, it would be impossible to count all of the positive changes in our existence, which are now taken for granted. In medicine, all one needs to do is to walk down the drug aisle of a supermarket or pharmacy and note the diverse over-the-counter preparations, which are available. Alternatively, one can observe the number of prescriptions prepared in the pharmacy, which are ready to be dispensed to the afflicted. In any case, numerous individuals are alive today who would have succumbed to their pathology yesterday.

Accordingly, the goal of this chapter is to consider the mechanisms through which the pharmacological efficacy of diverse substances are based on their interaction with GAPDH, their ability to alter GAPDH gene expression as well as their capacity to change GAPDH subcellular localization initiating a pharmacodynamics effect. In each instance, such changes provide a basic mechanism underlying neuropharmacology, cancer pharmacology as well as cardiovascular pharmacology. For that reason, the chapter is divided into three such sections, which not only describe relevant investigations but also include a specialized table delineating drug pharmacodynamics in relation to GAPDH structure and function. A generalized model figure is provided illustrating the dynamic intracellular results of drug–GAPDH interactions. Further, through analysis of their GAPDH interactions, these findings may provide new insight into our understanding of the pharmacological utility of readily available over-the-counter medicines.

## 1. GAPDH AND NEUROPHARMACOLOGY

As indicated in Table IV.1, considerable evidence exists interrelating the pharmacology of diverse neuropharmacological agents with changes in GAPDH structure and function.

Glyceraldehyde-3-Phosphate Dehydrogenase (GAPDH). http://dx.doi.org/10.1016/B978-0-12-809852-3.00015-7
**259**

**TABLE IV.1 GAPDH and Neuropharmacology[a]**

| Agent | Pathology | GAPDH Interaction | Result | References |
|---|---|---|---|---|
| CGP 3466 (Omigapil) | Parkinson's disease | Binds to GAPDH; site uncertain | Prevention of apoptosis | Kragten et al. (1998) |
| ONO-1603 | Dementia | Prevents increased GAPDH transcription and protein synthesis | Prevention of apoptosis | Katsube et al. (1989) |
| Rasagiline | Parkinson's disease | Uncertain | Prevention of GAPDH nuclear translocation | Maruyama et al. (2001) |
| Tetrahydroaminoacridine (THA; Tacrine) (Cognex) | Alzheimer's disease | Binding to GAPDH proapoptotic promoter | Prevention of apoptotic GAPDH expression | Tsuchiya et al. (2004) |
| Donepezil (Aricept) | Alzheimer's disease | Binding to GAPDH proapoptotic promoter | Prevention of apoptotic GAPDH expression | Tsuchiya et al. (2004) |
| R-(-)-Deprenyl (Selegiline; TCH3466; CGP 3466) (Omigapil) | Parkinson's disease | Binds to GAPDH; site uncertain | Prevents SNO GAPDH nitrosylation; nuclear translocation; apoptosis | Hara et al. (2006) |
| Cocaine | Addiction | Stimulation of SNO-GAPDH formation and nuclear translocation | Regulation of gene expression; induction of autophagy | Xu et al. (2013) and Guha et al. (2016) |

[a]In Chronological Order.

## 1.1 GAPDH and Cocaine

Throughout the long history of cocaine use, its central nervous system effects have been well established as has its symptomology as a drug of abuse. Recently, two intriguing studies indicate that its interaction with S-nitrosylated GAPDH may provide a basis for its significant cellular pathology (Xu et al., 2013; Guha et al., 2016). As related in Chapter 8, S-nitrosylation of GAPDH is not only a critical event whose formation results in a pleiotropic cascade within affected cells but also that it results in a defined pattern of downstream gene regulation. Each is of particular relevance in these studies.

These two reports provide complimentary information with respect to first, the role of SNO-GAPDH[149] in cocaine-induced gene regulation; and second, the specific cell death mechanism through which SNO-GAPDH[149] mediates cocaine cytotoxicity. In the first study (Xu et al., 2013), the definition of SNO-GAPDH[149] as a modulator of cocaine toxicity was examined as a function of behavioral or neurotoxic doses of cocaine. The use of such differential cocaine doses permitted a delineation of GAPDH effects specific to each drug exposure.

These studies yielded several salient findings: each exposure resulted in the formation of SNO-GAPDH[149]; knockout of n-nitric oxide synthase (nNOS) abolished the effect of cocaine on this signaling pathway; as defined by ChIP assays, differential effects on gene regulation were observed using the two different doses, i.e., behavioral doses induced CREB binding to the c-fos promoter while neurotoxic doses induced p53 binding to the PUMA promoter; the use of a GAPDH mutant (C150S) demonstrated the requirement for GAPDH[cys149]; and the use of the D1 antagonist SCH23390 to block both stimulation of SNO-GAPDH[149] formation and the synaptic actions required for the behavioral actions of cocaine. In toto, these studies not only established a role for SNO-GAPDH[149] in cocaine pharmacodynamics, but also the mediation of this signaling pathway by dopamine signaling and the differential effect of cocaine dose on GAPDH regulation of downstream gene expression. In addition, the selectivity of GAPDH regulation of gene expression as a function of dose indicates a complex GAPDH mechanism underlying cocaine toxicity.

In the subsequent study (Guha et al., 2016), the cell death mechanism through which SNO-GAPDH[149] mediated cocaine toxicity was examined. As indicated in Chapter 8, recent studies revealed a minimum of four such pathways: apoptosis, autophagy, parthanatos, and necroptosis. Each is characterized by its own individual pattern of gene expression and cell morphology, by the expression of specific biomarkers and by sensitivity to specific inhibitors.

The relevant cocaine-mediated cell death mechanism may be defined by examination of these parameters. Quantitation of increases in LC3-II (microtubule-associated protein 1 light chain-3), a biomarker of autophagy coupled with morphological observation indicated the role of this cell death mechanism in cocaine-mediated SNO-GAPDH signaling. The requirement for GAPDH[cys149] was established by substitution studies using the GAPDH mutant (C150S).

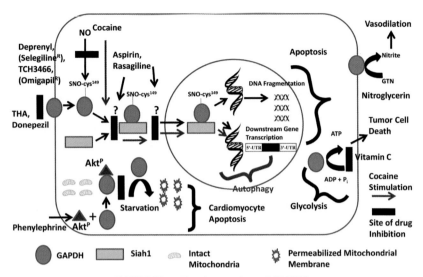

FIGURE IV.1   The pharmacology of GAPDH.

Lastly, as the use of specific inhibitors of the other three pathways failed to inhibit cocaine-associated cell death, these findings, these cumulative findings establish a new mechanism for cocaine-mediated cytotoxicity requiring the formation and use of SNO-GAPDH in an autophagic cell death mechanism. An illustration of this new mechanism is illustrated in Fig. IV.1.

## 1.2  GAPDH, R-(-)-Deprenyl (Selegiline), and Its Derivatives

Deprenyl and its derivatives are indicated for the treatment of Parkinson's disease. As indicated in Table IV.1, several detailed studies demonstrated their interaction with GAPDH as a basis mechanism potentially related to their clinical efficacy. Initial studies used CGP 3466, a deprenyl-related compound, to establish their binding to GAPDH (Kragten et al., 1998). As with many such investigations, affinity protocols were utilized as a general screening mechanism to determine drug–cellular protein interactions. In these studies, proteins that preferentially bound to immobilized GCP 346 were recovered than analyzed by SDS-PAGE.

Subsequent peptide analysis identified GAPDH as well as other proteins as CGP 3466 binding proteins. These included α- for β-actin, a mixture of α- and β-tubulin as well as α- and β spectin. To define the specific association of CGP 3466 with GAPDH, commercially available rabbit muscle GAPDH was used successfully in the affinity protocols. Further, it was demonstrated that NAD$^+$ could competitively inhibit GAPDH binding. This indicated not only that the GAPDH NAD$^+$ binding site was recognized by CGP 3466 but also it was in accord with a previous suggestion that the latter contained many of the requisite moonlighting GAPDH sequences responsible for its diverse interactions (Sirover, 2014).

The functional significance of CGP 3466–GAPDH binding was demonstrated through three sequential studies. CGP 3466 protected cells from apoptosis as did the introduction of antisense GAPDH sequences. Further, apoptosis induced by overexpression of GAPDH could be prevented by incubation with CGP 3466. In toto, these initial studies identified CGP 3466–GAPDH binding as well as its potential physiological significance, i.e., the prevention of apoptosis, which is associated with neuronal cell death that marks the progression of Parkinson's disease.

Subsequent studies defined the exact mechanism through which this GAPDH interaction prevented apoptosis (Hara et al., 2006). These investigations used both deprenyl (selegiline) and TCH346 (CGP3466). Mechanistic studies were performed using both a macrophage cell line in culture quantification of apoptotic cell death in cerebellar neurons and the use of 1-methyl-4-phenyl-1,2,3,6-tetrahydropyridine (MPTP) treatment to induce neuronal death in vivo.

Using all three experimental paradigms, it was demonstrated that both deprenyl and TCH346 prevented S-nitrosylation of GAPDH; that each prevented formation of the GAPDH–Siah1 protein complex; that deprenyl prevented apoptosis in vivo; and that it inhibited the formation of the GAPDH–Siah1 protein–protein complex in MPTP-treated animals. In toto, these studies represent not only a seminal finding (inhibition of SNO-GAPDH) but also that inhibition prevented all of the subsequent complex steps required in GAPDH mediation of programmed cell death as detailed in Chapter 8 and illustrated in Fig. IV.1. The adage that comes to mind is, "An ounce of prevention is worth a pound of cure." A further significance of these studies is that this effect could be observed at subnanomolar drug concentrations. In pharmacology, it is axiomatic that the smaller the effective dose that can be used, the lesser may be the side effects of the respective clinical protocol.

## 1.3 GAPDH and Antidementia Pharmacology

Early studies indicated the potential utility of the drug ONO-1603 in the treatment of dementia. Included among those investigations was the analysis of its effects on neuronal apoptosis in general and GAPDH expression in particular (Katsube et al., 1989). The model system used was the in vitro cultivation of cerebral cortical cells (CCCs) and cerebellar granular cells (CGCs) in culture, which exhibited a time-dependent increase in apoptosis as measured by the determination of cell viability, the extent of DNA fragmentation as well as by the overexpression of GAPDH. The latter was indicated by transcriptional increases as measured by Northern blot and by translational increases as measured by immunoblot quantitation of the 37 kDa protein.

These studies revealed that exposure of neuronal cells in culture to ONO-1603 delayed programmed cell death in a dose-dependent manner. In treated cells, there was an increase in the number of viable cells, a decrease in DNA fragmentation, as well as decreases in the extent of transcriptional and translational GAPDH expression. Pharmacokinetic analysis demonstrated that the maximal effect of ONO-1603 was observed when the drug was added at Day 7 of in vitro culture (termed 7 DIV), while less or no effect was detected if the drug was added at 9 or 11 DIV.

Analysis of GAPDH expression in untreated cells demonstrated that increases in GAPDH RNA were observed at 13 DIV in CCC and at 15 DIV in (CGCs). Addition of ONO-1603 at 7 DIV blocked this increase suggesting that intermediary steps were required for the observed change in GAPDH mRNA expression. Further, subcellular analysis revealed that the increases in the GAPDH protein were restricted to the particulate fraction with no change observed in the cell supernatant. Addition of the drug at 7 DIV blocked the particulate fraction increase. Accordingly, although these studies did not progress further with respect to drug development, they provided some intriguing findings with respect to the role of GAPDH in neuropharmacology. In particular, the temporal relationship with respect to the effect of ONO-1603 exposure in relation to changes in GAPDH mRNA and protein expression is intriguing.

## 1.4 GAPDH and the Pharmacology of Rasagiline

Its ability to function as a monoamine oxidase B inhibitor may be considered as an important indication for the use of rasagiline in the treatment of Parkinson's disease (Youdim et al., 2001). That being said, as with many other drugs, their examination provides evidence for additional mechanisms, which may underlie their therapeutic efficacy. In that regard, analysis of rasagiline and its derivatives demonstrated its role in the prevention of apoptosis induced not only by a variety of neurotoxins but also as a result of Parkinsonism-induced cell degradation. This included affects both on mitochondria to preclude diverse changes in its structure and function as well as prevention of nuclear events intrinsic to the initiation of programmed cell death (rev. in Mandel et al., 2005).

With respect to the latter, past studies indicated the significant role of GAPDH in neurotoxin-induced apoptosis (Maruyama et al., 2001). In those studies, programmed cell death caused by an endogenous dopaminergic neurotoxin, N-methyl(R)salsolinol, included GAPDH nuclear translocation. Intriguingly, as with the above-described effect of ONO-1603, there appeared to be a distinct temporal order of cellular changes following neurotoxin exposure. In particular, mitochondrial changes were observed 3 h thereafter, while nuclear accumulation of GAPDH was detected after 16 h thereafter in over 50% of the cells. Intriguingly, this percentage remained constant up to analysis at 24 h.

The effect of rasagiline was determined by preincubation of the cells with 1 µM rasagiline for 30 min prior to a neurotoxin exposure of 200 µM. That seemingly very short preincubation prevented nuclear GAPDH translocation (58% in neurotoxin-treated cells vs. 14% in rasagiline and neurotoxin-treated cells). Accordingly, as illustrated in Fig. IV.1, these studies demonstrate the ability of this Parkinson's disease drug to prevent toxin-mediated GAPDH nuclear accumulation.

## 1.5 GAPDH and the Pharmacology of Alzheimer's Disease

The interrelationship between Alzheimer's disease and the regulation, structure, and function of GAPDH has been considered in detail in Chapters 8 and 12. For that reason, as indicated in Table IV.1, this discussion is limited to the role of GAPDH in the pharmacodynamics of drugs used to treat Alzheimer's disease. In particular, these involve two such drugs, tetrahydroaminoacridine (THA, Tacrine, Cognex) and Donepezil (Aricept).

As discussed previously, a molecular genetic analysis was employed using luciferase reporter constructs in cytarabine-treated cells to select for those sequences, which enhanced GAPDH activity subsequent to exposure to this apoptotic stimulus. Using this protocol a proapoptotic promoter was located between positions −154 and −84 upstream from the GAPDH structural gene (Tsuchiya et al., 2004). The use of this reporter construct increased GAPDH expression 3- to 4-fold. The region contained a TATA-like box motif whose mutation diminished proapoptotic promoter activity.

To test the ability of antidementia drugs to affect promoter activity, cytarabine-treated cells containing the construct were exposed to deprenyl, THA, and donepezil. Significant depression of promoter function was observed with both THA and with donepezil (greater than 80% in each instance). In contrast, deprenyl exposure reduced activity by approximately 40%. In toto, these findings suggest that each Alzheimer's disease specific drug bound to the promoter region, reducing promoter activity and eliminating the apoptotic specific increase in GAPDH expression (Fig. IV.1).

## 2. GAPDH AND CANCER PHARMACOLOGY

As indicated in Table IV.2, several studies indicate not only the interrelationship between GAPDH and the pharmacodynamics of several well-known anticancer agents but also provide evidence for new roles of several well-established over-the-counter medicines. With respect to the former, GAPDH may be modified by either 3-bromopyruvate or saframycin (Ganapathy-Kanniappan et al., 2009; Xing et al., 2004, respectively). With respect to the latter, vitamin C modification of GAPDH may provide a new therapy for colon cancer (Yun et al., 2015), while vitamin D has the capacity to regulate GAPDH mRNA expression in poorly differentiated cells (Desprez et al., 1992).

**TABLE IV.2 GAPDH and Cancer Pharmacology[a]**

| Agent | Pathology | GAPDH Interaction | Result | References |
|---|---|---|---|---|
| Vitamin D | Mammary cancer | Increased mRNA expression | Decreased cancer cell growth | Chouvet et al. (1986) and Desprez et al. (1992) |
| Saframycin | Cancer cell lines | GAPDH–drug–DNA complex | GAPDH nuclear translocation; apoptosis | Xing et al. (2004) |
| Bromopyruvate | Hepatocellular carcinoma | Not determined | Apoptosis | Ganapathy-Kanniappan et al. (2009) |
| Vitamin C | Colorectal cancer | Glutathionylation of GAPDH[cys149] | Inhibition of glycolysis and tumorigenesis | Yun et al. (2015) |

[a]*Chronological Order.*

## 2.1 GAPDH and Vitamin C

Since the discovery by a British surgeon that limes could prevent scurvy in British seamen (known thereafter affectionately as "lymies"), vitamin C has been well recognized as an essential nutrient having numerous functions in vivo and is perhaps now the most widely used over-the-counter medication. Biochemical analysis demonstrated that, at low concentrations (0.1–0.5 mM), it specifically oxidizes GAPDH[cys149] forming a sulfenic acid derivative (Schmalhausen et al., 2003). This modification decreased GAPDH glycolytic activity but increased GAPDH acylphosphatase activity. The former was slightly decreased (<50% activity remaining after incubation with 0.5 mM ascorbate and 0.5 mM NAD[+] at 120 min). In contrast, the latter increased significantly at 20 min incubation under identical conditions. Ligand studies demonstrated the greatest effect of $PO_4^{--}$ on GADPH acylphosphatase activity. Mechanistic analysis indicated the role of $Fe^{++} \leftrightarrow Fe^{+++}$ interconversion as the basis for this alteration of GAPDH structure and function.

Intriguingly, recent cellular studies suggest the possibility not only that vitamin C may also be a potential colon cancer chemotherapeutic agent but also that GAPDH may be a major intracellular vitamin C target (Yun et al., 2015; commented on by Reczek and Chandel, 2015).In these studies, the effects of vitamin C were determined in wild-type colorectal cancer cells (CRC) as compared to that observed in CRC cells containing a *KRAS* or a *BRAF* mutation. The latter mutations may be of significance as over 50% of CRC cells contain such mutations, which may result in heightened tumor resistance to current therapeutic protocols.

These studies may be divided into two parts: first, cell culture and animal studies that demonstrated that vitamin C administration inhibited cell proliferation and tumor formation by *KRAS* or *BRAF* mutant CRC cells; and, second, a mechanistic analysis that established GAPDH as a unique vitamin C target in those mutant cells.

In the first part, sequential studies indicated that mutant CRC cells preferentially transport dehydroascorbate (DHA), the oxidized form of vitamin C into the cell, presumably due to upregulation of *GLUT1*; cell growth and colony formation was diminished by vitamin C treatment of *KRAS* and *BRAF* mutant cells; mutant cell viability was reduced in vitamin C-treated cells as defined by the appearance of the apoptotic marker annexin V; vitamin C-inhibited mutant tumor xenograft production in mice; analysis of metabolic changes in vitamin C-treated mutant CRC cells indicated increased levels of glycolytic intermediates upstream of GAPDH and diminished levels of such compounds downstream of GAPDH; and, lastly, ATP levels in vitamin C-treated mutant cells were significantly lower than in untreated cells indicating a block in glycolysis.

The second part analysis was based both on the previous studies described above and those in Chapter 8, which demonstrated the susceptibility of GAPDH$^{cys149}$ to oxidative modification. As vitamin C treatment of mutant CRC cells would increase the production of reactive oxygen species, it seemed reasonable to suggest that GAPDH$^{cys149}$ would be a preferred target. For that reason, GAPDH S-glutathionylation was analyzed by immunoprecipitation using both a GAPDH and an anti-S-glutathionylation specific antibody. GAPDH modification was 2- to 3-times higher in vitamin C-treated cells. Determination of GAPDH activity revealed that a 1-h incubation with vitamin C decreased GAPDH activity 50%. As GAPDH requires NAD$^+$, further studies revealed that its concentration diminished significantly after vitamin C treatment. Examination of poly (ADP-ribose) polymerase (PARP) demonstrated its activation as well. As NAD$^+$ is a PARP substrate, enhancement of its activity would diminish NAD$^+$ levels thereby further affecting GAPDH activity reducing tumor cell glycolysis. Of note, inhibition of PARP activity partially rescues vitamin C-treated CRC mutant cells.

In toto, these cumulative findings identify the interrelationship between GAPDH and the cancer preventive properties of vitamin C. As such, they provide compelling evidence highlighting this new and unique role for GAPDH in cancer prevention. Further, as noted in Chapter 8, given the role of SNO-modified GAPDH in programmed cell death, the observed vitamin C-dependent increase in the apoptotic marker annexin V suggests the possibility of a further role for S-glutathionylated GAPDH, i.e., apoptotic induction that, in conjunction with diminution of glycolysis, accelerates tumor cell death. In addition, given the ability of oxidized GAPDH to function as an acylphosphatase, it may be of interest to determine whether mutant CRC may develop the ability to utilize this GAPDH activity as a means to overcome the growth inhibitory effects of vitamin C. The role of acylphosphatase as a means to bypass GAPDH in anaerobic energy generation is discussed further in Section VI.

## 2.2 GAPDH and Vitamin D

The role of Vitamin D in basic cell functions and the pathologies, which arise from its deficiency, are well characterized. With respect to the former, its role in $Ca^{++}$ metabolism and bone formation is well established. Analysis of calcium metabolism identified an active form as 1,25-dihydroxyvitamin $D_3$ (1,25-$(OH)_2D_3$). With respect to the latter, its deficiency especially in young children may present as rickets, well recognized by its characteristic deficiencies in bone formation. As an aside, it may be an enduring, intriguing mystery as to whether or not Tiny Tim in the Charles Dickens' classic, "A Christmas Carol," suffered from rickets that was cured due to proper nutrition provided by the benevolence of a reborn Ebenezer Scrooge.

Past studies indicated that exposure to vitamin D inhibited the proliferation of breast cancer cells (Chouvet et al., 1986). Subsequently, the effect of vitamin D on GAPDH expression was examined in each breast cancer cell line that was characterized by differences in differentiation (Desprez et al., 1992). They observed that in BT-20, the poorly differentiated cell line, GAPDH mRNA was overexpressed as compared to the levels observed in MCF-7cells, which were highly differentiated. Further, the use of transcriptional and translational inhibitors (Actinomycin D and cycloheximide, respectively) indicated that the GAPDH mRNA increase in the former was resistant to Actinomycin D yet sensitive to cycloheximide. This raises the intriguing possibility that BT-20 GAPDH mRNA may be preexisting and subject to translational regulation. More recent studies indicate that vitamin D induced changes in GAPDH mRNA expression during hematopoiesis and myeloid cell differentiation (Piszczatowski et al., 2014). In toto, although these studies indicate overexpression of GAPDH in poorly differentiated cells (which are, for the most part, more aggressive tumor cells), which increase in gene expression does not translate into an increase in cell proliferation. The basis for this apparent inconsistency is unknown.

## 2.3 GAPDH and 3-Bromopyruvate

Past studies demonstrate the utility of the halogenated pyruvate, 3-Bromopyruvate (3-BrPA), as a potent inhibitor of glycolysis (Ko et al., 2004; Pereira da Silva et al., 2009; Ehrke et al., 2015). As such, it has been proposed as a cancer chemotherapeutic agent, given, as discussed in Chapter 11, tumor cells depend on glycolysis as their prime source of energy generation (rev. in Shoshan, 2012).

To identify the potential targets of 3-Bromopyuvate in cancer cells, hepatocellular cancer cells were incubated with radiolabeled 3-BrPA to determine its intracellular targets (Ganapathy-Kanniappan et al., 2009). Two-dimensional gel electrophoresis identified a unique radiolabeled 37 kDa protein as its major target, which was identified as GAPDH both by LCMS/MS and by immunoprecipitation analysis. GAPDH activity was decreased in a dose–response manner. As GAPDH is an abundant protein, the specificity of its modification was demonstrating by noting that other abundant proteins, i.e., β-actin or α-tubulin, were not modified.

The cellular consequences of 3-BrPA were examined not only by determining the time-dependent decrease in cell viability but also by quantitating 3-BrPA-induced cell death. The latter was demonstrated by flow cytometry and by annexin V fluorescence. It was considered that apoptosis was the result of ATP depletion. That being said, the site of 3-BrPA GAPDH modification is unknown and subcellular GAPDH localization studies were not performed. Given the role of GAPDH in programmed cell death, those studies may be of interest.

## 2.4 GAPDH and Saframycin

Exposure to the Saframycin A (SafA) antibiotics results in the inhibition of tumor cell growth (rev. in Aune et al., 2002). Mechanistic studies indicate not only do these antibiotics alkylate duplex DNA but also that they induce DNA–protein cross-links (Martinez et al., 1999). As the protein involved in cross-linking was unknown, studies were initiated to identify that protein and to determine its function (Xing et al., 2004). In those studies, an affinity chromatography protocol was developed in which duplex DNA was bound to an affinity resin, which was subsequently exposed to SafA. Protein samples were absorbed to the affinity column. Bound proteins were eluted by heat denaturation. A 37 kDa protein was detected, which bound to the SafA-duplex DNA affinity column and, which was identified as GAPDH.

The physiological relevance of this interaction was then determined. First, it was noted that the formation of the GAPDH–SafA-duplex DNA complex did not inhibit GAPDH glycolytic activity as defined by in vitro assay. Although it may be concluded that the latter binds to GAPDH at a site that does not interfere with catalysis, it was noted that the GAPDH concentration in vivo is c.70 μM, while the maximal concentration of SafA used in in vivo studies is between 1 and 100 nM. Accordingly, as GAPDH would always be in excess, it may not be unusual to observe no decrease in activity.

That being said, other studies demonstrated that SafA exposure resulted in nuclear translocation of GAPDH although most of the protein remained in the cytoplasmic fraction. This finding is reminiscent of those, which established the role of SNO-modified GAPDH in apoptosis, i.e., a small amount of GAPDH in the nucleus could have a profound effect (Hara and Snyder, 2006). Further,

RNA interference studies indicated that the ablation of cellular GAPDH rendered the recipient cells resistant to the antiproliferative effects of SafA. In toto, these findings highlight the specific role of GAPDH in the pharmacology of the Saframycin antibiotics.

## 3. GAPDH AND CARDIOVASCULAR PHARMACOLOGY

Recent evidence suggests a unique relationship between the action of cardiovascular drugs and GAPDH structure and function (Table IV.3).

## 3.1 GAPDH and Phenylephrine

Phenylephrine (PE) is well known for its use both in the treatment of hypotension and a nasal decongestant. Both indications are based on its vasoconstrictor activity. Also well known is its ability to prevent cardiac myocyte apoptosis through the prevention of mitochondrial-induced cell death (Naderi et al., 2010). Similarly, GAPDH may function to regulate mitochondrial membrane permeabilization (Tarze et al., 2007). A recent study provides evidence to indicate an interrelationship between PE protection of cardiac myocytes and this moonlighting GAPDH function (Yao et al., 2012).

In that study, apoptosis in cardiomyocyte cells was initiated by starvation. Mitochondrial membrane depolarization and nuclear apoptosis were monitored. Kinetic analysis indicated this change in mitochondrial function preceded apoptotic-associated nuclear events. Notably, addition of PE inhibited mitochondrial membrane depolarization and the associated increase in the formation of reactive oxygen species. It also increased the levels of bcl-2, inhibited Bax cytosolic→mitochondrial translocation, and decreased both cytochrome C and Caspase 3 activity. It was noted also that addition of PE increased GAPDH activity in a time-dependent manner (2–16 h incubation). However, as determined by immunoblot analysis, there did not appear to be an increase in GAPDH protein levels. Further, addition of iodoacetic acid (IAA), a known GAPDH inhibitor, prevented the protective effect of PE in cardiomyocytes. Overexpression of GAPDH by transfection of the GAPDH gene into cardiomyocytes eliminated starvation-induced apoptosis.

These studies present an apparent inherent contradiction with respect to GAPDH function, i.e., these findings indicate a protective role for GAPDH in mitigating mitochondria-induced apoptosis, while another analysis demonstrates conclusively that GAPDH participates actively in mitochondrial membrane permeabilization and other mitochondrial apoptotic events (Tarze et al., 2007). This apparent contradiction can be resolved by noting that reduction of Akt phosphorylation also reduced PE-induced protection of cardiomyocytes. As noted in Chapter 11, there is a distinct interrelation between GAPDH–Akt protein–protein interaction and the prevention of apoptosis and caspase-independent cell death (Fig. 11.2). Each involves Akt and phosphorylation. Accordingly,

**TABLE IV.3  GAPDH and Cardiovascular Pharmacology[a]**

| Agent | Pathology | GAPDH Interaction | Result | References |
|---|---|---|---|---|
| Heparin | Myocardial infarction; pulmonary thrombus (anticoagulant) | Formation of heparin–GAPDH amyloid structures | Prevention of neurodegenerative disease | Ávila et al. (2014) and Torres-Bugeau et al. (2012) |
| Phenylephrine | Hypotension | Increase of GAPDH activity; mechanism undetermined | Prevention of mitochondrial induced apoptosis | Yao et al. (2012) |
| Glycerol trinitrate (nitroglycerin) | Angina | Increase of GAPDH GTN reductase activity | Metabolism of GTN to nitrite | Seabra et al. (2013) |
| Aspirin | Myocardial infarction | Binding of aspirin metabolite salicylic acid; site undetermined | Inhibition of GAPDH nuclear translocation; inhibition of cell death | Choi et al. (2015) |

[a]Chronological order.

it may be reasonable to suggest that PE mediates a similar protective pathway in cardiomyocytes. In this instance, the Akt–GAPDH protein–protein interaction serves a protective function for normal cells while, as described in Chapter 11, the identical interaction is used to facilitate cancer development. The former is illustrated in Fig. IV.1 in which PE-induced cardioprotection involves downstream regulation of both Akt and GAPDH. In toto, they suggest a fundamental role for GAPDH in the pharmacology of phenylephrine.

## 3.2 GAPDH and Aspirin

As with vitamin C, aspirin is one of the most utilized over-the-counter medicines used routinely by millions of individuals on a daily basis. Further, as with vitamin C, its pharmacodynamics has not remained static as witnessed by its use as a preventive coronary therapy and its potential effect on reducing colon cancer incidence.

Recent studies also suggest not only that the role its metabolites may be also of significance to its pharmacology significance but also that their interactions with GAPDH may underlie their pharmacodynamics effects (Choi et al., 2015; Klessig, 2016). In these studies, HeLa cell-binding partners of salicylic acid (SA), aspirin's major metabolite, were identified by affinity-binding assays. Proteins that recognized SA were retained then recovered through the use of high SA concentrations in the eluant. Mass spectroscopic analysis identified GAPDH as such a binding partner. The specificity of SA–GAPDH binding was validated by physiochemical analysis using a recombinant human GAPDH protein. The binding efficacy of the recombinant protein to a series of natural and synthetic SA derivatives was also determined.

As described in Chapter 8, moonlighting GAPDH is deeply involved in apoptosis. In particular, its nuclear translocation is an a priori requirement for programmed cell death. For that reason, the effect of SA–GAPDH binding on GAPDH nuclear translocation and subsequent apoptosis was examined. Cells were exposed to the classical DNA damaging agent N-methyl-N-nitroso-N'-nitroguanidine (MNNG) following which GAPDH nuclear accumulation and subsequent apoptosis were determined.

As expected, following MNNG treatment, nuclear translocation of GAPDH was observed followed by apoptosis. Intriguingly, SA treatment prevented MNNG induced GAPDH nuclear accumulation and subsequent apoptosis. Notably, using the series of synthetic SA derivatives, the extent of inhibition of translocation and of apoptosis appeared proportional to the binding strength of the SA derivatives. In toto, these studies uncovered both a new aspect of aspirin pharmacology and of moonlighting GAPDH function.

## 3.3 GAPDH and Nitroglycerin

Glycerol trinitrate (GTN, aka nitroglycerin) is indicated for the treatment of angina. Administered sublingually, its metabolism results in the rapid release of nitric oxide in vivo. The latter, originally identified as endothelial relaxing factor (Furchgott, 1999), is an effective vasodilator providing the requisite relief of chest pain due to the constriction of coronary arteries.

Initial studies identified macrophage mitochondrial aldehyde dehydrogenase (mtALDH), functioning as a GTN reductase, as the enzyme primarily responsible for the bioactivation of GTN (Chen et al., 2002). That being said, evidence has also been presented indicating a role for GAPDH in the metabolism of this pharmacologically important compound. Historically, it was reported that GTN interacted with the active site GAPDH[cys149] converting it structurally to a sulfenic acid derivative that inhibited it glycolytic activity (You et al., 1975; Jakschik and Needleman, 1973; respectively). This modification of GAPDH structure unmasked its acylphosphatase activity (You et al., 1975).

Recently, examination of the role of GAPDH as a GTN reductase may indicate not only this moonlighting activity but also its physiological significance (Seabra et al., 2013). In these studies, it was demonstrated that GAPDH in vitro and in vivo could metabolize GTN. This enzymatic reaction was both time and GTN concentration dependent, requiring DTT and $NAD^+$. Of significance, the final product of the reaction was nitrite, which would be of significance pharmacologically. Disulfuram (Antabuse) was a competitive inhibitor similar to the result observed with mtALDH (Chen et al., 2002).

The physiological and clinical significance of these findings may reside in their analysis of GTN metabolism using intact erythrocytes. As noted, nitroglycerin is administered sublingually so that it may be readily absorbed into the blood stream, thereby avoiding first-pass metabolism. Initial studies indicating the functional diversity of GAPDH identified it as a membrane-associated protein (Kant and Steck, 1973; McDaniel et al., 1974; Yu and Steck, 1975; Kliman and Steck, 1980; Allen et al., 1987). Considering the abundance of GAPDH in vivo and this specific subcellular localization, as indicated in Fig. IV.1, it may be reasonable to suggest that this moonlighting GAPDH reaction may be a basic mechanism underlying the clinical efficacy of nitroglycerin.

## 3.4 GAPDH and Heparin: An Unexpected Surprise

One of the hallmarks of moonlighting proteins is the discovery of new functions for well-characterized proteins, which revolutionize our understanding of cell function. In that regard, heparin is now a century's old anticoagulant, which continues to provide a standard, effective, and, for the most part, safe treatment protocol for coronary pathologies. That being said, it is perhaps astonishing to consider not only that the use of this now classical anticoagulant may provide an intriguing therapy for the prevention of a neurodegenerative disease but also that its interaction with GAPDH may be intrinsic to that new activity.

As with many moonlighting studies, a recent report noted that, in vitro, heparin induced changes in GAPDH structure (Torres-Bugeau et al., 2012). Using a variety of elegant biophysical parameters, it was determined that GAPDH, normally present in solution as a soluble tetramer, could change its conformation to a fibril structure, which was amyloid-like in nature when incubated with heparin. Heparin-induced GAPDH aggregation resulted in a significant loss of GAPDH enzyme activity. This change in structure could be prevented by incubation with glyceraldehyde-3-phosphate which itself could be competed successfully with increased heparin concentrations.

The potential physiological significance of this finding relates to the effect of such heparin-induced GAPDH changes on the aggregation of α-synuclein. As previously discussed in Chapter 12, the latter is an important protein implicated in the etiology of Parkinson's disease, and it is present in Lewy bodies and binds GAPDH. As its aggregation may be one of the singular early events in Parkinson's disease development, inhibition or facilitation of that aggregation could influence positively or negatively the etiology of this disease. In that regard, incubation of α-synuclein with heparin-induced GAPDH structures facilitated the former's aggregation.

Although that finding would indicate a pathological effect of heparin-induced GAPDH aggregation, further cellular studies suggest a protective role mitigating dopaminergic cell death (Ávila et al., 2014). In particular, in human neuroblastoma cells, incubation with α-synuclein oligomers (α-syn$_{oli}$) reduced cell viability by 40%. In contrast, incubation of cells with α-syn$_{oli}$ plus heparin-induced GAPDH species (HI-GAPDH$_{ESS)}$ restored cell viability to control values. Further analysis revealed that incubation with α-syn$_{oli}$ increased liposome membrane permeability by 30%; that incubation with α-syn$_{oli}$ plus GAPDH increased this to 40%, while incubation with α-syn$_{oli}$ + HI-GAPDH$_{ESS}$ reduced leakage to 5%. Kinetic analysis indicated the latter effect was virtually immediate (5 min) declining further thereafter.

Based on those studies, it was hypothesized that, contrary to the expectation from the previous report that HI-GAPDH$_{ESS}$ may act as a scavenger protecting cells from the effects of exogenous α-syn$_{oli}$. Accordingly, as previous studies indicated that GAPDH may be secreted extracellularly (rev. in Sirover, 2014), unbeknownst to the clinicians administering heparin as a cardioprotective drug, it may very well be that, at the same time they were treating a coronary problem, they may also be aiding the patient by preventing α-syn$_{oli}$ neurocytotoxicity!

## 4. SUMMARY

It is hoped that, from this discussion, the reader will gain insight into a new aspect of moonlighting GAPDH. In particular, GAPDH function may provide fundamental mechanisms underlying not only the therapeutic efficacy of a series of structurally distinct pharmacological agents but also of cocaine, a major drug of abuse. In many cases, the interrelationship between GAPDH and drug pharmacodynamics appears to focus on the complex role of moonlighting GAPDH in apoptosis, autophagy, or other mechanisms of cell death (Fig. IV.1). As indicated, many of the discussed drugs appear to inhibit a unique particular step in GADPH-mediated programmed cell death thereby functioning as protective agents. GAPDH as a glycerol trinitrate reductase may be involved in nitroglycerin metabolism facilitating its action, and, as a heparin conjugate, providing protection against the development of an age-related neurodegenerative disease. Lastly, it appears that basic mechanisms underlying the action of the two most used drugs, aspirin and vitamin C, focus on GAPDH, i.e., the former by preventing apoptosis and the latter by inhibition of glycolysis in tumor cells.

## REFERENCES

Allen, R., Trach, K., Hoch, J., 1987. Identification of the 37-kDa protein displaying a variable interaction with the erythroid cell membrane as glyceraldehyde-3-phosphate dehydrogenase. J. Biol. Chem. 262, 649–653.

Aune, G., Furuta, T., Pommier, Y., 2002. Ecteinascidin 743: a novel anticancer drug with a unique mechanism of action. Anticancer Drugs 13, 545–555.

Ávila, C., Torres-Bugeau, C., Barbosa, L., et al., 2014. Structural characterization of heparin-induced glyceraldehyde-3-phosphate dehydrogenase protofibrils preventing α-synuclein oligomeric species toxicity. J. Biol. Chem. 289, 13838–13850.

Chen, Z., Zhang, J., Stamler, J., 2002. Identification of the enzymatic mechanism of nitroglycerin bioactivation. Proc. Natl. Acad. Sci. U.S.A. 99, 8306–8311.

Choi, H., Tian, M., Manohar, M., et al., 2015. Human GAPDH is a target of aspirin's primary metabolite salicylic acid and its derivatives. PLoS One. http://dx.doi.org/10.1371/journal.pone.0143447.

Chouvet, C., Vicard, E., Devonec, M., Saez, S., 1986. 1,25 Dihydroxyvitamin D, inhibitory effect on the growth of two human breast cancer cell lines (MCF-7, BT-20). J. Steroid Biochem. 24, 373–376.

Desprez, P., Poujol, D., Saez, S., 1992. Glyceraldehyde-3-phosphate dehydrogenase (GAPDH, E.C. 1.2.1.12.) gene expression in two malignant human mammary epithelial cell lines: BT-20 and MCF-7. Regulation of gene expression by 1,25-dihydroxyvitamin $D_3$ (11,25-(OH)2D3). Cancer Lett. 64, 219–224.

Ehrke, E., Arend, C., Dringen, R., 2015. 3-Bromopyruvate inhibits glycolysis, depletes cellular glutathione, and comprises the viability of cultured primary rat astrocytes. J. Neurosci. Res. 93, 1138–1146.

Furchgott, R., 1999. Endothelium-derived relaxing factor: discovery, early studies, and identification as nitric oxide. Biosci. Rep. 19, 235–251.

Ganapathy-Kanniappan, S., Geschwind, J., Kunjithapatham, R., et al., 2009. Glyceraldehyde-3-phosphate dehydrogenase (GAPDH) is pyruvylated during 3-bromopyruvate mediated cell death. Anticancer Res. 29, 4909–4918.

Guha, P., Harraz, M., Snyder, S., 2016. Cocaine elicits autophagic cytotoxicity via a nitric oxide-GAPDH signaling cascade. Proc. Natl. Acad. Sci. U.S.A. 113, 1417–1422.

Hara, M.R., Snyder, S.H., 2006. Nitric oxide-GAPDH-Siah: a novel cell death cascade. Cell. Mol. Neurobiol. 26, 527–538.

Hara, M., Thomas, B., Cascio, M., et al., 2006. Neuroprotection by pharmacologic blockade of the GAPDH death cascade. Proc. Natl. Acad. Sci. U.S.A. 103, 3887–3889.

Jakschik, S., Needleman, P., 1973. Sulfhydryl reactivity of organic nitrates: biochemical basis for inhibition of glyceraldehyde-3-phosphate dehydrogenase and monoamine oxidase. Biochem. Biophys. Res. Commun. 53, 539–544.

Kant, J., Steck, T., 1973. Specificity in the association of glyceraldehyde 3-phosphate dehydrogenase with isolated human erythrocyte membranes. J. Biol. Chem. 248, 8457–8464.

Katsube, N., Sunaga, K., Aishita, H., Chuang, D., Ishitani, R., 1989. ONO-1603, a potential antidementia drug, delays age-induced apoptosis and suppresses overexpression of glyceraldehyde-3-phosphate dehydrogenase in cultured central nervous system neurons. J. Pharmacol. Exp. Ther. 288, 6–13.

Klessig, D., 2016. Newly identified targets of aspirin and its primary metabolite, salicylic acid. DNA Cell Biol. 36, 163–166.

Kliman, H., Steck, T., 1980. Association of glyceraldehyde-3-phosphate dehydrogenase with the human red cell membrane: a kinetic analysis. J. Biol. Chem. 255, 6314–6321.

Ko, Y., Smith, B., Wang, Y., et al., 2004. Advanced cancers: eradication in all cases using 3-bromopyruvate to deplete ATP. Biochem. Biophys. Res. Commun. 324, 269–275.

Kragten, E., Lalande, I., Zimmerman, K., et al., 1998. Glyceraldehyde-3-phosphate dehydrogenase, the putative target of the antiapoptotic compounds CGP 3466 and R-(-)-deprenyl. J. Biol. Chem. 273, 5821–5828.

Mandel, S., Weinreb, O., Amit, T., Youdin, M., 2005. Mechanism of neuroprotective action of the anti-Parkinson drug rasagiline and its derivatives. Brain Res. Brain Res. Rev. 48, 379–387.

Martinez, E., Owa, T., Schreiber, S., Corey, E., 1999. Phthalascidin, a synthetic antitumor agent with potency and mode of action comparable to ecteinascidin 743. Proc. Natl. Acad. Sci. U.S.A. 96, 3496–3501.

Maruyama, W., Akao, Y., Youdim, M., Davis, B., Naoi, M., 2001. Transfection-enforced Bcl-2 overexpression of an anti-Parkinson drug, rasagiline, prevent nuclear accumulation of glyceraldehyde-3-phosphate dehydrogenase induced by an endogenous dopaminergic neurotoxin, N-methyl(R)salsolinol. J. Neurochem. 78, 727–738.

McDaniel, C., Kirtley, M., Tanner, M., 1974. The interaction of glyceraldehyde 3-phosphate dehydrogenase with human erythrocyte membranes. J. Biol. Chem. 249, 6478–6485.

Naderi, R., Imani, A., Faghihi, M., Moghimian, M., 2010. Phenylephrine induces early and late cardioprotection through mitochondrial permeability transition pore in the isolated rat heart. J. Surg. Res. 164, e37–e42.

Pereira da Silva, A., El-Bacha, T., Kyaw, N., et al., 2009. Inhibition of energy-producing pathways of HepG2 cells by 3-bromopyruvate. Biochem. J. 417, 717–726.

Piszczatowski, R., Rafferty, B., Rozado, A., Tobak, S., Lents, N., 2014. The glyceraldehyde 3-phosphate dehydrogenase gene (GAPDH) is regulated by myeloid zinc finger 1 (MZF-1) and is induced by calcitriol. Biochem. Biophys. Res. Commun. 451, 137–141.

Reczek, C., Chandel, N., 2015. Revisiting vitamin C and cancer. Science 350, 1317–1318.

Schmalhausen, E., Pleten, A., Muronetz, V., 2003. Ascorbate-induced oxidation of glyceraldehyde-3-phosphate dehydrogenase. Biochem. Biophys. Res. Commun. 308, 492–496.

Seabra, A., Ouellet, M., Antonic, M., Chrétien, M., English, A., 2013. Catalysis of nitrite generation from nitroglycerin by glyceraldehyde-3-phosphate dehydrogenase (GAPDH). Nitric Oxide 35, 116–122.

Shoshan, M., 2012. 3-bromopyruvate: targets and outcomes. J. Bioenerg. Biomembr. 44, 7–15.

Sirover, M., 2014. Structural analysis of glyceraldehyde-3-phosphate dehydrogenase functional diversity. Int. J. Biochem. Cell Biol. 57, 20–26.

Tarze, A., Deniaud, A., Le Bras, M., et al., 2007. GAPDH, a novel regulator of the pro-apoptotic mitochondrial membrane permeabilization. Oncogene 26, 2606–2620.

Torres-Bugeau, C., Ávila, C., Reisman-Vozari, R., et al., 2012. Characterization of heparin-induced glyceraldehyde-3-phosphate dehydrogenase early amyloid-like oligomers and their implication in α-synuclein aggregation. J. Biol. Chem. 287, 2398–2409.

Tsuchiya, K., Tajima, H., Yamada, M., et al., 2004. Disclosure of a pro-apoptotic glyceraldehyde-3-phosphate dehydrogenase promoter: anti-dementia drugs depress its activation in apoptosis. Life Sci. 74, 3245–3258.

Xing, C., LaPorte, J., Barbay, J., Myers, A., 2004. Identification of GAPDH as a protein target of the saframycin antiproliferative agents. Proc. Natl. Acad. Sci. U.S.A. 101, 5862–5886.

Xu, R., Serritella, A., Sen, T., et al., 2013. Behavioral effects of cocaine mediated by nitric oxide-GAPDH transcriptional signaling. Neuron 78, 623–630.

Yao, L., Wang, Y., Liu, X., et al., 2012. Phenylephrine protects cardiomyocytes from starvation-induced apoptosis by increasing glyceraldehyde-3-phosphate dehydrogenase (GAPDH) activity. J. Cell. Physiol. 227, 3518–3527.

You, K.-S., Benitez, L., McConachie, W., Allison, W., 1975. The conversion of glyceraldehyde-3-phosphate dehydrogenase to an acylphosphatase by trinitroglycerin and inactivation of this activity by azide and ascorbate. Biochim. Biophys. Acta 284, 317–330.

Youdim, M., Gross, A., Finberg, J., 2001. Rasagiline [*N*-Propargyl-1R(+)-aminoindan], a selective and potent inhibitor of mitochondrial monoamine oxidase B. Br. J. Pharmacol. 132, 500–506.

Yu, J., Steck, T., 1975. Associations of band 3, the predominant polypeptide of the human erythrocyte membrane. J. Biol. Chem. 250, 9176–9184.

Yun, J., Mullarky, E., Liu, C., et al., 2015. Vitamin C selectively kills *KRAS* and *BRAF* mutant colorectal cancer cells by targeting GAPDH. Science 350, 1391–1396.

## FURTHER READING

Anderson, R., Smit, M., Joone, G., Van Staden, A., 1990. Vitamin C and cellular immune functions: protection against hypochlorous acid-mediated inactivation of glyceraldehyde-3-phosphate dehydrogenase and ATP generation in human leukocytes as a possible mechanism of ascorbate-mediated immunostimulation. Ann. N. Y. Acad. Sci. 587, 34–48.

Endo, A., Hasumi, K., Sakai, K., Kanbe, T., 1985. Specific inhibition of glyceraldehyde-3-phosphate dehydrogenase by koningic acid (heptelidic acid). J. Antibiot. 38, 920–925.

Ishitani, R., Tajima, H., Takata, H., et al., 2003. Proapoptotic protein glyceraldehyde-3-phosphate dehydrogenase: a possible site of action of antiapoptotic drugs. Prog. Neuropsychopharmacol. Biol. Psychiatry 27, 291–301.

Nakazawa, M., Uehara, T., Nomura, Y., 1997. Koningic acid (a potent glyceraldehyde-3-phosphate dehydrogenase inhibitor)-induced fragmentation and condensation of DNA in NG108-15 cells. J. Neurochem. 68, 2493–2499.

Phadke, M., Krynetskaia, N., Krynetskiy, E., 2013. Cytotoxicity of chemotherapeutic agents in glyceraldehyde-3-phosphate dehydrogenase-depleted human lung carcinoma A549 cells with the accelerated senescence phenotype. Anticancer Drugs 24, 366–374.

Pradhan, L., Mondal, D., Chandra, S., Ali, M., Agarwal, K., 2008. Molecular analysis of cocaine-induced endothelial dysfunction: role of endothelin-1 and nitric oxide. Cardiovasc. Toxicol. 8, 161–171.

Steinritz, D., Weber, J., Balszuweit, F., Thiermann, H., Schmidt, A., 2013. Sulfur mustard induced nuclear translocation of glyceraldehyde-3-phosphate dehydrogenase (GAPDH). Chem. Biol. Interact. 206, 529–535.

Voskresenskiy, A., Sun, L., 2008. The housekeeping gene (GA3PDH) and the long interspersed nuclear element (LINE) in the blood and organs of rats treated with cocaine. Ann. N. Y. Acad. Sci. 1137, 309–315.

Section V

# The Unique Role of Sperm-Specific GAPDH

---

*There is an exception to every rule.*

An Old Proverb

The "central dogma" of moonlighting proteins is the assertion that, from a single protein with a well-established, classical activity and subcellular distribution, it is possible to derive a moonlighting protein with many separate activities and diverse intracellular localizations. As noted, GAPDH may be such a quintessential protein. This is demonstrated by the depth of diverse GAPDH functions described in the previous chapters in which its normal moonlighting activities and differential subcellular locations are considered. It is also indicated by those chapters in which its pathological activities are enumerated.

That being said, the goal of this chapter is to consider what may be an intriguing reversal of that "central dogma," i.e., the presence of a unique spermatogenic GAPDH protein, which has a restricted "housekeeping" function, may not exhibit any moonlighting activity yet, is essential for perhaps one of the most critical tasks in Nature, i.e., the propagation of the species. This includes its unique localization (being covalently linked to the fibrous sheath in the principal piece of the flagellum), its genetic organization, its structural properties as compared to somatic GAPDH, the novel temporal sequence of its mRNA and protein regulation during spermatogenesis, its association with other "housekeeping" proteins to provide the requirement for rapid and efficient ATP generation, and, perhaps intriguingly, its expression in tumor cells.

## 1. IDENTIFICATION OF SPERMATOZOIC GAPDH

As with many organs and tissues, spermatozoic GAPDH, along with other glycolytic enzymes, was identified and its activity characterized (Harrison, 1971; Brooks, 1976). The latter studies indicated, with respect to male gonadal distribution, that the level of spermatozoic glycolytic enzymes was significantly higher than that observed for other gonadal locales. This selective increase in glycolytic protein expression provides presumably the basis for the spermatogenic focus on anaerobic cytoplasmic energy production as contrasted with aerobic mitochondrial ATP production (Mann and Lutwak-Mann, 1981).

Glyceraldehyde-3-Phosphate Dehydrogenase (GAPDH). http://dx.doi.org/10.1016/B978-0-12-809852-3.00016-9
**279**

Intriguingly, it was noted also that analysis of spermatozoic glycolytic enzymes was technically difficult with variable results obtained by standard extraction protocols (Harrison, 1971). A large quantity of those proteins was contained in sperm cytoskeletal structures as contrasted with those proteins released by standard cell disruption procedures. Accordingly, a more rigorous procedure would be necessary to release those proteins into soluble form for reliable quantitation. This technical requirement was the first suggestion of a distinctive characteristic of spermatogenic glycolytic enzymes in general and GAPDH-S in particular as somatic GAPDH is readily assayed in crude cell extracts or in subcellular fractions (rev. in Sirover, 1999).

## 2. LOCALIZATION OF SPERMATOZOIC GAPDH

To determine the specific distribution of GAPDH-S, it was necessary first to establish reliable protocols for its isolation (Westhoff and Kamp, 1997). Using boar spermatozoa, a variety of conventional protocols failed to liberate enzymatically active GAPDH-S. These included altering pH, increasing ionic strength, addition of ATP, the use of detergents, or digestion with chymotrypsin. In contrast, mild digestion with trypsin released enzymatically active GAPDH-S. Quantitation following trypsin digested indicated that 90% of GAPDH-S was bound in some manner to a spermatogenic cytoskeletal structure. As subsequent studies have demonstrated, mild trypsin digestion is now a standard protocol in GAPDH-S investigations.

The recovery of boar GAPDH-S in sufficient quantity permitted the production of an anti-GAPDH-S polyclonal antibody. Control immunoblot analyses demonstrated the immunological specificity of the antibody as well as the requirement of mild trypsin digestion to release the bound GAPDH-S into soluble form. Subsequent immunogold labeling of boar sperm flagella demonstrated fibrous sheath labeling. No labeling of any other structure was detected. Therefore, Westhoff and Kamp proposed that GAPDH-S contains a specific peptide linker that functions to bind the protein to the fibrous sheath of the principal piece of the flagellum. Further, it was hypothesized that such GAPDH-S binding functions to provide the requisite energy for dynein ATPases to facilitate sperm motility. As illustrated in Fig. V.1, the distance between the mitochondria and the flagellum may be such so as to preclude the efficient use of ATP generated during aerobic respiration for flagellum function. In addition, the associate of GAPDH with dynein may not be unusual given the role of each in intracellular membrane trafficking (Tisdale et al., 2009).

Further studies using fluorescence analysis demonstrated GAPDH-S localization to the principal piece of the mouse tail (Bunch et al., 1998). Subsequent immunoblot studies using isolated mouse fibrous sheath localized GAPDH-S to that structure. Surprisingly, the GAPDH-S $M_r$ of 69 kDa was much higher than that observed for its somatic counterpart (see below). Although these studies

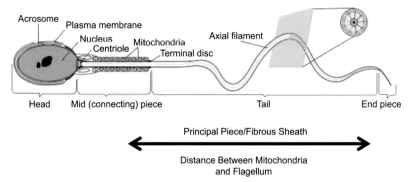

**FIGURE V.1**    Spatial relationships in mammalian spermatozoa. *Adapted from Wikipedia.*

suggested the generality of GAPDH-S fibrous sheath localization, examination of human and rat GAPDH-S yielded different findings (Liu et al., 2013). In those studies, immunoanalysis suggested a diverse spermatogenic localization. The latter could not be due to antibody cross-reactivity as the GAPDH-S antibodies did not recognize somatic GAPDH in a wide variety of tissues. No subcellular localization studies were performed, so it is unclear if the human and rat GAPDH-S proteins localize to the fibrous sheath.

## 3. GENETIC ANALYSIS OF GAPDH-S

In literature, an exceptional mystery begins with an event that perplexes both the characters in the literary composition and the audience as well. Whether in a novel, a play, or a movie, the plot develops and, as events unfold, the cloud of mystery starts to dissipate and its explanation comes into view.

## 3.1 Isolation of the GAPDH-S Gene

Such was the case of the mystery protein detected in the spermatogenic fibrous sheath (Fenderson et al., 1988). A monoclonal antibody, designated ATC, was prepared by immunizing Fisher rats against an extract from rat adult testis cells. It was used then to screen for immunoreactive proteins in testis. Such a protein was identified and localized to the fibrous sheath. It exhibited apparent molecular weight of 67,000. The unknown protein had a wide phylogenetic distribution although no immunoreactive proteins were detected in humans. It was noted also that the unknown protein was difficult to extract, present in insoluble spermatogenic fractions.

Subsequently, monoclonal antibody ATC was used to immunoscreen a mixed mouse germ cell cDNA library in λgt11 (Welch et al., 1992). That analysis yielded two clones. Sequence analysis demonstrated their similarly to somatic cell GAPDH. However, there were significant differences between the gene encoding the ATC identified protein and that well known for somatic cell

GAPDH. Perhaps the most notable was the presence of an N-terminal domain of 105 amino acids that did not match either somatic cell GAPDH or any other known sequence.

Using start and stop codons contained within the cDNA sequence, the ATC monoclonal antibody-isolated GAPDH cDNA encoded a protein of some 438 amino acids with a calculated $M_r$ of 47.4 kDa (substantially different from the 67 kDa size after SDS-PAGE). The sequence was highly enriched for proline (49.5%) and contained two polyproline domains (20 and 17 residues, respectively). However, it did conserve the GAPDH cofactor and substrate-binding sites. Calculated sequence similarities for the ATC identified cDNA were 64%, 61%, 72%, 71%, and 69% for *Escherichia coli*, *Saccharomyces cerevisiae*, *Drosophila melanogaster*, chicken, mouse, and human somatic cell GAPDH, respectively. Northern blot analysis revealed a 1.5-kb mRNA that not only corresponded to the calculated amino acid sequence but also was significantly larger than the 1.3-kb somatic cell GAPDH transcript. Considering all the available evidence, it was concluded that the cDNA recognized by monoclonal ATC was a unique GAPDH species, which could be termed GAPDH-S.

Further sequence comparisons between GAPDH-S and somatic GAPDH demonstrated both similarities and differences between each protein (Kuravsky and Muronetz, 2007). Similarities included those sequences involved in the glycolytic enzyme reaction, the AP-2-binding motif which may be related to membrane trafficking and that domain responsible to cell cycle–related protein degradation. A noticeable difference is the lack of the nuclear export signal in the GAPDH-S protein.

## 3.2 Genomic Characterization of the GAPDH-S Gene

The next step was to examine the genomic organization of GAPDH-S using the above identified GAPDH-S cDNA as the start of the screening mechanism (Welch et al., 1995). In contrast to the novelty of the GAPDH-S cDNA and protein, it appears that the mouse GAPDH-S gene is relatively unremarkable. It contains requisite exons and introns. It does not contain a TATA box, which indicates some variability in the transcription initiation site. This may provide the foundation for the observation of heterogeneity in the size of testicular GAPDH mRNA (Mezquita et al., 1998). That being said, Southern blot analysis demonstrated that GAPDH-S is a single copy gene. This is in direct contrast to that genomic organization observed for the somatic GAPDH gene. Although the latter is encoded by a single structural gene, in spite of one functional gene, somatic mammalian cells contain a plethora of GAPDH genomic sequences (Piechaczyk et al., 1984; Fort et al., 1985).

Further studies demonstrated the presence of GAPDH-S in both human and rat spermatozoa (Welch et al., 2000, 2006, respectively). Their genomic organization was similar. In the former, chromosomal analysis demonstrated that the GAPDH-S gene was localized to chromosome 19. This is in contrast to somatic

GAPDH which localizes to chromosome 12 in humans and to chromosome 6 in mice (Bruns and Gerald, 1976; Bruns et al., 1978, respectively). Both human and rat GAPDH-S are characterized by the addition of the N-terminal proline-rich domain although there may be some variance in its length, i.e., the human GAPDH-S is 30 amino acids shorter. Rat GAPDH-S is 94% similar to its mouse counterpart and 83% similar to human GAPDH-S. In toto, this suggests a conservation of GAPDH-S in mammalian spermatozoa.

## 3.3 Developmental Regulation of the GAPDH-S Gene

As spermatogenesis is highly programmed, the temporal expression of the GAPDH-S gene was examined within that complex sequence of selective gene activation. In the mouse, analysis of GAPDH-S mRNA indicated strict transcriptional regulation. Little GAPDH-S mRNA was detected in steps 4–6. In contrast, significant mRNA levels were observed during steps 7–15 with diminution of mRNA levels at step 16 (Mori et al., 1992). Using a different developmental classification, mouse GAPDH-S mRNA was minimal at day 18. In contrast, 2 days later, at day 20 considerable upregulation of the RNA was detected (Welch et al., 1992). Further mouse studies demonstrated the temporal relationship between the synthesis of GAPDH-S mRNA and the appearance of immunoreactive GAPDH-S protein (Bunch et al., 1998). In those studies, significant GAPDH-S mRNA was detected at step 9 which diminished significantly at steps 12–13 of mouse development. In contrast, little immunoreactive protein was observed at step 9, while considerable protein was detected at steps 12–13. It was suggested that translational regulation of GAPDH-S was responsible for the delay in the synthesis of the GAPDH-S protein.

A different classification was used to examine the developmental regulation of the rat GAPDH-S gene (Welch et al., 2006). In that study, postnatal GAPDH-S mRNA was detected from days 29–50. GAPDH-S protein levels were minimal in stage IX and stage XIII spermatids. In contrast, significant GAPDH-S protein was observed in stage V and stage VI tubules with considerable reduction in stage VII. Using boar spermatozoa, it was demonstrated that the synthesis of GAPDH-S mRNA was detected in round spermatids while the GAPDH-S protein could only be observed in condensing spermatids (Feiden et al., 2008). These studies provide further documentation of GAPDH-S translational regulation.

## 3.4 Requirement of the GAPDH-S Gene for Male Fertility

Genetic analysis indicated the requirement of GAPDH-S function for sperm motility (Miki et al., 2004). In these studies, transgenic mice which were homozygous deficient in the GAPDH-S gene were constructed. Termed Gapds$^{-/-}$, they contained defects in exons 5 and 6, which specifies amino acids within the GAPDH-S catalytic site and the NAD$^+$-binding site, respectively. As expected, determined by immunocytochemistry, no spermatogenic GAPDH-S protein was detected. Further, no GAPDH enzyme activity was observed, which demonstrated the absence of the somatic GAPDH isozyme. Structural analysis demonstrated the integrity of the fibrous sheath.

Intriguingly, quantitation of ATP levels indicated that the Gapds$^{-/-}$ sperm contained approximately 10% of the normal level of ATP even though mitochondrial oxygen consumption was normal. This diminution in ATP levels may be considered perplexing as, indicated in Fig. V.1, although there is considerable distance between sperm mitochondria and the fibrous sheath, the former would be presumably intact and able to synthesize considerable ATP.

In spite of this defect in ATP production, the number of sperm and their morphology were similar to wild type as was mating behavior, testis weight, and morphology. Body weights were similarly equivalent. In contrast, analysis of the transgenic sperm revealed that they exhibited significant defects in motility. This observation may provide the physiological significance for the location of GAPDH-S in the fibrous sheath, i.e., the mitochondria are simply too far away and the ATP they produce cannot be efficiently transported to the flagellum; hence the need for GAPDH-S within that locale. This supposition is in accord with diminished sperm function and reduced sperm motility in normal animals using the inhibitor α-chlorohydrin to reduce GAPDH activity (Terrell et al., 2011).

Significantly, homozygous Gapds$^{-/-}$ mice were infertile. In contrast, heterozygous male Gapds$^{+/-}$ mice and homozygous female Gapds$^{-/-}$ mice are fertile as are mice lacking the sperm-specific cytochrome c (c$_t$) protein (Narisawa et al., 2002). Further, it has been proposed that the presence of anti-GAPDH-S antibodies may be a biomarker of infertility (Fu et al., 2016). In toto, these cumulative results strongly suggest that GAPDH-S is a critical spermatogenic protein whose depletion results in severe and specific effects on male fertility.

## 4. PURIFICATION AND PROPERTIES OF GAPDH-S

The isolation and characterization of GAPDH-S were facilitated by the observation that mild trypsin treatment of isolated sperm cytoskeletal structures released bound GAPDH-S (Westhoff and Kamp, 1997). As the latter contains c.90% of GAPDH-S activity, the use of this protocol was especially critical for any successful purification of GAPDH-S as well as the subsequent analysis of its catalytic activity. That being said, the GAPDH-S protein so isolated was not that which functions in vivo, i.e., it is reasonable to suggest that cleavage of the N-terminal proline-rich domain changes the conformation and perhaps the properties of the GAPDH-S protein. Nevertheless, with that caveat in mind, the use of that protocol resulted in the provision of considerable information with respect to GAPDH-S structure and function.

Initial studies reported the purification of the abovementioned active fragment of human GAPDH-S (Shchutskaya et al., 2008). The protein so isolated,

$M_r$ of 40 kDa, was comparable in size to the calculated 44.5 kDa of GAPDH-S minus the N-terminal domain. Peptide analysis indicated that it spanned the entire protein. Investigation of its catalytic properties indicated a higher specific activity and higher $K_m$ as compared to somatic GAPDH. However, its pH optimum and $K_m$ for $NAD^+$ was similar. Intriguingly, molecular modeling of its homotetrameric form suggested that the four N-terminal domains may reside opposite each other, at the distal parts of the oligomer. As such, they would be exposed which would facilitate binding to the fibrous sheath. In that regard, it was suggested that a GAPDH-S-specific cys[21] is present in that domain and could be involved in a disulfide bond with a corresponding cysteine in the fibrous sheath. This represents an attractive possibility.

## 4.1 Isolation and Characterization of Recombinant GAPDH-S

Although the above studies are informative, detailed analyses required the preparation of the recombinant protein. As indicated in Table V.1, presented in chronological order, four laboratories were successful in preparing recombinant GAPDH-S. Each has not only provided significant information with respect to its structure and function but also encountered the same difficulty, i.e., the impossibility of preparing a full-length recombinant protein containing the N-terminal domain that was not insoluble.

In the first study, a recombinant rat GAPDH-S was produced in *E. coli* (Frayne et al., 2009). As expected, expression of full-length GAPDH-S was unsuccessful with the majority of GAPDH-S in an insoluble form. Accordingly, a construct in which that domain was omitted was used. However, although expression was observed in *E. coli*, it appears that a large proportion of that

**TABLE V.1** Purification and Properties of Recombinant Sperm-Specific GAPDH

| Species | Vector | Oligomeric Structure | Unique Finding | References |
|---------|--------|---------------------|----------------|------------|
| Rat | *Escherichia coli* | Heterotetramer | *E. coli* GAPDH/ GAPDH-S protein complex | Frayne et al. (2009) |
| Human | *E. coli* | Homotetramer | GAPDH-S stability | Elkina et al. (2010) |
| Human | *E. coli* | Homotetramer | Increase in catalytic efficiency | Chaikuad et al. (2011) |
| Human | Baculovirus | Homotetramer | Contraceptive test | Lamson et al. (2011) |

protein was also insoluble. That being said, it was stated that a small amount was soluble and sufficient for purification.

Surprisingly, a 36-kDa protein copurified with the rat protein. Analysis by SDS-PAGE clearly indicated a higher $M_r$ for the rat protein. Visualization suggested a proportionality between the two with the rat protein present to a smaller extent. The 36-kDa protein was identified as the *E. coli* GAPDH monomer. Gel filtration analysis indicated its physical association into a complex with a molecular mass indicative of a tetrameric protein. Analysis of the crystal structure indicated a 1:3 GAPDH-S/*E. coli* GAPDH ratio in the heterotetramer. Sequence analysis indicated a 65% identity across the entire molecule increasing to 85% in the dimerization domain. Thus, this presents sufficient similarity between each protein to form the heterotetramer extending from the respective primary sequences to their quaternary structure.

In the second study, a recombinant human GAPDH-S was expressed in *E. coli* (Elkina et al., 2010). As with each other study, expression of the full-length GAPDH-S yielded insoluble protein which required either guanidine hydrochloride or urea for extraction. Accordingly, a truncated sequence was used which lacked the 68 N-terminal amino acids and encoded a protein of 339 amino acids. The recombinant protein produced by that construct (termed dN-GAPDS) exhibited an $M_r$ of 38 kDa which was quite comparable to the expected $M_r$ of 36.9 kDa.

Although the catalytic properties of dN-GAPDS were comparable to that of somatic GAPDH, there were noticeable and potentially physiologically significant, differences between the two proteins. Thermal inactivation studies demonstrated a 25% reduction in dN-GAPDS activity when incubated for 10 min at 55°C. In contrast, somatic GAPDH retained only c.10% of its activity at that interval and temperature. In addition, no further decrease in dN-GAPDS was observed when the protein was incubated for a total of 20 min. At 70°, dN-GAPDS retained c.40% of its activity at 5 min, while somatic GAPDH lost 100% of its activity at c.5 min. These findings demonstrate a fourfold lower inactivation rate for dN-GAPDS. Further studies demonstrated that dN-GAPDS was significantly resistant to denaturation by guanidine hydrochloride with a 20-fold lower inactivation rate constant.

The comparison of the respective protein structure provided a possible explanation for these results. As discussed, although there are two additional cysteines in dN-GAPDS, cys[94], and cys[150], their spatial relationship would preclude their contribution to the enhanced stability of dN-GAPDS. Further analysis would also tend to exclude enhanced hydrogen bonding. In contrast, the number of salt bridges in dN-GAPDS exceeds those found in somatic GAPDH. Further analysis using molecular modeling suggested the role of three "buried" salt bridges and specific proline residues as contributors to GAPDS stabilization (Kuravsky et al., 2014). Mutational analysis indicated not only the contribution of P164 and P326 but also the salt bridge D311–H124 to GAPDS stability. In toto, these studies provide the data and foundation to indicate that, considering its function and localization, there may be an inherent logic to this physical difference between GAPDH-S and its somatic counterpart.

In the third study, a detailed crystallographic study was undertaken using a truncated human recombinant GAPDH-S which was termed hGAPDS$_{\Delta N}$ (Chaikuad et al., 2011). Two crystal structures were analyzed. In one, NAD$^+$ and inorganic phosphate were bound. In the second, NAD$^+$ and glycerol were bound. In each instance, the structures were similar to other GAPDH structures obtained from diverse sources. The recombinant protein was present as a homotetramer and its structure was compared in detail to human somatic GAPDH. It was concluded that there is broad overlap in common structural features.

That being said, it was noted that there was a specific kinetic difference between hGAPDS$_{\Delta N}$ and its human somatic counterpart. Although the K$_m$, V$_{max}$, and k$_{cat}$ are similar, the hGAPDS$_{\Delta N}$ K$_m$ for NAD$^+$ (35 µM) is threefold lower than that observed for its human somatic counterpart (100 µM). This difference translates into a comparable increase in catalytic efficiency of hGAPDS$_{\Delta N}$. In this regard, it was noted that the structural differences between human GAPDH-S and human GAPDH are located on the protein surface, which may result in different electrostatic potentials. The latter may provide the basis for the observed kinetic differences.

In the fourth study, baculovirus infection of insect cells was used to produce a recombinant GAPDH-S (Lamson et al., 2011). As with the other studies, a truncated GAPDH-S construct was used (residues 69–408). The protein produced exhibited an M$_r$ of 40 kDa and a native M$_r$ of c.130 kDa. The former is in accordance with the calculated M$_r$, while the latter is indicative of a tetramer. There did not appear to be any formation of a heterotetramer nor was there any difference in specific activity as compared to other sources. Using the recombinant protein as a testing vehicle for potential contraceptive agents, it was noted that a preliminary scan indicated the potential utility of three such agents.

## 4.2 Mechanisms of GAPDH-S Reactivation

A recent study indicated the role of the chaperone, TRiC, in maintaining GAPDH-S structure and function (Naletova et al., 2011). The former is a large protein with a number of subunits localized in reticulocytes and in testis. In these studies, the ability of TRiC to restore enzyme activity after denaturation was examined using several different proteins. These included somatic GAPDH, the previously described dN-GAPDS (Elkina et al., 2010; see above), and a new GAPDH-S construct termed ChBD-GAPDS. The latter was constructed by the N-terminal addition of a chitin-binding domain to dN-GAPDS. The rationale for its construction may be the continuing concern with respect to the utility of studies using truncated GAPDH-S proteins lacking the N-terminal proline-rich domain.

Proteins were denatured by incubation with 4 M guanidine hydrochloride. Notably, incubation of somatic GAPDH for 15 min resulted in a complete loss of activity. In contrast, a 24h incubation was required to eliminate activity in the case of the dN-GAPDS and the ChBD-GAPDS recombinant proteins. Reactivation was then tested by incubation with TRiC.

These studies demonstrated a TRiC-dependent reactivation of both dN-GAPDS and the ChBD-GAPDS recombinant proteins. This reactivation was increased fivefold for both recombinant proteins by the addition of $Mg^{++}$-ATP. It was time dependent with a maximal extent observed at 100min. In contrast, there was no effect of TRiC or $Mg^{++}$-ATP on the renaturation of somatic GAPDH. Thus, these results suggest not only that TRiC may play a role in the stabilization of GAPDH-S but also that the addition of N-terminal amino acids does not, in this instance, affect enzyme kinetics. Further, as the stability of GAPDH-S may be a rate-limiting factor in sperm motility and, ultimately, fertility, mechanisms to maintain GAPDH-S may be of significance.

## 4.3 Mechanisms of GAPDH-S Catalysis

Historically, as a putative "housekeeping protein," GAPDH was used to study basic mechanisms of oligomeric protein catalysis. As each of the monomers in the GAPDH tetramer binds $NAD^+$, the binding of one $NAD^+$ molecule to one of the monomers can influence subsequent binding of additional $NAD^+$ molecules to other monomers in a positive or negative manner, termed positive or negative cooperativity.

Mutational analysis was used to examine the role of $NAD^+$ binding cooperativity in catalysis by GAPDH-S (Kuravsky et al., 2015). Five proteins were used: somatic GAPDH, dN-GAPDS, and three mutant dN-GAPDS proteins in which their novel GAPDS salt bridges were individually altered. As described previously, it was considered that these salt bridges may contribute to the enhanced stability of the spermatogenic protein. In two, glutamic acid was changed to glutamine and, in the third, termed dN-GAPDS (D311N), in which $asp^{311}$ has been changed to $asn^{311}$ to alter the salt bridge which may connect the $NAD^+$-binding site with the catalytic domain. By definition, substitution of the 311 aspartic acid residue with a 311 asparagine residue would destroy the D311–H124 salt bridge.

Analysis of cooperativity indicated that the control somatic GAPDH exhibited its characteristic negative cooperativity, while dN-GAPDS displayed positive cooperativity as did the two glutamate to glutamine mutants. In contrast, the dN-GAPDH (D311N) mutant bound $NAD^+$ noncooperatively. Its $K_m$ was increased 2.5-fold as was its specific activity. It also exhibited a slight increase in thermal stability. As such, these findings highlight further the significance of the D311–H124 salt bridge in GAPDH-S structure and function.

The nature of GAPDH-S' stability and activity was examined further focusing on intersubunit and interdomain interactions (Makshakova et al., 2015). In these studies, five different GAPDH proteins were analyzed, GAPDH, GAPDH-S and three mutants, E244Q, E96Q, and D311N each of which would affect one of the three unique GAPDH-S salt bridges. Using these proteins, a detailed analysis was presented. For example, it was noted that intersubunit interactions are stronger in GAPDH-S than GAPDH possibly due to the involvement of glu$^{352}$ in the former as opposed to gln$^{280}$ in the latter. Subunit contact in GAPDH-S as compared to GAPDH involves the substitution of arg$^{265}$ and his$^{275}$ for gly$^{193}$ and leu$^{203}$, respectively. Mutant analysis demonstrated that glu244gln destabilized the intersubunit salt bridge resulting in a diminution of intersubunit energy; that glu96gln affects interdomain interaction energy and that asp311asn diminishes intersubunit energy. In toto, these cumulative studies represent an elegant analysis of GAPDH-S (rev. in Muronetz et al., 2015). They also present an exceptional model for the use of recombinant proteins and mutational studies as a means to assess the importance of specific amino acids or domains in protein function.

## 5. FORMATION OF A GLYCOLYTIC GAPDH-S FIBROUS SHEATH PROTEIN COMPLEX

As described above, it was demonstrated that not only was GAPDH-S localized in the fibrous sheath but also that genetic studies showed that GAPH-S$^{-/-}$ mice had 90% decreases in their sperm ATP levels. Accordingly, it seemed reasonable to suggest a sperm-specific localization of glycolytic proteins in the fibrous sheath responsible for the ATP production for sperm motility and, ultimately, fertility.

One of the original studies to suggest that possibility examined the localization of boar GAPDH-S, triosephosphate isomerase (TIM), pyruvate kinase (PK), and phosphoglycerate kinase (PGK) (Westhoff and Kamp, 1997). Following trypsin digestion, GAPDH-S, TIM, and PK copurified with a defined stoichiometry between GAPDH-S and TIM as defined by activity measurements. This suggested that they were present in a multiprotein complex. Subsequently, the coordinate temporal expression of boar GAPDH-S and PK was determined as was the necessity of mouse PGK2 for sperm function and male fertility (Feiden et al., 2008; Danshina et al., 2010, respectively).

Further studies in the mouse demonstrated that additional glycolytic enzymes were physically associated with the fibrous sheath (Krisfalusi et al., 2006). This included aldolase 1 and lactate dehydrogenase A (LDHA). As previously noted, the inclusion of LDH in the fibrous sheath multienzyme protein complex would permit the required regeneration of NAD$^+$ for continued glycolytic ATP production (Westhoff and Kamp, 1997).

In summary, these cumulative findings suggest a model in which a glycolytic multienzyme complex is formed late in spermatogenesis, which is responsible for the generation of some 90% of the ATP generated in vivo. This is illustrated in Fig. V.2, which denotes the cyclic nature of NAD$^+$ reduction and NADH oxidation (Fig. V.2). Accordingly, it is reasonable to suggest that other enzymes may be associated with this complex. Finally, although recent

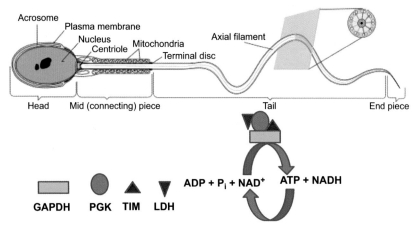

**FIGURE V.2** Fibrous sheath localization of glycolytic enzymes in mammalian spermatozoa. *Adapted from Wikipedia.*

findings indicated that PGK2 was essential for sperm function, it was noted that its deletion was not a total bar to motility and fertility (Danshina et al., 2010). As a possible explanation, it was suggested that acylphosphatase may provide a means to bypass PGK2 to produce 3-phosphoglycerate although with limited efficacy.

## 6. ROLE OF GAPDH-S IN HUMAN PATHOLOGY

Given the localization and restricted focus of GAPDH-S, it seems counterintuitive to suggest its role in human pathology apart from infertility. That being said, recent findings suggest that aberrant GAPDH-S expression may be observed during somatic cell tumorigenesis as well as its dysfunction in instances of diabetes.

### 6.1 Expression of GAPDH-S in Melanoma

As previously discussed, somatic GAPDH may play a fundamental role in tumorigenesis. Recent evidence suggests that GAPDH-S may also be involved (Sevostyanova et al., 2012). In these studies, melanoma cell lines were examined for GAPDH-S expression through immunoblot analysis using GAPDH-S-specific antibodies. Those studies revealed the presence of a 37-kDa GAPDH-S protein which, it was suggested, may be a GAPDH-S fragment having lost the N-terminal proline-rich domain. Alternatively, GAPDH-S may be synthesized as the full-length protein with a cleavage of the N-terminal domain. Localization studies revealed a cytoplasmic localization. Intriguingly, immunoprecipitation analysis demonstrated its physical association with somatic GAPDH with a 1:1 stoichiometry. The formation of this heterotetramer is consistent with previous studies in *E. coli* with a recombinant GAPDH-S protein (Frayne et al., 2009).

Although it is clear that GAPDH-S is expressed in melanoma cells, its role remains uncertain. It was noted that there appears to be a cancer-related dysregulation of testis genes, termed cancer/testes-associated (CTA) genes. However, their role in tumorigenesis is unknown. The observation that GAPDH-S forms a heterotetramer with somatic GAPDH in melanoma cells suggests its role in anaerobic energy generation, a known characteristic of cancer cells. Whether it functions also in moonlighting GAPDH activities is uncertain. As recent evidence suggests its role in sperm–egg interactions (Petit et al., 2014; Margaryan et al., 2015) as well as the potential interaction of the N-proline region of GAPDH-S with SH3 domains (Tatjewski et al., 2016), analysis of the functional diversity of GAPDH-S may be of interest.

## 6.2 Role of GAPDH-S in Diabetes

As previously described (Chapter 14), the structure and function of somatic GAPDH may be affected by hyperglycemia and metabolic syndrome. In addition, as indicated in Chapter 8, somatic GAPDH is involved fundamentally both as a target for oxidative stress as well as a mediator of the cellular response to that physiological and environmental challenge. Accordingly, it may come as no surprise that the structure and function of GAPDH-S may also be dependent on its interaction with reactive oxygen species (ROS).

Initial studies considered the effect of ROS on normal GAPDH-S (Elkina et al., 2011). In these investigations using human and equine sperm, two parameters were examined: sperm motility and GAPDH-S activity. It was determined that sperm characterized by low motility exhibited significantly lower GAPDH-S activity than did the high mobility samples. Examination of activity as a function of ROS exposure indicated demonstrable effects of superoxide anion, hydroxyl radical, and hydrogen peroxide on catalysis. Kinetic analysis indicated a significantly faster effect of $H_2O_2$ (5 min vs. 2 h). The effect of ROS could be ameliorated by glutathione.

The potential relationship between these findings and diabetes was examined in clinical studies using samples from normal and diabetic patients (Liu et al., 2015). As indicated, there was a coordinate temporal relationship between GAPDH-S activity, sperm motility, and infertility in patients as a function of time after diagnosis. The first two parameters diminished progressively, while the third increased progressively. It was also determined that addition of exogenous GAPDH-S to sperm samples could protect the latter against ROS damage. Accordingly, these cumulative findings suggest a new pathology may be associated with diabetes as well as its interrelationship with oxidative stress.

## 7. SUMMARY

It is hoped that, from this discussion, the reader will gain insight into the fundamental role of GAPDH-S in sperm structure and function. As summarized in Table V.2, GAPDH-S may be a uniquely designed protein through which an organism provides the necessary means to ensure the propagation of the species. This involves its restricted function, localization, and protein–protein

**TABLE V.2  Comparative Properties of Sperm-Specific and Somatic Mammalian GAPDH**

| Property | Sperm-Specific GAPDH | Somatic Cell GAPDH | References |
|---|---|---|---|
| Chromosomal localization | Chromosome 19 | Chromosome 12 (human); chromosome 6 (mouse) | Welch et al. (2000), Bruns and Gerald (1976) and Bruns et al. (1978) |
| GAPDH mRNA | 1.5 kb | 1.3 kb | Welch et al. (1992) and rev. in Sirover (1999) |
| $M_r$ | 58 kDa | 37 kDa | Shchutskaya et al. (2008) and rev. in Sirover (1999) |
| Amino acid length | 438 | 333 | Welch et al. (1992) |
| N-terminal proline-rich domain | Present | Absent | Welch et al. (1992) |
| Distribution | Fibrous sheath | Cytoplasm, nucleus, membrane | Westhoff and Kamp (1997), rev. in Tristan et al. (2011) and Sirover (2012) |
| Thermoinactivation ($70°$, $k_{obs}$) | $0.96\,min^{-1}$ | $0.226\,min^{-1}$ | Elkina et al. (2010) |

interactions that distinguish GAPDH-S from its somatic counterpart. In particular, the addition of the N-terminal proline-rich domain permitting covalent attachment to the fibrous sheath is remarkable. Further, examination of its genomic structure, temporal developmental regulation, and catalytic properties demonstrates its physiological significance. That being said, the mechanisms that underlie its translational regulation are unclear. Their characterization may further illuminate the complexity inherent in this seemingly straightforward role of GAPDH-S in Nature.

## REFERENCES

Brooks, D., 1976. Activity and androgenic control of glycolytic enzymes in the epididymis and epididymal spermatozoa of the rat. Biochem. J. 156, 527–537.

Bruns, G., Gerald, P., 1976. Human glyceraldehyde-3-phosphate dehydrogenase in man-rodent somatic cell hybrids. Science 192, 54–56.

Bruns, G., Pierce, P., Regina, V., Gerald, P., 1978. Expression of GAPDH and TPI in dog-rodent hybrids. Cytogenet. Cell Genet. 22, 547–551.

Bunch, D., Welch, J., Magyar, P., Eddy, E., O'Brien, D., 1998. Glyceraldehyde-3-phosphate dehydrogenase-S protein distribution during mouse spermatogenesis. Biol. Reprod. 58, 834–841.

Chaikuad, A., Shafqat, N., Al-Mokhtar, R., et al., 2011. Structure and kinetics of human sperm-specific glyceraldehyde-3-phosphate dehydrogenase-GAPDS. Biochem. J. 435, 401–409.

Danshina, P., Geyer, C., Dai, Q., et al., 2010. Phosphoglycerate kinase 2 (PGK2) is essential for sperm function and male fertility in mice. Biol. Reprod. 82, 136–145.

Elkina, Y., Atroshchenko, M., Bragina, E., Muronetz, V., Schmalhausen, 2011. Oxidation of glyceraldehyde-3-phosphate dehydrogenase decreases sperm motility. Biochemistry (Mosc) 76, 268–272.

Elkina, Y., Kuravsky, M., El'darov, M., et al., 2010. Recombinant human sperm-specific glyceraldehyde-3-phosphate dehydrogenase: structural basis for enhanced stability. Biochim. Biophys. Acta 1804, 2207–2212.

Feiden, S., Wolfrum, U., Wegener, G., Kamp, G., 2008. Expression and compartmentalization of the glycolytic enzymes GAPDH and pyruvate kinase in boar spermatogenesis. Reprod. Fertil. Dev. 20, 713–723.

Fenderson, B., Toshimori, K., Muller, C., Lane, T., Eddy, E., 1988. Identification of a protein in the fibrous sheath of the sperm flagellum. Biol. Reprod. 38, 345–357.

Fort, P., Marty, L., Piechaczyk, M., et al., 1985. Various rat adult tissues express only one major mRNA species from the glyceraldehyde-3-phosphate-dehydrogenase multigenic family. Nucleic Acids Res. 13, 1431–1443.

Frayne, J., Taylor, A., Cameron, G., Hadfield, A., 2009. Structure of insoluble rat sperm glyceraldehyde-3-phosphate dehydrogenase (GAPDH) via heterotetramer formation with *Escherichia coli* GAPDH reveals target for contraceptive design. J. Biol. Chem. 284, 22703–22712.

Fu, J., Yan, R., Luo, Y., et al., 2016. Anti-GAPDHS antibodies: a biomarker of immune infertility. Cell Tissue Res. http://dx.doi.org/10.1007/s00441-016-2361-6.

Harrison, R., 1971. Glycolytic enzymes in mammalian spermatozoa: activities and stabilities of hexokinase and phosphofructokinase in various fractions from sperm homogenates. Biochem. J. 124, 741–750.

Krisfalusi, M., Miki, K., Magyar, P., O'Brien, D., 2006. Multiple glycolytic enzymes are tightly bound to the fibrous sheath of mouse spermatozoa. Biol. Reprod. 75, 270–278.

Kuravsky, M., Barinova, K., Aleksandra Marakhovskaya, A., et al., 2014. Sperm-specific glyceraldehyde-3-phosphate dehydrogenase is stabilized by additional proline residues and an interdomain salt bridge. Biochim. Biophys. Acta 1844, 1820–1826.

Kuravsky, M., Muronetz, V., 2007. Somatic and sperm-specific isoenzymes of glyceraldehyde-3-phosphate dehydrogenase: comparative analysis of primary structures and functional features. Biochemistry (Mosc) 72, 744–749.

Kuravsky, M., Barinova, K., Asryants, R., Schmalhausen, E., Muronetz, V., 2015. Structural basis for the NAD binding cooperativity and catalytic characteristics of sperm-specific glyceraldehyde-3-phosphate dehydrogenase. Biochimie 115, 28–34.

Lamson, D., House, A., Danshina, P., et al., 2011. Recombinant human sperm-specific glyceraldehyde-3-phosphate dehydrogenase (GAPDHS) is expressed at high yield as an active homotetramer in baculovirus-infected insect cells. Protein Expr. Purif. 75, 104–113.

Liu, J., Sun, C., Zhang, C., Wang, X., Li, J., 2013. Location and characterization of GAPDS in male reproduction. Urol. Int. 90, 449–454.

Liu, J., Wang, Y., Gong, L., Sun, C., 2015. Oxidation of glyceraldehyde-3-phosphate dehydrogenase decreases sperm motility in diabetes mellitus. Biochem. Biophys. Res. Commun. 465, 245–248.

Makshakova, O., Semenyuk, P., Kuravsky, M., et al., 2015. Structural basis for regulation of stability and activity in glyceraldehyde-3-phosphate dehydrogenases. Differential scanning calorimetry and molecular dynamics. J. Struct. Biol. 190, 224–235.

Mann, T., Lutwak-Mann, C., 1981. Male Reproductive Function and Semen. Springer Verlag, Berlin.

Margaryan, H., Dorosh, A., Capkova, J., et al., 2015. Characterization and possible function of glyceraldehyde-3-phosphate dehydrogenase-spermatogenic protein GAPDHS in mammalian cells. Reprod. Biol. Endocrinol. 13, 15.

Mezquita, J., Pau, M., Mezquita, C., 1998. Several novel transcripts of glyceraldehyde-3-phosphate dehydrogenase expressed in adult chicken testis. J. Cell. Biochem. 71, 127–139.

Miki, K., Qu, W., Goulding, E., et al., 2004. Glyceraldehyde-3-phosphate dehydrogenase-S, a sperm-specific glycolytic enzyme, is required for sperm motility and male fertility. Proc. Natl. Acad. Sci. U.S.A. 101, 16501–16506.

Mori, C., Welch, J., Sakai, Y., Eddy, E., 1992. In situ localization of spermatogenic cell-specific glyceraldehyde-3-phosphate dehydrogenase (Gapd-s) messenger ribonucleic acid in mice. Biol. Reprod. 46, 859–868.

Muronetz, V., Kuravsky, M., Barinova, K., Schmalhausen, E., 2015. Sperm-specific glyceraldehyde-3-phosphate dehydrogenase-an evolutionary acquisition of mammals. Biochemistry (Mosc) 80, 1672–1689.

Naletova, I., Popova, K., Eldarov, M., et al., 2011. Chaperonin TRiC assists the refolding of sperm-specific glyceraldehyde-3-phosphate dehydrogenase. Arch. Biochem. Biophys. 516, 75–83.

Narisawa, S., Hecht, N., Goldberg, E., et al., 2002. Testis-specific cytochrome *c*-null mice produce functional sperm but undergo early testicular atrophy. Mol. Cell. Biol. 22, 5554–5562.

Petit, F., Serres, C., Auer, J., 2014. Moonlighting proteins in sperm–egg interactions. Biochem. Soc. Trans. 42, 1740–1743.

Piechaczyk, M., Blanchard, J., Marty, L., et al., 1984. Posttranscriptional regulation of glyceraldehyde-3-phosphate dehydrogenase gene expression in rat tissues. Nucleic Acids Res. 123, 6951–6963.

Sevostyanova, I., Kulikova, K., Kuravsky, M., Schmalhausen, E., Muronetz, V., 2012. Sperm-specific glyceraldehyde-3-phosphate dehydrogenase is expressed in melanoma cells. Biochem. Biophys. Res. Commun. 427, 649–653.

Shchutskaya, Y., Elkina, Y., Kuravsy, M., Bragina, E., Schmalhausen, E., 2008. Investigation of glyceraldehyde-3-phosphate dehydrogenase from human sperms. Biochemistry (Mosc) 73, 185–191.

Sirover, M., 1999. New insight into an old protein: the functional diversity of mammalian glyceraldehyde-3-phosphate dehydrogenase. Biochim. Biophys. Acta 1432, 159–184.

Sirover, M., 2012. Subcellular dynamics of multifunctional protein regulation: mechanisms of GAPDH intracellular translocation. J. Cell. Biochem. 113, 2193–2200.

Tatjewski, M., Gruca, A., Plewczynski, D., Grynberg, M., 2016. The proline-rich region of from human sperm may bind SH3 domains, as revealed by a bioinformatics study of low-complexity protein segments. Mol. Reprod. Dev. 83, 144–148.

Terrell, K., Wildt, D., Anthony, N., et al., 2011. Glycolytic enzyme activity is essential for domestic cat (*Felis catus*) and cheetah (*Acinonyx jubatus*) sperm motility and viability in a sugar-free medium. Biol. Reprod. 84, 1198–1206.

Tisdale, E., Azizi, F., Artalejo, C., 2009. Rab2 utilizes glyceraldehyde-3-phosphate dehydrogenase and protein kinase Cι to associate with microtubules and to recruit dynein. J. Biol. Chem. 284, 5876–5884.

Tristan, C., Shahani, N., Sedlak, T., Sawa, A., 2011. The diverse functions of GAPDH: views from different subcellular compartments. Cell. Signal. 23, 317–323.

Welch, J., Barbee, R., Magyar, P., Bunch, D., O'Brien, D., 2006. Expression of the spermatogenic cell-specific glyceraldehyde-3-phosphate dehydrogenase (GAPDS) in rat testis. Mol. Reprod. Dev. 73, 1052–1060.

Welch, J., Brown, P., O'Brien, D., Eddy, E., 1995. Genomic organization of a mouse glyceralde-hyde- 3-phosphate dehydrogenase gene (Gapd-s) expressed in postmeiotic spermatogenic cells. Dev. Genet. 16, 179–189.

Welch, J., Brown, P., O'Brien, D., et al., 2000. Human glyceraldehyde-3-phosphate dehydroge-nase-2 gene is expressed specifically in spermatogenic cells. J. Androl. 21, 328–338.

Welch, J., Schatte, E., O'Brien, D., Eddy, E., 1992. Expression of a glyceraldehyde-3-phosphate dehydrogenase gene specific to mouse spermatogenic cells. Biol. Reprod. 46, 869–878.

Westhoff, D., Kamp, G., 1997. Glyceraldehyde-3-phosphate dehydrogenase is bound to the fibrous sheath of mammalian spermatozoa. J. Cell Sci. 110, 1821–1829.

## FURTHER READING

Kuravsky, M., Aleshin, V., Frishman, D., Muronetz, V., 2011. Testis-specific glyceraldehyde-3-phosphate dehydrogenase. BMC Evol. Biol. 11, 160.

# Section VI

# Discussion

*To everything*
*Turn, Turn, Turn,*
*There is a Season,*
*Turn, Turn, Turn*
*And a Time for every purpose under Heaven.*

Ecclesiastes by way of Pete Seeger.

As indicated in this work, there is a breadth and depth of GAPDH function which, at times, may be considered breathtaking. Surely, those of us present at the beginning of this journey could not have appreciated the scope and the magnitude of its activities, both in normal cell function and in cell pathology. For those reasons, it seemed appropriate that this was the time and the place, both scientifically and personally, to present not only our knowledge of moonlighting GAPDH activity but also to indicate how those functions reveal basic parameters intrinsic to cell, tissue, and organ viability as well as to consider how those activities relate to the etiology of human disease.

Many themes are interwoven into the analysis of moonlighting GAPDH activity presented in this work. In these chapters, the role of protein–protein and protein–nucleic interactions are, in many instances, a priori requirements for moonlighting GAPDH activities. This is illustrated in Chapter 2 describing the role of GAPDH in transcriptional gene expression, its role in hypoxia (Chapter 9), and its interaction with age-related neurodegenerative disorders (Chapter 12). Similarly, the subcellular localization of GAPDH and the changes observed therein are representatives of its functional diversity. This is noted first in Chapter 1 in which immunocytochemical analyses were used to note, at that time, "peculiar" intracellular changes in the localization of a classical housekeeping protein.

Another theme involves the pleiotropic nature of moonlighting GAPDH function. This is exemplified by the consequences of a single GAPDH posttranslational modification. As related in Chapter 8, a single change, S-nitrosylation of its active site cysteine, results not only in the complex pattern of protein and gene expression, which characterizes programmed cell death, but also may be involved in heme regulation. Similarly, as described in Chapter 5, GAPDH phosphorylation is involved in changes in cytoskeletal structure facilitating membrane transport. Further, these small changes in protein structure, reflected in the presence of GAPDH isozymes, may provide the basis for new GAPDH

Glyceraldehyde-3-Phosphate Dehydrogenase (GAPDH). http://dx.doi.org/10.1016/B978-0-12-809852-3.00017-0

**297**

activities. This is illustrated in Chapter 4 in which its membrane fusogenic activities are based on the presence of a specific, basic GAPDH isozyme that lacks glycolytic activity.

Another theme is the complexity of its moonlighting activities. This is described in Chapter 3 in which its posttranscriptional role in determining mRNA stability is considered. As discussed, binding of GAPDH to 3′-UTR sequences may have three different effects depending on the mRNA in question and, therefore, distinct cellular consequences related to the function of the respective mRNA. This is also emphasized in Section IV in which the pharmacology of moonlighting GAPDH is described, highlighting both the number of drugs, which target GAPDH and the corresponding indications of treatment.

Finally, another major theme developed in this work is the importance of moonlighting GAPDH activity with respect to significant cell functions and pathology. Each of the chapters and sections describe cellular parameters and pathological conditions, which are not inconsequential. Indeed, they run the gamut of existential cell functions to pathological states, which threaten that existence. Apart from the chapters considered above, the critical nature of GAPDH function is illustrated by the unique role of sperm-specific GAPDH (Section V), its role in determining the effects of both excitatory and inhibitory neurotransmitters (Chapter 7), its facilitation of cancer development (Chapter 11) as well as its duality of function in infection and immunity (Chapter 13). All in all, considering the past three decades, it has been an interesting time.

# Index